STRUCTURAL
INVESTIGATION
OF POLYMERS

Ellis Horwood Series in
POLYMER SCIENCE AND TECHNOLOGY

Series Editors: T. J. KEMP, Professor of Chemistry, University of Warwick
J. F. KENNEDY, Professor of Chemistry, University of Birmingham

This series, which covers both natural and synthetic macromolecules, reflects knowledge and experience from research, development and manufacture within both industry and academia. It deals with the general characterization and properties of materials from chemical and engineering viewpoints and will include monographs highlighting polymers of wide economic and industrial significance as well as of particular fields of application.

STRUCTURAL INVESTIGATION OF POLYMERS

G. BODOR D.Sc
Head of Polymer Physics Section
Research Institute for the Plastics Industry
Budapest, Hungary

Translation Editor

J. HODGKINSON
Imperial College of Science, Technology and Medicine, London

ELLIS HORWOOD
NEW YORK   LONDON   TORONTO   SYDNEY   TOKYO   SINGAPORE

English Edition first published in 1991
a coedition between
ELLIS HORWOOD LIMITED
Market Cross House, Cooper Street,
Chichester, West Sussex, PO19 1EB, England

A division of
Simon & Schuster International Group
A Paramount Communications Company
and
AKADÉMIAI KIADÓ
Publishing House of the
Hungarian Academy of Sciences, Budapest, Hungary

© 1991
Translation: © 1991 Ellis Horwood Limited

*Distributed in East European countries, China, Northern Korea, Cuba, Vietnam and
Mongolia by* ACADEMIA, Publishing House of the Czech and Slovak Academy of
Sciences, Vodičkova 40, 11229 Praha 1, The Czech and Slovak Federal Republic

Printed in Hungary by Akadémiai Kiadó és Nyomda Vállalat, Budapest

British Library Cataloguing in Publication Data

Bodor, G. (Geza)
    Structural investigation of polymers.
    1. Polymers. Structure & properties
    I. Title
    547.7

ISBN 0–13–852989–2

Library of Congress Cataloging-in-Publication Data

Bodor, G.
    Structural investigation of polymers/G. Bodor; translation editor: J. Hodgkinson.
        p. cm. – (Ellis Horwood series in polymer science and technology)
    Includes bibliographical references and index.
    ISBN 0–13–852989–2
    1. Polymers. 2. Conformational analysis. I. Title. II. Series. QD381. B568 1990
    547.7–dc20   90–40591
                                                                                        CIP

Preface

This book is based on the author's lectures, given at the Technical University of Budapest. It has benefited from many other sources, in particular the lectures given by Prof. E. W. Fischer at the Polytechnic Institute of Brooklyn in 1964/65, the excellent book by F. Billmeyer, and of course experimental results from the Department of Polymer Micromorphology at the Polymer Research Institute where the author has been a coworker since 1952.

The first version of this book was written in 1968–69, for special postgraduate courses, the second was printed in proper book form by Műszaki Könyvkiadó, Budapest in 1982. In the intervening years polymer characterization techniques have developed so rapidly that this revised edition has been prepared in English at the suggestion of many colleagues from abroad.

The book is intended for graduate students working in the field of polymers, for research workers in the polymer industry and for those, who are simply interested in polymer characterization.

The structure of polymers is described and closely related to their morphology, and the methods of investigation are outlined sufficiently to put relevant questions to the experts in these methods. The author would be pleased, if scientists, working in the field of polymer analysis could also make use of this presentation, at least on related subjects.

Finally, the author wishes to thank Dr. György Bánhegyi who not only translated the text from Hungarian but also contributed important comments and advice during the translation. The manuscript was prepared by Mrs. Irén Szlancsók, to whom the author would like to express thanks for her patience and diligence.

Géza Bodor

Acknowledgements

The permission of the following publishers to reproduce Tables and Figures is gratefully acknowledged.

Academic Press (London)
Vonk C. G. (1982): Synthetic Polymers in the Solid State, in Small Angle Light Scattering. Ed. by O. Glatter and O. Kratky. Fig. 7.112.

American Chemical Society (Washington)
Dorman D. E., Otocka E. P. and Bovey F. A. (1972): *Macromolecules* **5**, 574. Fig. 8.35.
Ferguson R. C. (1967): *Polymer Preprints* **8**, 1026. Fig. 8.43.
Johnson L. F., Heatley F. and Bovey F. A. (1970): *Macromolecules* **3**, 175. Fig. 8.44.
Slichter W. P. (1961): *Rubber Chem. Technol.* **34**, 1574. Fig. 8.50.

American Institute of Physics (New York)
Anderson F. R. (1964): *J. Appl. Phys.* **35**, 64. Fig. 4.31.
Slichter W. P. and David D. D. (1964): *J. Appl. Phys.* **35**, 3103. Figs 8.47, 8.48.

Bell Laboratories (Murray Hill)
Bovey F. A. (1972): Resolution NMR of Macromolecules. Figs 8.15, 8.17, 8.18, 8.27, 8.28–8.31, 8.38–8.40.

Butterworth Scientific Ltd. (Surrey)
Dlugosz J. and Michie R. I. C. (1960): *Polymer* **1**, 41. Fig. 4.26.
Wlochowitz A. and Eder M. (1984): *Polymer* **25**, 1268. Fig. 6.19.
Powles J. G., Strangede J. H. and Sandiford D. J. H. (1963): *Polymer* **4**, 401. Fig. 8.49/a.

Powles J. G., Hunt B. I. and Sandiford D. J. H. (1964): *Polymer* **5,** 505.
Fig. 8.49/*b*.
Cornell University Press (Ithaca, N.Y.)
Paul J. Flory (1962): Principles of Polymer Chemistry. Figs 2.2, 2.3.
Hüthig and Wepf Verlag (Mainz)
Turner-Jones A., Aizlewood J. M. and Beckett D. R. (1964): *Die Makrom. Chemie* **75,** 134. Fig. 7.34.
Interscience Publishers (New York)
Mark H. (1943): The Investigation of High Polymers with X-Rays in the Chemistry of Large Molecules.
eds: Burk R. E. and Grummit O. Fig. 7.25.
Optical Society of America (New York)
Colthup N. B. (1950): *Journal of the Optical Society of America* **40,** 397.
Fig. 8.1.
Pergamon Press Ltd. (Oxford)
Margerison D. and East G. C. (1967): Introduction to Polymer Chemistry. Figs 2.10, 2.12, 2.27–2.30, 2.35, 2.44, 2.46, 2.49–2.56.
Tables 2.9–2.13, 2.16–2.21.
The Biochemical Journal (London)
P. Andrews (1964): *Biochem. J.* **91,** 222. Fig. 3.16.
The Faraday Society (London)
Schmidt J. A. S. (1955): *Discussions Faraday Soc.* **19,** 207. Fig. 8.51.
The Institute of Physics (London)
Farrow G. and Ward I. M. (1960): *British J. Appl. Phys.* **11,** 543. Fig. 8.46. Table 8.12.
Prof. Dr. J. Springer (Berlin)
Einführung in die Theorie der Lichtstreuung verdünnter Lösungen Grosser Moleküle, (1970) Figs 2.31–2.34, 2.37–2.40.
Springer Verlag (Heidelberg)
Schurz J. (1974): Physikalische Chemie der Hochpolymeren.
Figs 8.52–8.55.
The Society of Polymer Science, Japan
Inoue Y., Nishioka A. and Chujo R. (1971): *Polymer Journal* **4,** 535.
Fig. 8.41.
Varian/James Shoolery (Palo Alto)
A Basic Guide to NMR, 1978. Figs 8.19–8.24.
J. Wiley and Sons (New York)
Fischer T., Kinsinger J. B. and Wilson C. W. (1966): *J. Pol. Sci. (Part B)* **4,** 379. Fig. 8.45.
Geil P. H. (1960): *J. Pol. Sci.* **44,** 449. Fig. 4.20.

Keller A. (1959): *J. Pol. Sci.* **36,** 361. Fig. 4.25.

Starkweather H. W. and Brooks R. E. (1959): *Appl. Polym. Sci.* **1,** 236. Fig. 6.10.

Staverman A. J., Pals D. T. F. and Kruissink Ch. A. (1957): *J. Pol. Sci.* **23,** 57. Figs 2.23, 2.24. Table 2.15.

Stehling, F. C. (1964): *J. Pol. Sci. (Part A)* **2,** 1815. Fig. 8.42.

Shyluk S. (1962): *J. Pol. Sci.* **62,** 318. Figs 3.11, 3.7, 3.8.

Wilson C. W. and Santee E. R. (1965): *J. Pol. Sci. (Part C)* **8,** Fig. 8.34.

I am very much obliged to

Prof. F. A. Bovey (Murray Hill)

Prof. J. Peterman (Hamburg)

Dr. V. B. F. Mathot (Geleen)

for their manuscripts and for the permission to use them in my work.

Contents

List of symbols used

Symbol	Meaning
A	orientation parameter
A	first virial coefficient
A	mass number of the atom
A_m	meridian reflection in X-ray diffraction
A_2	second virial coefficient according to Eq. 2.80
A_3	third virial coefficient according to Eq. 2.80
$A\%$	absorbance in $\%$
a	exponent of the SMH-equation
a	lattice constant
a^*	lattice constant of the reciprocal space
a	particle diameter
a	specific absorbance known earlier as the extinction coefficient
B	second virial coefficient according to Eq. 2.2.79
B	compressibility (see K) B is referred to in the text as bulk modulus
B	temperature coefficient
B	line halfwidth
b	lattice constant
b^*	lattice constant of the reciprocal space
C	third virial coefficient
C	sound velocity
C_p	heat capacity of the sample at constant pressure
$C(x)$	Cauchy (or Lorentz) function
c	concentration, g/cm^3
c	speed of light in vacuum
c_n	propagation velocity of light in a medium of refractivity n
c	lattice constant
c^*	lattice constant of the reciprocal space
c_m	mass concentration, g/kg
D	optical density (absorbance)

D	crystalline particle size
D	crystalline defect factor
D_{hkl}	size of crystallite normal to the hkl plane, nm
d	interplanar distance between atomic planes
d	sample thickness, cm
E	electrical field strength
E	tensile or Young's modulus
E_D	activation energy on the surface of the nucleus
E	sonic Young' modulus
E_{lat}^0	lateral modulus of the fully oriented fibre
E_{or}	sonic modulus of oriented sample
E_u	sonic modulus of the unoriented sample
$E(t)$	stress relaxation modulus
$F_{(p)}$	mass fraction of solution in summative fractionation
f	orientation factor
f	Cabannes correction factor
f_x	fraction of chains of length x remaining in solution
f_{cr}	crystalline orientation
f_{am}	amorphous orientation
$f_{spec, am}$	specific amorphous orientation
$f(M_i)$	molecular mass distribution function
G	free enthalpy, Gibbs potential
G	relative scattered intensity
G	growth rate constant of nucleation
G	shear modulus
G	gauche conformation, when the R substituent is in *trans* position with respect to the chain (see Fig. 8.37)
G'	gauche conformation when both the R substituent and the chain carbon atom are in *gauche* position with respect to the chain
$G(t)$	shear relaxation modulus
G^x	dynamic shear modulus
G'	in-phase component of the G^x modulus (storage modulus)
G''	90° out-of-phase component of the dynamic modulus (loss modulus)
G_{el}	elastic part of the free enthalpy
G^*	free enthalpy needed for the formation of a nucleus of critical size
$G(x)$	Gaussian function
g	measure of branching
H	enthalpy
$\bar{H}$	average end-to-end distance
H_v	small angle light scattering, vertical primary beam, horizontal analyzer
H_0	magnetic field strength
H_1	magnetic field strength, perpendicular to H_0
H_{loc}	local magnetic field (in NMR spectroscopy)
h	chain end-to-end distance
h	scattering vector of the reciprocal space

h	Planck constant (6.6254×10^{-34} Js)
h	heterotactic (atactic) configuration (mr)
I	scattered or transmitted intensity
I	identity period
I	spin quantum number
I_0	intensity of incident light
I_0	maximum value of the function
I_s	energy of scattered light
$I(\Theta)$	intensity of scattered X-rays in SAXS
$I_{\Theta,u}$	scattered intensity of unpolarized light under angle Θ
$I_{\Theta,H}$	scattered intensity of horizontally polarized light under angle Θ
$I_{\Theta,v}$	scattered intensity of vertically polarized light under angle Θ
$I_a(d)$	intensity of the amorphous ring in amorphous sample
$I(M_j)$	integrated molecular mass distribution function
$I(s)$	intensity of the coherent X-ray scattering at point s
$I_c(s)$	scattered intensity originating from the crystalline parts
i	isotactic configuration (mm)
J	coupling constant of the protons
2_{τ}	coupling constant of geminal protons
3_{τ}	coupling constant of vicinal protons
K	constant of the SMH equation
K	compressibility
K_{cr}	compressibility of the crystalline parts
K_{am}	compressibility of the amorphous parts
k	Boltzmann constant ($1.380\,45 \times 10^{-23}$ J.K^{-1})
k'	Huggins constant
L	long period
l	length of capillary
M	molecular mass of polymer
M	macroscopic magnetic moment
M_0	molecular mass of the base molecule
$\bar{M}_n$	number average molecular mass
$\bar{M}_m$	mass average molecular mass
$\bar{M}_v$	viscosity average molecular mass
$\bar{M}_z$	z-average molecular mass
M_c	critical molecular mass below which the rubber elastic properties are not apparent
m	total mass
m	magnetic quantum number
m	meso-configuration
m	line shape function
mm	isotactic configuration
mr	heterotactic (atactic) configuration
m_L	mass of the liquid polymer during crystallization
m_0	total mass of polymer during crystallization
N_1, N_2	number of molecules of solvent and polymer
N	number of macromolecules

N_0	number of basic molecules in the polymer
N	nucleation constant, the number of nuclei formed during unit time interval
N_{Av}	Avogadro number (6.0247×10^{23})
N_v	number of molecules per unit volume
n	number of moles of a given component, $n = N/N_{Av}$
n	exponent of the Avrami equation
n	order of reflection in X-ray diffraction
n	refractive index
n	population difference between two states of spin
n_{equ}	equilibrium value of the population difference
O	orientation parameter
P	degree of polymerization
P	polydispersity
$P(\Theta)$	Debye scattering function
p	number of nuclei per unit volume
p	induced oscillating dipole moment
p_1	vapour pressure of solution
p_1^0	vapour pressure of solvent
Q	ratio of experimental and theoretical intensities
R	gas constant $(8.314 \, J/mole.K)$
R	volume ratio of precipitated and diluted phases
R_0	radius of anisotropic sphere in SALS experiments
R_θ	reduced scattered intensity
r	racemic configuration
rr	syndiotactic configuration
S	entropy
S	intensity ratio measured at equator and meridian
$\bar{S}$	radius of gyration, scattering mass radius
s	distance of chain segment from centre of mass of chain
s	parameter of the reciprocal lattice $(h/2\pi)$
s	syndiotactic configuration (rr)
T	absolute temperature
T	*trans* conformation
T	transmission, I/I_0
T_m	melting point
T_{am}	size of amorphous area
T_{cr}	size of crystalline area
T_f	flow point
T_{crit}	critical temperature of phase separation, at infinite molecular mass it is equal to the Θ temperature
T_g	glass temperature
T_G	area under Gaussian curve
t	time, usually in sec units
t_1	flow time of solvent
t_2	flow time of solution
t_1	spin–lattice (longitudinal) relaxation time

t_2	spin–spin (transverse) relaxation time
$t_{1/2}$	half-time (in crystallization)
t_{corr}	correlation time (time needed for 1 radian rotation)
u	atomic vibrational amplitude
V	specific volume of sample
V_{am}	specific volume of an amorphous polymer
V_{cr}	specific volume of fully crystalline polymer
V_e	effective (apparent) volume of dissolved material
V_v	small angle light scattering, vertically polarized light, vertical analyzer
$V(x)$	Voigt function
v	volume fraction
v_1	volume fraction of solvent
v_2	volume fraction of polymer
v	volumetric flow rate
$\bar{v}_1$	molar volume of solvent
$\bar{v}_2$	molar volume of polymer
$\boxed{v_2}$	specific volume of polymer
$v_{2,crit}$	critical volume fraction at phase separation point
$W(x)$	Voigt approximation function
$w(M)$	differential molecular mass distribution function
w	half of the line halfwidth (halfwidth $= 2w = \beta$)
x	number of segments in a polymer (the volume of the segments is equal to that of the solvent molecules)
x	crystalline fraction
Y_1	molar fraction of solvent
Y_2	molar fraction of polymer
Z	atomic number
z	asymmetry coefficient
α	Flory linear expansion coefficient (degree of perturbation)
α	polyarizability of the molecule
α	line-shape parameter in the Voigt approximation function
β	linewidth (radian)
β_m	experimental linewidth
β_{inst}	instrument linewidth
β	integrated linewidth
β	magnetic moment of electron
Γ	optical path difference
Γ_2	second virial coefficient in Eq. 2.81
Γ_3	third virial coefficient in Eq. 2.81
γ	gyromagnetic ratio of nucleus
γ	shear deformation (strain)
Δ	depolarization value
Δ	optical birefringence
ΔE	energy difference
ΔH_b	heat of evaporation, J/g
ΔH_f	heat of freezing, J/g

ΔH_{cr}	melting heat of fully crystalline polymer, J/g
ΔH_m	melting heat, J/g
ΔS^*	configurational entropy change
Δp_1	change in partial vapour pressure
ΔT_b	boiling point elevation
ΔT_f	freezing point depression
ΔV	voltage difference
δ	chemical shift ($\delta = 10 - \tau$)
δ	phase angle
δ	parameter characterizing the ratio of diffusion and growth constants, and the branching in fibril formation
δ	solubility parameter
δ_v	half-width of the NMR resonance line, Hz
ε	extensional deformation (strain)
ε	dielectric constant
ε	parameter used for characterizing the degree of coiledness
Θ	Bragg angle, half of the angle between incident and scattered radiation (angle of observation is 2Θ)
Θ	angle of observation in light scattering measurements (because of historical reasons it is different from the previous definition)
Θ	relaxation time
Θ	theta temperature
χ_1	entropy parameter
λ_0	wavelength of light in vacuum
λ'	wavelength of light in solution
η	viscosity
$[\eta]$	limiting viscosity number (known earlier as the intrinsic viscosity)
$[\eta]_\Theta$	limiting viscosity number calculated from the Flory–Fox equation
μ	chemical potential
μ_E	excess chemical potential
μ	magnetic moment of nucleus
μ	linear absorption coefficient
ν	frequency, (cycle/sec, Hz)
ν	Poisson's ratio
ν^*	wavenumber (cm^{-1})
ν_0	frequency of Larmor precession, Hz
π	osmotic pressure
ρ	density
ρ_1	density of solvent
ρ_2	density of polymer
ρ_{am}	density of fully amorphous polymer
ρ_{cr}	density of fully crystalline polymer
ρ_g	apparent density of coils
ρ_L	density of liquid phase during crystallization
ρ_S	density of solid phase during crystallization
$\rho(a)$	diameter distribution function
$\rho_{\text{spherical equivalent}}$	the density of that sphere the viscosity of which would be equal to that of the polymer coil

σ	extensional stress
σ	coefficient in the Brönsted–Schulz equation, describing the degree of separation
σ	standard deviation, in GPC measurements the width of the chromatogram
σ_{mol}	linewidth of a monomodal peak in GPC measurement
σ_e	surface energy
σ_{inst}	instrument zone-width in GPC measurements
τ	turbidity coefficient
τ	chemical shift in NMR measurements
τ	time period
τ	shear stress
ϕ	volume fraction of dissolved parts
ϕ	final average volume of crystallized parts
χ_1	Flory–Huggins polymer–solvent interaction parameter
$\chi_{1,crit}$	critical value of the interaction parameter, where phase separation appears (around 0.5)
ψ_1	entropy parameter
ω_0	radial frequency of Larmor precession motion

INTRODUCTION

Low molecular inorganic materials can be definitely identified by determining their chemical composition and allotropic modification.

For low molecular organic materials containing tertiary carbon, some further data are required to describe steric isomers usually available from tables in standard references.

In the case of macromolecules (polymers), in addition to chemical composition, the structure of molecular chain, molecular mass, distribution of molecular mass, distribution of amorphous and crystalline areas, orientation of molecules, size and distribution of crystalline areas and crystalline structure formed within these areas are required for characterization. All these factors together determine the properties of polymers and by changing any of them the final properties of the polymer can be influenced in a predictable way.

Sorting out and measuring those properties characteristic of polymers has not yet been solved in all cases. Characterization of branched and cross-linked polymers is especially difficult, since the latter ones are insoluble so several methods are inapplicable in their cases. Because of this, the description of linear macromolecules will be the main subject treated.

PART ONE

CHARACTERIZATION OF POLYMER MOLECULES

CHAPTER 1

Chemical composition and configuration of polymer chains

Position and arrangement of atoms in a polymeric chain can be divided into two categories:

— Position fixed by chemical bonds, for example *cis*- or *trans*-isomers or stereo-isomers (D- or L-forms). This kind of order which can be changed only by disrupting chemical bonds and is known as configuration.

— Structural differences originating from rotations around single bonds. These structures, including the various forms of polymer chains in solution, are called conformations.

In solution or in the melt where the mobility of the molecules is relatively high, several different conformations can follow each other within a short period. By cooling the melt, molecular mobility becomes more and more hindered. The mechanical properties of the solid product will depend on the molecular mass of the polymer chain and on the crystallinity. If the molecular

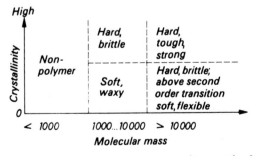

Fig. 1.1. Dependence of polymer properties on molecular mass and crystallinity.

mass exceeds about 10,000, and the material is crystallizable, the product will be fairly hard and tough. If no crystalline order has been formed, the material will be rigid and brittle below its secondary transition temperature, and soft and flexible above this point. If the molecular mass is lower than about 10,000, crystallizable polymers yield a rigid product, while non-crystalline types give soft, wax-like ones (Fig. 1.1).

1.1. HEAD-TO-TAIL LINKAGE

From a structural point of view polyethylene is the simplest polymer. In this case it is of no importance which end of the molecule is linked to the nearest ethylenes during chain formation. In the formation of other polyolefins, however, there are still more possibilities for linkage. As an example, take polypropylene. If the methyl groups follow each other regularly (Fig. 1.2*a*, the head-to-tail scheme), the structure will be different from that produced when the methyl groups follow each other immediately in the chain (Fig. 1.2*b*, the head-to-head linkage). Of course head-to-head and head-to-tail linkages may occur in the same molecule (Fig. 1.2*c*).

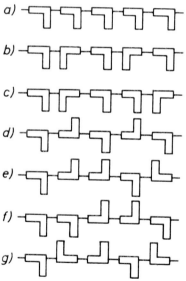

Fig. 1.2. Linkage possibilities of monomers containing asymmetric carbon atoms. *a* Head-to-tail structure, isotactic order; *b* head-to-head structure; *c* mixed structure; *d* head-to-tail structure, alternating, or syndiotactic order; *e* head-to-tail structure, with disordered atactic steric order; *f* head-to-tail structure, with steric order alternating after every second monomer; *g* mixed structure, with disordered atactic steric order.

1.2. STEREOREGULARITY

The structure of molecular chains containing asymmetric carbon atoms cannot be defined by the mode of linkage (head-to-head or head-to-tail) only. Properties are over-ridingly influenced by the sequence of steric activity of the asymmetric centres following each other in the same mode of linkage (e.g. head-to-tail). Some possibilities are: regular (d-d-d or l-l-l) (Fig. 1.2a), alternating (d-l-d-l) (Fig. 1.2d) or random sequence (Fig. 1.2e). Study of the spatial arrangement of side groups and synthesis of polymers with a given spatial configuration has become a central theme for polymer chemical research in the past few years. Natta et al. (1955) synthesized a whole series of polymers with regular configuration (stereospecific polymers) by means of stereoselective catalyst systems. If the activity of the asymmetric centres in the chain is uniform (see Fig. 1.2a), the polymer is referred to as isotactic, if it is alternating (see Fig. 1.2d), the chain is syndiotactic, and if it is random (see Fig. 1.2e), the polymer is atactic. Polymer chains with asymmetric carbon atoms can be synthesized in a variety of forms (as e.g. d-d-l-l-...) (Fig. 1.2f), but the description and nomenclature will be omitted here.

The dependence of physical properties on structure can be characterized by the fact that the melting point of atactic polypropylene is around 60–70°C, whereas that of isotactic polypropylene is 175°C.

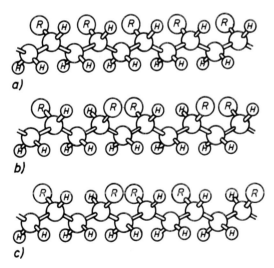

a)

b)

c)

Fig. 1.3. Steric structure of molecular chains containing asymmetric carbon atom. a Isotactic, b syndiotactic, c atactic.

Schematic representation as shown in Fig. 1.2 helps in understanding the situation, but it can be misleading. It gives the impression that stereospecific polymers could be transformed into each other simply by rotations around single covalent bonds. In fact such a transformation is impossible because of the tetrahedral configuration of carbon atoms (Fig. 1.3).

1.3. CIS–TRANS ISOMERY

Other kinds of rotational isomers can be formed if one of the initial compounds shows *cis-trans* isomery. An example for this is the copolymer of styrene with the esters of maleic or fumaric acid. From the two dicarboxilic acids, maleic acid shows *cis-* while fumaric acid *trans*-structure:

$$CH-COOCH_3 \qquad\qquad\qquad HC-COOCH_3$$
$$\| \qquad\qquad\qquad\qquad\qquad\qquad \|$$
$$CH-COOCH_3 \qquad\qquad\qquad CH_3COO-CH$$

Maleic acid dimethyl ester   Fumaric acid dimethyl ester

If rotation is free in the resulting polymer, both compounds yield the same product. If, however, rotation is impossible or greatly hindered, two polymer products are possible with the following, quite different structures:

Styrene-maleic ester copolymer

Styrene fumaric ester copolymer

These two products exhibit different physical properties: if the composition is similar, the maleic product is soft and rubber-like, whereas the fumaric is rigid and brittle. Differences are due to the fact that rotation is not free around the chain because of the steric hindrance of the ester groups.

If there are double bonds in the main chain of the polymer, the resulting *cis-trans* isomers can significantly modify its properties. Natural rubber and guttapercha are classical examples for such differences but in the case of other diene-polymers with rubber-like qualities (as polybutadiene, polyisoprene, polychloroprene) *cis-trans* isomery plays an equally important role.

Cis-trans isomers are also present in polymeric chains containing cycloaliphatic rings.

An important example is the Kodel polyester fibre, which is poly (1,4-cyclohexyl-terephthalate) with the following repeat unit:

Because of the 1,4-disubstituted cyclohexane ring the melting points of the resulting *cis-* and *trans*-isomers are 260°C and 320°C respectively.

1.4. BRANCHING

Polymerization of a bifunctional monomer (as e.g. vinyl derivatives) results in a regular chain, without branching, at least in principle. In practice, however, for several reasons various branching processes occur, the product therefore contains several different side groups. In polyethylene, e.g. one can find methyl groups and often longer side chains comparable with the main chain itself. Appearance of side chains is determined by several factors, mostly by production technology. Side chains strongly influence polymer properties. Highly branched polyethylene is e.g. softer, less crystallizable, its melting point is lower, and is more soluble than its linear or slightly branched counterpart.

1.5. THE EFFECT OF COPOLYMERIZATION ON THE STRUCTURE OF POLYMERS

During copolymerization four basic structural types of macromolecule can be formed:

(i) Alternating copolymers in which different monomeric units follow each other regularly, and can be characterized by the following regular structure:

A-B-A-B-A-B-.

(ii) Statistical or *random copolymers* in which monomeric units follow each other irregularly, and various monomer sequences are possible, e.g.:

A-A-B-A-A-A-B-B-A

Properties of such copolymers are usually close to the average of those of the homopolymers prepared from the monomeric components. Structure of such random linear copolymers is usually looser, and dye or solvent molecules can penetrate them relatively easily. Regular linear copolymers sometimes form new types of crystals, the structure of which is not characteristic of any of the components.

(iii) Block copolymers contain the components in uniform segments of various length that link to each other, e.g.:

A-A-A-A-A-A-A-B-B-B-B-B-B-B-A-A-A-A-A

Block copolymers are not synonymous with bulk-polymerized products. The latter term refers to a solvent-free polymerization technique.

(iv) In the case of *graft copolymers* the second component is subsequently polymerized onto the synthesized polymeric chain of the first component. A typical structure is:

```
A-A-A-A-A-A
  |        |
  B        B
  |        |
  B        B
  |        |
  .        .

  .        .
```

A very interesting technique of graft–copolymer synthesis is the irradiation of the base polymer with X- or γ-rays. In such cases free radicals are formed on the polymeric chain, and by adding a monomer (polymerizable by a free radical mechanism) grafts of a desired length can be synthesized.

Properties of block and graft copolymers are usually not the average of the components but some of them add to each other. If, for example, one of the components is watersoluble and the other one is soluble in apolar solvents, the block or graft copolymer can dissolve in both types of solvent. The reason of this is that the soluble part can solubilize the other. Such phenomena appear if the two components are otherwise compatible. In such cases micellar structures appear in the solution.

CHAPTER 2

Molecular mass of polymers

The size of a polymer molecule can be characterized by its molecular mass (before the introduction of the SI system, by molecular weight) or the degree of polymerization. Degree of polymerization (P) gives the number of basic monomeric units in the macromolecule:

$$P = \frac{N_0}{N} \tag{2.1}$$

where N is the number of macromolecules formed, N_0 is that of the original basic units.

Similarly one can speak about the degree of polycondensation if the macromolecule is formed by polycondensation, but frequently the term: degree of polymerization (stepwise polymerization) is used in the case of polycondensates as well.

Molecular mass of a polymer can be calculated if the molecular mass of the basic molecule, M_0 is multiplied by the degree of polymerization:

$$M = PM_0. \tag{2.2}$$

In the case of polycondensates elimination of by-products (e.g. water) takes place in parallel with the formation of the macromolecule; so when calculating the molecular mass, the mass of the structural repeat unit has to be multiplied by the degree of polymerization.

The lengths of the chains are determined by stochastic processes during both stepwise polymerization and polycondenzation. If the reaction proceeds stepwise, the length of the chain will be determined by the local occurrence of the reactive groups at the ends of the growing chains. In the case of radical polymerization the length of the chain will be determined by the time available for chain growth before dezactivation of the growing macroradical sets on. In any event, polymerization yields chains of different length, the length distribution of which can be determined by statistical methods used for each type of polymerization. This means that a polymer system can be characterized by molecular mass distribution and by its measurable averages rather than molecular mass.

As molecular masses of polymer samples are always averages, different methods yield different results depending on the method of averaging. Methods based on counting the molecules in a given mass of material yield the so-called number average, which is usually designated by $\bar{M}_n$. In many cases the number average is close to the maximum of the mass distribution curve.

Other methods measure quantities proportional to the mass of molecules, in this case the mass-averaged molecular mass in obtained. Since weighing is proportional to the mass, and molecules with higher molecular mass will achieve higher weight factors in the averaging procedure, then the mass average is higher than (or in the case of uniform molecules at best equal to) the number average. The term weight average molecular weight, $\bar{M}_w$, has been used extensively in the literature but here $\bar{M}_m$, mass averaged molecular mass is preferred. The $\bar{M}_m/\bar{M}_n$ ratio can be used to characterize the spread of the molecular mass distribution curve. This ratio for most polymers varies between 1.0 and 50. A single molecular mass average characterizes the size of the particles in such a system only where they are uniform, i.e., if the system is monodisperse.

In order to show how averages depend on the measuring method, let us consider the following two systems:
— System I: 90 molecules with a molecular mass of 1000;
                10 molecules with a molecular mass of 10,000;
— System II: 10 molecules with a molecular mass of 1000;
                90 molecules with a molecular mass of 10,000.
Number average is the ratio of the total mass and the number of the molecules:

$$\bar{M}_n = \frac{\sum_i N_i M_i}{\sum_i N_i} \tag{2.3}$$

where N_i is the number of the i-th type of molecule. Mass average is obtained by the multiplication of mass fraction and molecular mass:

$$\bar{M}_m = \sum_i m_i M_i \tag{2.4}$$

where the mass fraction can be given as:

$$m_i = N_i M_i / \sum_i N_i M_i \tag{2.5}$$

and from this it follows that:

$$\bar{M}_m = \frac{\sum_i N_i M_i^2}{\sum_i N_i M_i}. \tag{2.6}$$

Certain methods measure the so called z-average which can be defined as:

$$\bar{M}_z = \frac{\sum_i N_i M_i^3}{\sum_i N_i M_i^2}. \tag{2.7}$$

These three averages can be related to the first, second and third moment of the molecular mass distribution function. Table 2.1 summarizes the various averages of the two systems mentioned.

Table 2.1

Various molecular mass averages for the systems given in the text

m	System I	System II
M_n	1,900	9,100
M_m	5,737	9,902
M_z	9,257	9,990

In general the following system of inequalities is valid for the various average values:

$$\bar{M}_z > \bar{M}_m > \bar{M}_n. \tag{2.8}$$

Practical polymer samples contain molecular chains of different size, that is, they are polydisperse, therefore molecular mass distribution is more precise than simple averages. This is particularly important if the distribution function exhibits several peaks, that is, if the system is bi- or polymodal. The relationship between various average values and the distribution curve is shown in Fig. 2.1.

The most important methods of molecular mass (or degree of polymerization) determination are the following:

— methods based on the measurement of colligative properties;
— sedimentation and diffusion methods;
— light scattering;
— viscometry;
— chemical methods.

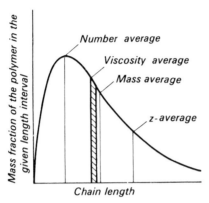

Fig. 2.1. Distribution of polymer molecules with respect to chain length (molecular mass) and various averages.

Methods listed under the first three headings are regarded as absolute methods, because they make possible the direct determination of the molecular mass of macromolecules by measuring experimental constants (as for example, density, refractive index, gas constant etc.).

Viscometry is only a relative method since the average molecular mass can be determined only if an empirical relationship based on absolute methods is available.

Chemical molecular mass determination can only be considered as an approximation, as it is based on end-group determination. This method works only if there is no branching during polymerization or polycondensation and if the end-group can be precisely characterised by analytical methods. However, if these conditions are fulfilled, the method is absolute. Molecular mass determination is always performed in solution and the derived mathematical equations are strictly valid in ideal solutions only. But ideal solutions can be obtained at infinite dilutions only, therefore molecular mass determination is based on values extrapolated to infinite dilution. Extrapolation is performed from values measured at different concentrations by determining the $c \rightarrow 0$ limit.

2.1. SOLUBILITY OF POLYMERS

Macromolecular materials can be divided into two main subgroups: amorphous and partially crystalline. The difference between them is based on X-ray study of the molecule. This structural difference is important from the point of view of solubility, since in the case of partially crystalline materials the crystalline areas have to melt before dissolution. In general partially crystalline materials dissolve at temperatures close to their melting point, for example, polyethylene dissolves in various solvents only when heated and poly(tetrafluoro ethylene) (melting point 327 °C) does not dissolve at all in any known solvents. If, however, there are specific interactions taking place between the polymer and the solvent (e.g. hydrogen bonds) dissolution can occur at lower temperatures. (Energy needed for dissolution is supplied by chemical energy.) For example nylon-6 and nylon-66 dissolve in cold formic acid. Interaction between amorphous polymers and liquids is similar to that arising between two liquids. It is well known that in processes occurring spontaneously free enthalpy must decrease. Solubility occurs only if ΔG, the free enthalpy of dissolution is negative. It is known from thermodynamics that:

$$\Delta G = \Delta H - T \Delta S \tag{2.9}$$

where G is the free enthalpy or Gibbs-potential (Guggenheim 1949), H is enthalpy, and T absolute temperature, S is entropy.

In American literature G is sometimes designated as F, unlike in European convention.

ΔS, the entropy of dissolution is normally large and, in the case of non-aqueous solutions, positive. The sign of ΔG is therefore usually determined by ΔH, the enthalpy mixing. If there exists an exothermic interaction between liquid and polymer, ΔH is negative, the system heats up, and the polymer dissolves. If the chemical effect is endothermic, ΔH is positive and the system cools down. In such cases the magnitude of ΔH determines the sign of the free enthalpy change, that is, whether the polymer dissolves or not.

In order to study the conditions for polymer dissolution molecular interactions present in polymers have to be reviewed.

2.1.1. Chemical bonds and intermolecular forces in polymers

The nature of bonds between the atoms in a molecule is studied by quantum chemistry. In the following some relevant definitions will be needed from this field.

2.1.1.1. Primary bonds

(i) Ionic bonds are formed by electron transfer between atoms. An electron octet of the negative ion is formed by taking up electrons from another atom. Electrostatic charges are characteristic of this type of bond. A typical example is table salt (NaCl).

(ii) Covalent bonds are those valence bonds which are formed by electrons distributed between two or more atoms forming a stable electronic cloud. This kind of bond occurs, for example, in the methane molecule and is typical for polymers.

(iii) Coordinative bonds are similar to covalent ones, since electron octets are formed by sharing electrons, but in this case both electrons are furnished by the same atom, the so-called donor. Such products are exemplified by the addition products of borontrifluoride.

Table 2.2

Some typical bond length and bond energy values

Bond	Bond length nm	Dissociation energy kJ/mole
C—C	0.154	349
C=C	0.134	613
C—H	0.110	416
C—N	0.147	307
C=N	0.145	895
C—O	0.146	361
C=O	0.121	752
C—F*	0.132–0.139	433–517
C—Cl	0.177	340
N—H	0.101	391
O—H	0.096	466
O—O	0.132	147

* Bond length decreases, and bond energy increases if further fluorine atoms are present.

After Billmeyer 1984.

(iv) Metallic bonds, which are virtually absent in polymers, are formed by electrons distributed freely among the atoms forming a mobile electronic cloud. The number of electrons in such systems is not sufficient for the formation of stable nucleus–electron octet arrangements. Good electronic conductivity is characteristic of this type of bond. Of all the above types this is the least understood and studied.

Table 2.2 lists some typical bond energies and lengths.

2.1.1.2. Secondary bonds

The existence of secondary intermolecular forces became known from the studies of van der Waals on real gases and they are often called van der Waals forces.

2.1.1.2.1. Dipolar forces

If in a molecule there are charges of equal magnitude but of the opposite sign, the molecule is called polar, and it has a dipole moment. At large distances this system behaves in a neutral manner. At molecular distances, however, charge separation becomes significant and plays an important role in intermolecular attraction. Interaction energy depends on the magnitude of dipoles, and in this kind of interaction thermal motion always acts against orientation, consequently the dipolar forces strongly depend on temperature.

2.1.1.2.2. Induction forces

Polar molecules influence their originally apolar neighbours with no dipole moment. The electric field of the molecular dipole causes a slight displacement of the nuclei and electrons in the neighbourhood of the dipolar molecule, and this leads to the formation of induced dipoles. Intermolecular interaction between a permanent and an induced dipole is called induction force. The magnitude of electrical and nuclear displacements is characterized by the molecular polarizability. Interaction energy in this case is always small and independent of temperature.

2.1.1.2.3. Dispersion forces

The fact that intermolecular forces also exist among apolar molecules indicates that there is a third type of intermolecular forces besides those already mentioned above.

Existence of such forces is shown by the slight temperature dependence of intermolecular forces even if dipolar interaction dominates over induction. Every molecule possesses a time dependent dipole moment, the average of which is zero for apolar molecules. This dipole moment is the result of the instantaneous position of electrons and nuclei.

These fluctuations disturb the electronic clouds of the neighbouring atoms giving rise to attractions that are referred to as dispersion forces. This type of interaction exists between all molecules and plays a dominant role among intermolecular interactions unless there are very strong dipoles present. In apolar systems only dispersion forces occur and are independent of temperature.

Interaction energy between dipolar molecules is inversely proportional to the sixth power of intermolecular distances. Similarly to primary forces, repulsion becomes dominant if atoms are getting much closer to each other than the equilibrum bond distances, i.e. at distances in the order of 0.3–0.5 nm.

Typical values of secondary intermolecular interaction energies range between 8 and 40 kJ/mole depending on the polarizability and dipole moment of the molecules.

2.1.1.2.4. Hydrogen bond

The hydrogen bond in which the hydrogen atom is bound to two other atoms, is especially important in polymers. In biopolymers it is believed to play a decisive role in biological processes.

The classical theory of chemical bonds regards hydrogen as being capable of participating in only one bond, which is partly ionic and partly covalent in character. If, however, hydrogen is bound to an electro-negative atom, and electron density around the proton is low, this proton can attract other electron-rich groups by electrostatic forces. A hydrogen bond thus appears between two functional groups within the same molecule or between two different molecules. One of these groups is usually acidic in character (proton donor). Such as hydroxyl, carboxyl, amine or amide groups. The other group must be basic in character. This group can contain oxygen (as in carbonyl or in hydroxyl groups) or nitrogen (as in amines or in amides) or sometimes halogen. All these considered the hydrogen bond is the strongest if formed among secondary forces. Some of the typical hydrogen bond lengths (between the bridging atoms) and dissociation energies are shown in Table 2.3.

Association of polar molecules in liquids (such as water, alcohol and hydrogen-fluoride), dimerization of simple fatty acids and important

Table 2.3

Some typical hydrogen bond lengths and
bond energies

Bond	Bond length nm	Dissociation energy kJ/mole
C—H—N	—	13
O—H—N	0.28	—
O—H—O	0.26—0.28	13—26
O—H—Cl	0.31	—
N—H—N	0.31	13—21
N—H—O	0.29—0.30	17
N—H—Cl	0.32	—
N—H—F	0.28	—
F—H—F	0.24	30

After Billmeyer 1984.

structural characteristics of polar polymers (as, for example, polyamides, cellulose and proteins) can be explained by the existence of hydrogen bonds.

Secondary forces usually do not influence significantly the formation of stable chemical compounds, rather, they lead to the formation of agglomerates of discrete molecules in the liquid or solid state. A number of physical properties (such as volatility, viscosity, surface tension, frictional properties, miscibility or solubility), can, however, be mainly explained by intermolecular forces.

Cohesion energy is by definition that amount of energy needed to part a molecule from its neighbourhood in the liquid or solid state and take it to infinite distance. This is close to the heat of evaporation or sublimation at a constant volume and its value can be estimated from thermodynamic data. Cohesion energy per unit volume—also called specific cohesion energy or cohesion energy density (CED)—is characteristic of the particular material. Some relevant data are shown in Table 2.4, reflecting the effect of chemical structure on this property.

2.1.2. Dissolution, miscibility. The relationship between volatility and molecular mass

The tendency for evaporation in liquids is a function of the total kinetic energy, and therefore highly depends on temperature. Boiling point depends on the ratio of cohesion energy to translational kinetic energy, so that, within a homologous series, it is a function of molecular mass. In substances with high molecular mass, cohesion energy can exceed that of primary bonds, and the molecule decomposes before evaporation. This point is reached by polymers far below the usual molecular mass range.

Melting point is also a function of cohesion energy but here other important factors must be taken into account. Such as the effect of molecular order or entropy. It is known from thermodynamics that chemical changes occur spontaneously, only if free enthalpy (the Gibbs-potential) decreases. This can happen if, during the configurational change, ΔH is overcompensated by the $T \cdot \Delta S$ term.

In general, a material with a high melting point also has a high boiling point, but this interrelation is not strict. In the case of symmetric molecules having low fusion entropy, the melting point is lower than when molecules with similar molecular mass exhibit a lower degree of symmetry.

2.1.2.1. Effect of polarity

Molecules having strongly polar groups attract their neighbours significantly. This is reflected in higher boiling and melting points, and other properties of the material also prove the presence of high cohesion energy density. Miscibility and solubility are also properties determined by intermolecular interactions. Solution or mixing energy (which can be either positive or negative) is the difference between the cohesion energy of the mixture and that of the pure components. A highly negative heat of solution promotes dissolution, whereas positive mixing energies decrease solubility. In the case of a positive mixing energy, the enthalpy of the system increases, and the reaction is endothermic, heat is consumed; whereas in the case of a negative mixing energy ($\Delta H < 0$), enthalpy decreases, and the process is exothermic, heat evolves. Limited solubility appears in the case of endothermic dissolution only, since $\Delta G = 0$, then:

$$H = T \Delta S \tag{2.10}$$

and since $\Delta S > 0$, then ΔH must be positive, so that heat is consumed during mixing. The miscibility of exothermic solutions is usually unlimited.

2.1.3. Types of polymer and intermolecular force

Some general conclusions can be drawn from the data shown in Table 2.4. Properties of polymers highly depend on secondary intermolecular interaction energies. If intermolecular attractive forces are weak so that the

Table 2.4

Cohesion energy density of linear polymers

Polymer	Repeat unit	Cohesion energy density, J/cm^3
Polyethylene	$-CH_2-CH_2-$	260
Polyisobutylene	$-CH_2-C(CH_3)_2-$	275
Polyisoprene	$-CH_2-C(CH_3)=CHCH_2-$	283
Polystyrene	$-CH_2CH(C_6H_5)-$	312
Poly(methyl methacrylate)	$-CH_2C(CH_3)(COOCH_3)-$	350
Poly(vinyl acetate)	$-CH_2CH(OCOCH_3)-$	370
Poly(vinyl chloride)	$-CH_2CHCl-$	385
Poly(ethylene terephthalate)	$-CH_2CH_2OCOC_6H_4COO-$	470
Poly(hexamethylene adipamide)	$-NH(CH_2)_6NHCO(CH_2)_4CO-$	770
Poly(acrylonitrile)	$-CH_2CHCN-$	1,000

After Billmeyer 1984.

cohesion energy is low, and molecular chains are flexible enough, the material can be easily distorted under the action of external forces. Materials with these properties are known as elastomers. If cohesion energy density is somewhat higher, the chain becomes more rigid—these are typical thermoplasts. If cohesion energy is high, the material becomes resistant to shear and tension which is characteristic of fibres.

Polyethylene stands out in the Table with its relatively low cohesion energy, but, because of its symmetrical chain structure, it belongs to the thermoplastic group.

2.1.4. Polymer solubility criteria. Solubility parameter

Dissolution of polymers is a two-step process; first the polymer swells, and the swollen polymer then dissolves. This second step can be promoted by stirring. It there are too many cross-links or hydrogen bonds in the polymer, which cannot be broken by the solvent, the process stops in the swelling stage.

Some empirical rules of polymer solubility became known even before the advance of theoretical studies.

1. The "similar dissolves similar" principle is applicable to polymers as well. It is frequently true that the polymer dissolves best in its own monomer or in a structurally similar solvent.

2. Increasing molecular mass decreases solubility. This is the principle of solution fractionation.

3. With increasing melting points solubility decreases. As a result, crystalline polymers dissolve to a lesser extent and they frequently become soluble only close to their melting point.

2.1.4.1. Solubility parameter

It has already been mentioned that dissolution of a material is possible only if the free enthalpy of the solution decreases with respect to the sum of the pure components. It has also been established that if the polymer–solvent interaction is strong, and ΔH negative, dissolution is always possible. If there are only dispersion interactions between the components ΔH is usually positive and its magnitude decides whether the material is soluble or not.

According to Hildebrand and Scott (1950), in such cases the mixing heat per unit volume can be expressed as:

$$\Delta H = v_1 v_2 (\delta_1 - \delta_2)^2 \tag{2.11}$$

where v is the volume fraction, indices 1 and 2 refer to the solvent and the polymer respectively, and δ is the solubility parameter, which is the square root of the cohesion energy density (CED).

In the case of low molecular mass materials, CED is the evaporation heat per unit volume at a given temperature, but for polymers, where the molecular chain cannot be evaporated, determination of CED is somewhat more difficult. According to the method of Gee (1946), swelling of slightly crosslinked polymers was studied in different solvents. Eq. (2.11) shows a minimum if δ_1 and δ_2 are close to each other, then ΔG is maximized. Following this train of thought swelling of the network may be compared in different solvents, and it is assumed that the CED value of the polymer is equal to that of the solvent which causes maximum swelling.

Small (1953) developed a method to calculate CED for polymers from group increments and—similarly to parachor—liquid data are used to establish these increments and to calculate the cohesion energy density of polymers.

In the history of polymer chemistry solubility of poly-acrylonitrile has proven to be an interesting problem. This polymer has been regarded as cross-linked for a long time, since it could not be dissolved in any known solvent at that time. The polymer itself has been known since 1925, moreover it has been utilized by industry in the production of oil-resistant acrylonitrile–butadiene rubbers, but its solvent was found by Rein one and a half decades later. So

Table 2.5

Solubility parameters of some solvents and polymers

Solvent	$(J/cm^3)^{1/2}$ δ_1	Polymer	$(J/cm^3)^{1/2}$ δ_2
n-Hexane	14.80	Polyethylene	16.2
Decalin	16.00	Polystyrene	18.9
Cyclohexane	17.00	Poly(methyl methacrylate)	18.6
Carbon tetrachloride	17.60	Poly(vinyl chloride)	19.45
		Poly(ethylene terephthalate)	21.9
2-Butanone	18.52	Nylon 6/6	27.8
Benzene	18.75	Poly(acrylo nitrile)	26.3
Chloroform	18.90		
Tetrahydrofuran	19.45		
Acetone	20.00		
Methanol	29.70		
Cyclohexanone	32.80		
Dimethyl formamide	25.00		

After Billmeyer 1984.

fibres were produced from this material in Germany as late as 1943, and the first pilot plant was built in 1945 in the USA.

Solubility of poly-acrylonitriles has been analyzed by Walker (1952) in terms of cohesion energy density and polarity. As shown in Table 2.5 the solubility parameter of poly-acrylonitrile is 26.3 $(J/cm^3)^{1/2}$. From the solvents mentioned in Table 2.5, the solubility parameter of dimethyl formamide stands closest to this value (25.0 $(J/cm^3)^{1/2}$) and in fact it is a good solvent for poly-acrylonitrile, but other solvents or solvent mixtures with slightly different solubility parameters cannot dissolve this polymer.

Solubility parameter values shown in Table 2.5. are the square roots of cohesion energy density (CED) given in Table 2.4. (In earlier Tables these values are given in $(cal/cm^3)^{1/2}$ units.)

Such tables frequently help in solving solubility problems. It is generally true that if there is no special chemical interaction between the solvent and the polymer, and their solubility parameters are far apart, then the polymer will not dissolve in the solvent. The reserve is not always true; even if the difference is minimal, the polymer will not necessarily dissolve. (There are polymers which do not dissolve in their own monomers, so during bulk polymerization they will precipitate.)

Sometimes a polymer can be dissolved in mixed solvents if the solubility parameter of the mixture approximates that of the polymer. Polystyrene ($\delta = 18.9$) e.g. dissolves in a mixture of acetone (20) and cyclohexane (17) but does not dissolve in either of the pure solvents.

These considerations cannot be applied to crystalline polymers, polar polymers in apolar solvents, or to apolar polymers in polar solvents, as solubility parameters are calculated from dispersion forces only.

There is no direct relation between the rate of dissolution and the principles presented above, since even if dissolution is possible from a thermodynamic point of view, diffusion of the solvent into the polymer influences the rate of dissolution. Solvents with smaller, more mobile molecules dissolve the polymer faster even if they are not very good solvents

Table 2.6

Most important solvents for some polymers

Polymer	Solvent
Poly(acrylonitrile)	Dimethyl formamide, nitro phenol, ethylene carbonate
Polyethylene	Aromatic and chlorinated solvents, above 70 °C
Poly(methyl methacrylate)	Esters, chlorinated and aromatic hydrocarbons
Polystyrene	Aromatic and chlorinated hydrocarbons
poly(tetrafluoro ethylene)	Not known
Poly(vinyl acetate)	Aromatic and chlorinated hydrocarbons, low molecular mass esters, methanol
Poly(vinyl alcohol)	Water (if the polymer is fully hydrolized)
Poly(vinyl chloride)	Tetrahydrofuran, toluene, dioxane (at lower molecular mass)
Poly(vinylidene chloride)	Tetrahydrofuran (warm), dioxane (warm, at lower molecular mass)
Natural rubber	Hydrocarbons
Butadiene–styrene copolymer (G.R.S.)	Chlorinated and nitrated hydrocarbons
Polychloroprene (Neoprene)	Chlorinated hydrocarbons, phenol, formic acid, trifluoro ethanol
Poly(ethylene terephthalate)	Benzyl alcohol (warm), nitrobenzene (warm)
Linear polyesters (alkyd resins)	Hydrocarbons, chlorinated hydrocarbons, low molecular mass alcohols, ketones, esters
Polypropylene	Above 135 °C decaline, tetraline, paraffins
Cellulose nitrate	Low molecular mass alcohols, ketones, esters
Cellulose acetate	Ketones, low molecular mass esters

After Billmeyer 1984.

thermodynamically. To achieve fast and effective dissolution, a mixture of both thermodynamically and kinetically good solvents is frequently used.

The more important solvents of some polymers are listed in Table 2.6.

2.1.5. Conformation of macromolecular chains in solution

As mentioned previously various molecular arrangements resulting from rotations around single bonds of the polymeric chain are called conformations. In solution the polymer forms a random coil. The effective volume of this coil far exceeds many times that of the polymeric segments because of the great variety of different conformations. The density of segments in a typical polymer solution is around 0.01 g/cm^3. The size of the random coil is highly influenced by the polymer–solvent interactions. In thermodynamically good solvents, where there is a strong interaction between the polymer and the solvent, the coil is extended. In poor solvents the reserve is true.

In the following section the conformational properties of macromolecules in ideal and real solutions will be treated.

2.1.5.1. The random coil

Fully extended chains can never be found in practice in solutions of polymers. Chains are always coiled to some extent, and this is characterized by the root-mean-square of the chain-end distances, the so-called average chain-end distance:

$$\bar{H} = (\bar{h^2})^{0.5}. \tag{2.12}$$

Another possible characteristic feature is the radius of gyration, which is the root-mean-square of distances between the mass centre of the molecule and the segments:

$$\bar{S} = \overline{(s^2)}^{0.5}. \tag{2.13}$$

This value is used to characterize the degree of intramolecular twisting, and can be calculated from light scattering or gel permeation chromatographic measurements. It can also be obtained from small angle X-ray scattering data and is frequently referred to as the "scattering mass radius".

In the case of linear polymers with a random distribution of end-to-end distances there is a very simple relationship between these two data:

$$\bar{h}^2 = 6\bar{s}^2$$

or

$$(\bar{H})^2 = 6(\bar{S})^2. \tag{2.14}$$

The simplest model of a polymer chain consists of x linear elements (with length a) joined without any restriction on the angle between the consecutive bonds. The probability (W) that this system will exhibit a given end-to-end distance, r, can be determined using classical probability theory. In the calculations it is assumed that the polymer consists of a very high number of point-like (i.e. extensionless) segments with a distance of x. The mass points can be placed on any point of a spherical surface with radius x around each other.

The result of this calculation is that the average end-to-end distance is proportional to the square root of the number of segments:

$$\bar{H}_0 = (\bar{h}_0^2)^{0.5} = ax^{0.5} \tag{2.15}$$

where the subscript 0 refers to the unperturbed molecules without interactions. Since the number of segments is in direct proportion to the molecular mass, the average end-to-end distance will be proportional to $M^{0.5}$. The expression:

$$\bar{h}_0^2/M \quad \text{or} \quad (\bar{H}_0)^2/M \tag{2.16}$$

is characteristic of the chain structure of the polymer which, in turn, is independent of the molecular mass and environmental factors (e.g. solvent) as well.

Spatial distribution of end-to-end distances can be described by the Gaussian distribution function shown in Fig. 2.2. According to the diagram, the probability of finding the other end at a given distance from the first segment decreases with the distance, the most probable situation is therefore when the two ends of the same chain are close to each other.

The probability of the occurrence of the other chain end within a spherical shell with a thickness of dr at a distance r relative to the first segment can be described by the function shown in Fig. 2.3. This exhibits a maximum at a non-zero radius, which can be explained by the fact that $\bar{H} = (\bar{h}_0^2)^{0.5}$ is also not zero. This Gaussian distribution function gives non-zero values even at distances longer than the total length of the straight chain without folding, that is, it does not describe the realistic situation. In some cases, however, where more realistic functions are available (e.g. for the description of rubber elasticity) much better approximations are available.

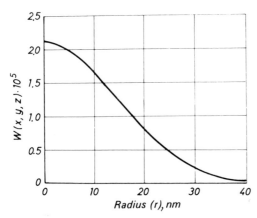

Fig. 2.2. Probability distribution of chain end-to-end distances in a polymer consisting of 10^4 segment of length 0.25 nm. $W(xyz)$ is given in nm^{-3}; r in nm units.

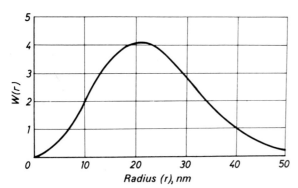

Fig. 2.3. Radial distribution of end-to-end distances in the same polymer shown in the previous Figure.

End-to-end distances calculated by finer methods (e.g. by the restriction of segmental angles) give similar results. These calculations are complicated by the determination of rotational hindrance. Actual molecular dimensions of polymers can be best determined experimentally.

2.1.5.2. Non-linear chains

Branched chains exhibit more chain-ends than their linear counterparts, consequently the radius of gyration is more adequate for their characterization. Assuming equal numbers of segments the volume of the branched molecule is smaller than that of the linear one. This volume contraction is usually described by the so-called g-factor:

$$g = \bar{s}^2 \text{(branched)}/\bar{s}^2 \text{(linear)}. \tag{2.17}$$

The value of g can be calculated by statistical methods for various types, and degrees, of branching. In the case of random branching, e.g. five nodes per molecule with three functional branching decreases the g value to 0.70. By measuring the contraction of the radius of gyration the number of branches can be determined.

2.1.5.3. Excluded volume

A certain portion of the random conformations does not exist, since if one segment is present at a particular point, it excludes other segments from that volume occupied by the first. Calculations under these conditions cannot yet be performed exactly.

Because of these interactions the average end-to-end distance will exceed the value indicated by calculations. The ratio of real and ideal (unperturbed) molecular dimensions has been named the Flory expansion coefficient:

$$(\bar{h}^2)^{0.5} = \alpha(\bar{h}_0^2)^{0.5} \tag{2.18}$$

or

$$\bar{H} = \alpha \bar{H}_0.$$

2.2. THERMODYNAMICS OF POLYMER SOLUTIONS

Studies of macromolecular solutions yield data concerning the shape and size of macromolecules. First we shall deal with phenomena related to the equilibrium of two phases: one is the pure solvent, and the polymer solution.

If these two phases are in equilibrium, the free enthalpy change associated with dilution can be given as:

$$\Delta G_1 = kT \ln [p_1/p_1^0] \tag{2.19}$$

where ΔG_1 is the free enthalpy change of the solvent dilution which evolves if a single molecule of the pure solvent (vapour pressure p_1^0) is added to an infinite amount of solution (vapour pressure p_1), k the Boltzmann-constant, and T the absolute temperature.

This expression describes all of the so-called colligative properties, such as vapour pressure depression, freezing point depression, boiling point elevation and osmotic pressure. All these properties cannot, however, be calculated a priori using thermodynamic principles alone; at least one has to be measured as a function of concentration, and the rest may then be calculated.

(i) *Ideal solutions*. The simplest case of mixing is the admixture of components 1 and 2 with molecules of similar size and force field. In this case the vapour pressure of both components will be proportional to the mole fraction of the given component, in accordance with the Raoult law. Such that:

$$p_1 = p_1^0 \left[\frac{N_1}{N_1 + N_2} \right] = p_1^0 Y_1 \tag{2.20}$$

where N_1 and N_2 are the number of molecules and Y_1 is the molar fraction of component 1. The free enthalpy change per molecule in this case is given by:

$$\Delta G_1 = kT \ln Y_1 . \tag{2.21}$$

The total free enthalpy difference of mixing is as follows:

$$\Delta G = N_1 \Delta G_1 + N_2 \Delta G_2,$$

$$\Delta G = kT(N_1 \ln Y_1 + N_2 \ln Y_2). \tag{2.22}$$

In the case of ideal mixing no energetic change occurs during mixing, that is, $\Delta H = 0$. Since

$$\Delta G = \Delta H - T \Delta S, \tag{2.9}$$

it follows that:

$$\Delta S = - k(N_1 \ln Y_1 + N_2 \ln Y_2). \tag{2.23}$$

As this expression is positive for all compositions, the second law of thermodynamics suggests that spontaneous mixing should occur at any molar ratio.

(ii) *Real solutions*. In practice very few liquids can be described by the Raoult law exactly. Three main types of liquid mixture may be distinguished:

1. *athermic solutions* where $\Delta H = 0$, but entropy cannot be calculated from Eq. (2.23).

2. *regular solutions* where ΔS is ideal, but ΔH is finite.

3. *irregular solutions* where both ΔH and ΔS deviate from the ideal values.

In practice it has been found that for molecules of similar size ΔS is nearly ideal if $\Delta H = 0$, therefore athermic solutions are nearly ideal.

If force fields around the molecules of the components are dissimilar, ΔH is finite, and has to be calculated from the cohesion energy density values mentioned earlier.

2.2.1. Entropy and heat of mixing in polymer solutions

Polymer solutions exhibit extraordinarily high deviations from the Raoult law and approximate ideal behaviour only at extremely low concentrations. If the concentration of the solution exceeds a few per cent, the deviation from ideal behaviour becomes so great that the Raoult law cannot be used even to estimate the thermodynamic properties of polymer solutions (Fig. 2.4).

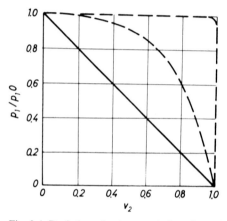

Fig. 2.4. Deviation of polymer solutions from the ideal. (After Gee and Treloar 1942.)

In this Figure activity (p_1/p_1^0) is plotted versus v_2, the volume fraction of the polymer. The upper curve is a theoretical one for the benzene-natural rubber system; the central curve represents experimental points, whereas the lower curve has been calculated using the approximation that the activity of the solvent is equal to its volume fraction (after Gee-Treloar).

2.2.2. Entropy of mixing. The Flory–Huggins theory

Deviation of polymer solutions from ideal behaviour is mainly due to low mixing entropy. This low mixing entropy is, in turn, the consequence of the range of difference in molecular dimensions between polymer and solvent.

To describe this situation Flory (1942), and Huggins (1942), developed theories independently. Since these theories differ from each other only slightly, they are usually referred to as the Flory–Huggins theory.

As a model experiment, let us assume two liquids with molecules of similar size. Their mixture can be represented in two dimensions by a rectangular network with the molecules of the two components in the nodes (Fig. 2.5). If there were only one pure material present, a single configuration of the network would exist, since the exchange of two identical molecules does not lead to a new configuration.

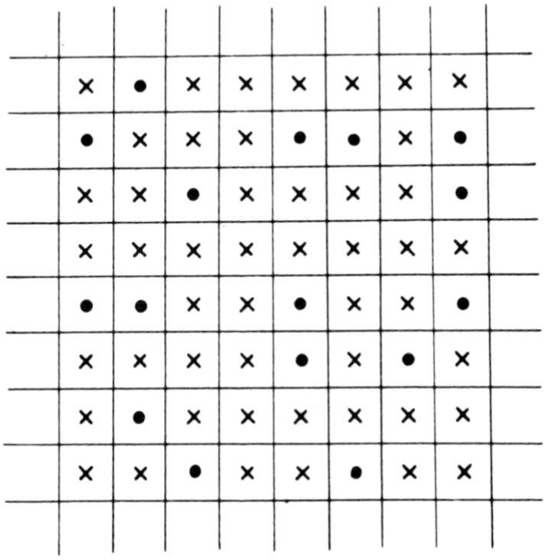

Fig. 2.5. Two-dimensional lattice model for the distribution of the components in a two-component non-polymeric mixture. (After Billmeyer 1984.)

Nevertheless, if two types of molecule are present, a wide variety of configurations is possible.

Let us assume that we have N_1 and N_2 molecules of components 1 and 2 respectively. Let us also assume a model network with $N_1 + N_2$ nodes. Any of the molecules can occupy any of the nodal points with equal probability. If the molecules can occupy the nodes in W different ways then the thermodynamic probability of the system, from which the entropy can be calculated, is equal to W. If the $N_1 + N_2$ molecules are distributed over the network, the first place can be filled with any of the molecules, that is, it can be filled in $N_1 + N_2$ different ways. The second node can be filled in $N_1 + N_2 - 1$ ways and so on so that the nodes can be filled in $(N_1 + N_2)!$ different ways. One has to take into account, however, that the exchange of molecules of component 1 does not give a new configuration, and, similarly, the molecules of component 2 are freely interchangable, thus the number of different configurations can be given as:

$$W = \frac{(N_1 + N_2)!}{N_1! N_2!}.$$ (2.24)

Before mixing the pure liquids can fill the network in a single way, so the thermodynamic probability of the pure components is equal to 1.

Since according to the Boltzmann formula, the entropy of mixing can be calculated as:

$$\Delta S = k [\ln W_{\text{final}} - \ln W_{\text{initial}}]$$ (2.25)

the entropy difference can be determined.

The logarithm of the initial thermodynamic probability is equal to zero. In the calculation of final probability, N_1 and N_2 are large numbers so the Stirling-approximation can be used:

$$\ln N! = N \ln N - N$$ (2.26)

and so

$$\Delta S = -k \left(N_1 \ln \frac{N_1}{N_1 + N_2} + N_2 \ln \frac{N_2}{N_1 + N_2} \right).$$ (2.27)

where $N_1/(N_1 + N_2)$ is the mole fraction of component 1, denoted as Y_1, whereas the mole fraction of the other component is Y_2. Such that:

$$\Delta S = -k(N_1 \ln Y_1 + N_2 \ln Y_2).$$ (2.23)

In the case of polymers, one of the mixing components consists of chain molecules with x (a large number) segments. All of these segments occupy one place in the aforementioned network. The arrangement is shown in Fig. 2.6.

The value of x is equal to the number of segments (monomeric units) of the polymer which are assumed to have the same volume as the solvent molecules themselves. In other words x is equal to the ratio of the volumes occupied by the polymer molecule and the solvent molecule (that is, the ratio of the molar volumes):

$$x = \bar{v}_2 / \bar{v}_1.$$ (2.28)

One can see from Fig. 2.6 that qualitatively the entropy of mixing of the polymer solution must be much lower than that of normal solutions, since the number of different occupational configurations is much lower. If a macromolecular segment is fixed at a given node, the neighbouring segment can occupy only certain nodes, and the position of the next segment is also limited and so on. To cope with this situation a method has been developed

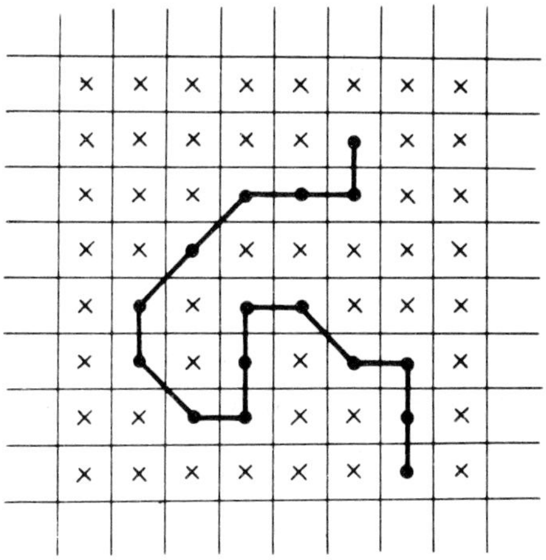

Fig. 2.6. Two-dimensional lattice model for the distribution of polymer segments in solution. (After Billmeyer 1984.)

by Flory and Huggins to calculate the mixing entropy of polymer solutions with the following result:

$$\Delta S = -k(N_1 \ln v_1 + N_2 \ln v_2) \tag{2.29}$$

where indices 1 and 2 refer to the solvent and polymer respectively, and v_1 and v_2 are the volume fractions defined as:

$$v_1 = N_1/(N_1 + xN_2), \tag{2.30}$$

$$v_2 = xN_2/(N_1 + xN_2). \tag{2.31}$$

2.2.3. Heat of mixing and free enthalpy of polymer solutions

As we have seen in the study of conditions for polymer dissolution, the mixing enthalpy (ΔH) can be described by the Hildebrand–Scott formula:

$$\Delta H = v_1 v_2 (\delta_1 - \delta_2)^2. \tag{2.11}$$

Flory rearranged this equation into a form similar to the van Laar equation applied to the mixing heat of two-component systems. Let us introduce the following notation:

$$k\,T\chi_1 = \frac{(\delta_1 - \delta_2)^2}{N_1/v_1} \tag{2.32}$$

where N_1/v_1 is the number of molecules in a unit volume of the solvent; $k\,T\chi_1$ the energy difference evolved when a single solvent molecule is placed in a large volume of pure polymer ($v_2 = 1$).

Putting this expression into the mixing enthalpy equation (Eq. (2.11)) we obtain the following result:

$$\Delta H = \chi_1 k\,T N_1 v_2 \tag{2.33}$$

where χ_1 denotes the interaction energy per solvent molecule divided by $k\,T$, and is known as the Flory–Huggins polymer–solvent interaction parameter.

Taking into account the Flory–Huggins mixing entropy (Eq. (2.29)) and using the mixing enthalpy according to Eq. (2.33) the free enthalpy of mixing of the polymer solution can be calculated from Eq. (2.9) as:

$$\Delta G = k\,T(N_1 \ln v_1 + N_2 \ln v_2 + \chi_1 N_1 v_2). \tag{2.34}$$

This equation is very important, since others can be derived from it which are related to experimentally measurable quantities. The equation can also be expressed using the molar quantities of the components ($n = N/N_{Av}$):

$$\Delta G = R\,T(n_1 \ln v_1 + n_2 \ln v_2 + \chi_1 n_1 v_2). \tag{2.35}$$

In the course of this derivation it has been assumed that the configurational mixing entropy is equal to the total mixing entropy. It has to be noted, however, that entropic effects due to the possible orientation of the components, or to intermolecular interactions, have been neglected. Because of this, in future sections the entropy change originating from configurational changes only will be denoted as ΔS^* (Flory 1953) in order to distinguish it from the experimentally measurable total entropy change, ΔS.

2.2.4. Partial molar quantities

It is well known from previous work (Guggenheim 1949) that the chemical potential of a given component is equal to the partial differentiation of the free enthalpy with respect to the number of moles of the given component. Partial molar quantities are denoted by a bar over the quantity. Chemical potential difference between the solvent and solution can be obtained from Eq. (2.34) by differentiating with respect to N_1. It has to be taken into account, however, that v_1 and v_2 are themselves a function of N_1. The result is:

$$\Delta \bar{G}_1 = kT\left[\ln(1-v_2)+\left(1-\frac{1}{x}\right)v_2+\chi_1 v_2^2\right]. \tag{2.36}$$

If Eq. (2.36) is multiplied by the Avogadro number the chemical potential per mole is obtained:

$$\mu_1 - \mu_1^0 = RT\left[\ln(1-v_2)+\left(1-\frac{1}{x}\right)v_2+\chi_1 v_2^2\right]. \tag{2.37}$$

Eq. (2.37) thus gives the free enthalpy difference if an athermal polymer solution is made from an amorphous polymer.

If the entropy given by Eq. (2.29) is multiplied by the Avogadro number and differentiated with respect to N_1, the molar configurational entropy of the solvent is obtained:

$$\overline{\Delta S_1^*} = -R\left[\ln(1-v_2)+\left(1-\frac{1}{x}\right)v_2\right]. \tag{2.38}$$

By comparing Eqs (2.37) and (2.38), the difference in the chemical potential can be expressed as:

$$\mu_1 - \mu_1^0 = -T\overline{\Delta S_1^*} + RT\chi_1 v_2^2 \tag{2.39}$$

where the final term is the relative partial molar enthalpy difference, or heat of dilution:

$$\overline{\Delta H_1} = RT\chi_1 v_2^2. \tag{2.40}$$

From this equation a useful formula can be derived for the osmotic pressure. There is a well-known relationship between osmotic pressure and chemical potential difference:

$$\pi \bar{v}_1 = -(\mu_1 - \mu_1^0) \tag{2.41}$$

where π is the osmotic pressure, and $\bar{v}_1$ the molar volume of the solvent. And so:

$$\pi = -\frac{RT}{\bar{v}_1}\left[\ln(1-v_2)+\left(1-\frac{1}{x}\right)v_2+\chi_1 v_2^2\right]. \tag{2.42}$$

Expanding the logarithmic expression into a Taylor series:

$$\ln(1-a) = -[a + a^2/2 + a^3/3 + \ldots], \tag{2.43}$$

$$\mu_1 - \mu_1^0 = -RT\left[v_2/x + \left[\frac{1}{2} - \chi_1\right]v_2^2 + v_2^3/3 + \ldots\right], \tag{2.44}$$

$$\pi = \frac{RT}{\bar{v}_1}\left[v_2/x + [1/2 - \chi_1]v_2^2 + v_2^3/3 + \ldots\right]. \tag{2.45}$$

This expression can be simplified if the solution concentration, c, is introduced. If we denote the specific volume of the polymer by $\boxed{v_2}$, and, since from Eq. (2.31) the volume fraction of the polymer is v_2. And so:

$$v_2 = c\boxed{v_2}. \tag{2.46}$$

The ratio of the molar volumes of the polymer and the solvent is equal to x according to Eq. (2.28):

$$x = \bar{v}_2/\bar{v}_1. \tag{2.28}$$

The ratio of the molar volume, $\bar{v}_2$ and the specific volume is equal to the molecular mass of the polymer:

$$\frac{\bar{v}_2}{\boxed{v_2}} = M\left[\frac{\text{ml/mole}}{\text{ml/g}} = \frac{\text{g}}{\text{mole}}\right]. \tag{2.47}$$

Using this notation:

$$\frac{v_2}{\bar{v}_1 x} = \frac{v_2}{(\bar{v}_2/\bar{v}_1)\bar{v}_1} = \frac{c\boxed{v_2}}{\bar{v}_2} = \frac{c}{M}, \tag{2.48}$$

and introducing into Eq. (2.45):

$$\pi = \frac{RTc}{M} + RT\left[\frac{\boxed{v_2}}{\bar{v}_1}^2(1/2 - \chi_1)c^2 + (\boxed{v_2}^3)\frac{c^3}{3\bar{v}_1} + \ldots\right]. \tag{2.49}$$

The first term of the right hand side of the equation is the ideal or van't Hoff formula, which is valid for infinite dilution. Higher terms indicate deviations according to the previous theory. The coefficients are called virial coefficients. These play an important role in experimental techniques, and, if their values are known experimentally, important conclusions can be drawn regarding the thermodynamic state of the polymer–solvent system.

2.2.5. Experimental results in determining the entropy of mixing in polymer solutions

The Flory–Huggins theory has been checked—among others—using the rubber–benzene system (Gee 1946, 1947). The free enthalpy change of the system has been determined from vapour pressure measurements, and the heat of dissolution has been measured by calorimetry. The concentration dependence of the latter has been described previously as:

$$\Delta \bar{H}_1 = R T \chi_1 v_2^2. \tag{2.40}$$

The result was that the entropy of mixing calculated from the enthalpy of mixing and the heat of dissolution was in good agreement with theoretical values for very low concentrations (Fig. 2.7).

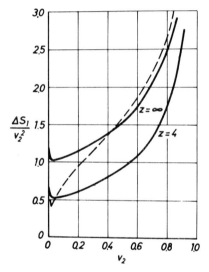

Fig. 2.7. The dependence of mixing entropy on the volume fraction of the polymer component. The dashed line indicates experimental results, full lines are theoretical curves calculated with different χ_1 values. Coordination numbers are 4 and infinity respectively. (After Gee 1946, 1947.)

The concentration dependence of partial molar dilution heats predicted by Eq. (2.40) could not be verified experimentally. This is shown in Fig. 2.8, where $\Delta \bar{H}_1/v_2^2$ is plotted versus v_2.

Other systems studied also show considerable deviations from this theory. For example, according to Eq. (2.40) χ_1 should be independent of v_2. However, Fig. 2.9. shows experimental values determined for several systems.

It is clear that, in practice, the Flory–Huggins interaction parameter varies as a function of v_2, the volume fraction of the polymer.

In spite of all these discrepancies the Flory–Huggins theory was a great step forward compared to the classical theory of ideal solutions in understanding the behaviour of polymer solutions. It has highlighted the principal differences between the thermodynamic treatment of ideal and polymer solutions.

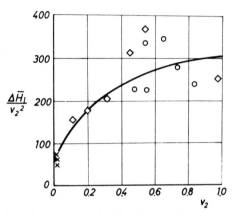

Fig. 2.8. Heat of dilution plotted versus the volume fraction of the polymer. Experimental data are taken from a natural rubber-benzene system. (After Gee 1946, 1947.)

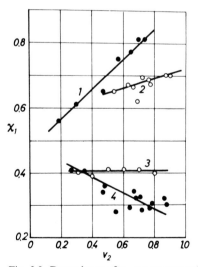

Fig. 2.9. Dependence of χ_1 on concentration. Curve *1* poly (dimethyl siloxane) in benzene (Newing 1950); Curve *2* polystyrene in benzene (Bawn 1950); Curve *3* natural rubber in benzene (Gee 1946, 1947); Curve *4* polystyrene in toluene. (Bawn 1950.)

2.2.6. Dilute solutions. The Flory–Kriegbaum theory

The network model applied to the Flory–Huggins theory has neglected the fact that, in highly dilute solutions, the polymer molecules are arranged so that there are polymer-free solvent regions, and other regions where the polymer molecules occur almost exclusively like clouds. Flory and Kriegbaum (1950) developed a model in which isolated segment clouds are assumed at great distance from each other with solvent in between. These clouds are nearly spherical, the segment-density reaches its maximum in the centre, and decreases monotonically, according to a Gaussian distribution function, with increasing distance from the centre. If a molecule is present in certain segments, all other molecules are excluded from them. This excluded volume effect, mentioned above, results in interactions among the molecules that can be characterized thermodynamically. The Flory–Kriegbaum equation describes these interactions by deviation from ideal solutions. Because of these interactions an excess chemical potential appears, which was investigated in detail by Flory and Kriegbaum (1950). The steps of the derivation will not be dealt with here, but the excess chemical potential can be obtained from Eq. (2.38) if x is chosen to be infinite:

$$(\mu_1 - \mu_1^0)_E = R\,T[\ln(1 - v_2) + v_2 + \chi_1 v_2^2] \tag{2.50}$$

where subscript E refers to the excess chemical potential.

By performing the series expansion the following expression results:

$$(\mu_1 - \mu_1^0)_E = -R\,T\left[\left[\frac{1}{2} - \chi_1\right]v_2^2 + v_2^3/3 + \dots\right] \tag{2.51}$$

which contains all terms appearing in Eq. (2.44) with the exception of the ideal term, and is therefore frequently referred to as the excess chemical potential.

If the polymer concentration in the given unit volume is low, it is sufficient to take into account the quadratic term only, terms of higher order being neglected.

Independent of these considerations the following general expression can be written for dilute systems by neglecting higher order terms:

$$(\mu_1 - \mu_1^0)_E = R\,T(\varkappa_1 - \psi_1)v_2^2, \tag{2.52}$$

where $\varkappa_1$ and ψ_1 are heat- and entropy parameters, and so:

$$\Delta \bar{H}_1 = R\,T\varkappa_1 v_2^2 \tag{2.53}$$

and

$$\Delta \bar{S}_1 = R\psi_1 v_2^2. \tag{2.54}$$

Within the limits of the idealized model the χ_1 parameter of Eq. (2.51) can be related to the parameters of Eq. (2.52) by comparing the two equations:

$$\varkappa_1 - \psi_1 = \chi_1 - \frac{1}{2}. \tag{2.55}$$

Frequently an ideal Θ-temperature is used which is defined as:

$$\Theta = \varkappa_1 T/\psi_1, \tag{2.56}$$

or

$$\varkappa_1 = \frac{\Theta}{T}\psi_1. \tag{2.56a}$$

By substitution:

$$\psi_1 - \varkappa_1 = \psi_1 \left(1 - \frac{\Theta}{T}\right). \tag{2.57}$$

So the excess chemical potential can be given as:

$$(\mu_1 - \mu_1^0)_E = -RT\psi_1 \left(1 - \frac{\Theta}{T}\right)v_2^2. \tag{2.58}$$

At $T = \Theta$ the chemical potential of segment–solvent interaction is equal to zero, according to Eq. (2.58). Putting it another way, Θ is the temperature where the excess chemical potential and the deviation from ideal behaviour vanishes. In this case, of course, the free energy of inter-segmental interactions is also zero.

The effect of the excluded volume decreases with decreasing temperatures and vanishes at $T = \Theta$. At this point the inter-segmental attraction and the excluded volume effect cancel each other. Below the Θ temperature molecular attraction dominates, and the excluded volume becomes "negative". If the temperature is further decreased the inter-segmental interaction becomes so strong that the polymer precipitates.

The actual conformation of the polymer chains can be regarded as a balance between the expansion forces originating from the excluded volume effect and the contraction forces being due to the less probable conformations of the segments. The free enthalpy term of the latter can be calculated similarly to the rubber elasticity of polymeric systems. The molecular dimensions are compared to the statistical end-to-end distances by the expansion factor:

$$\bar{H} = \alpha \bar{H}_0. \tag{2.18}$$

The elastic part of free enthalpy in this case is equal to:

$$\Delta G_{el} = k T \left[\frac{3}{2} (\alpha^2 - 1) - \ln \alpha^3 \right]. \tag{2.59}$$

2.3. MEASUREMENT OF OSMOTIC PRESSURE

2.3.1. Theory of osmotic pressure determination

The basic principle behind osmotic pressure determination (and of all related, so-called colligative properties) is the Raoult law which states that the vapour pressure of the solution is proportional to that of the solvent and the mole fraction of the solvent molecules:

$$p_1 = p_1^0 Y_1. \tag{2.20}$$

Since with solutions of polymers the mole fraction of the solvent is always less than unity, it follows that the vapour pressure of such solutions is smaller than that of the solvent, and the proportionality factor is the mole-fraction of the solvent.

The basic equation for osmotic pressure has already been derived:

$$\pi = \frac{R T}{M} c + \frac{R T \boxed{v_2}^2}{\bar{v}_1} \left(\frac{1}{2} - \chi_1 \right) c^2 + \frac{R T \boxed{v_2}^3}{3\bar{v}_1} c^3, \tag{2.49}$$

which after rearrangement becomes:

$$\frac{\pi}{R T c} = \frac{1}{M} + \frac{\rho_1}{\rho_2^2 M_1} \left(\frac{1}{2} - \chi_1 \right) c + \frac{\rho_1}{3\rho_2^3 M_1} c^2 \tag{2.50}$$

where ρ_1 and ρ_2 are the densities of the solvent and polymer respectively, and the following relations have been included:

$$\bar{v}_1 = \frac{M_1}{\rho_1}$$

and

$$\boxed{v_2} = \frac{1}{\rho_2}. \tag{2.60}$$

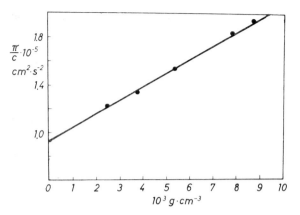

Fig. 2.10. Dependence of osmotic pressure on concentration. Data obtained from polystyrene dissolved in toluene. (After Margerison and East 1967.)

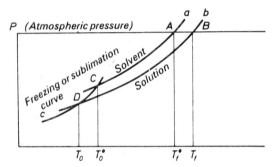

Fig. 2.11. The effect of vapour pressure depression on boiling and freezing points.

By plotting π/c, the osmotic pressure measured at low concentrations versus c, curves as seen in Fig. 2.10. can be obtained. The intercept of the π/c axis at $c = 0$ gives RT/M, whereas the slope gives the second virial coefficient.

In general:

$$Q = Ac + Bc^2 + Cc^3 + \ldots. \tag{2.61}$$

Similar expressions are valid for the partial vapour tension of the solvent in equilibrium with the solution, for vapour pressure depression (Eq. 2.62), which can be converted into temperature difference (ΔT_p, Eq. 2.63), for freezing point depression, and for boiling point elevation (Fig. 2.11) (Eq. 2.64):

$$\Delta p_1 = \frac{p_1^0 \bar{v}_1}{N_{Av}} N_v + B' N_v^2 + C' N_v^3 + \ldots, \tag{2.62}$$

$$\Delta T_p = \frac{R T^2 \bar{v}_1}{\Delta H_v N_{Av}} N_v + B' N_v^2 + C' N_v^3 + \ldots, \tag{2.63}$$

$$\Delta T_f = \frac{R T^2 \bar{v}_1}{\Delta H_f N_{Av}} N_v + B'' N_v^2 + C'' N_v^3 + \ldots, \tag{2.64}$$

$$\Delta T_b = \frac{R T^2 \bar{v}_1}{\Delta H_b N_{Av}} N_v + B''' N_v^2 + C''' N_v^3 + \ldots \tag{2.65}$$

where N_v is the number of molecules per unit volume, N_{Av} the Avogadro number, ΔH_f heat of solidification, ΔH_v the heat of vapourization, ΔH_b the boiling heat in J/g units, A, B and C denote the first, second and third virial coefficients respectively.

The N_v/N_{Av} term in Eq. (2.62, 2.65) can be expressed as:

$$N_v/N_{Av} = c/\bar{M}_n. \tag{2.66}$$

Since the definition of the number average molecular mass is:

$$\bar{M}_n = \frac{\Sigma N_i M_i}{\Sigma N_i} \tag{2.3}$$

and the total number of molecules per unit volume is given by:

$$N_v = \Sigma N_i = \frac{\Sigma N_i M_i}{\bar{M}_n}. \tag{2.67}$$

The concentration may be expressed in g/cm^3 units:

$$c[g/cm^3] = \frac{\Sigma N_i M_i}{N_{Av}} \left[\frac{gmole^{-1} \cdot cm^{-3}}{mole^{-1}} \right] \tag{2.68}$$

and so

$$N_v = \frac{c N_{Av}}{\bar{M}_n}. \tag{2.66}$$

Thus when measuring colligative properties the number average molecular mass results.

In the low concentration limit:

$$\lim_{c \to 0} \frac{\pi}{c} = \frac{R T}{\bar{M}_n}, \tag{2.69}$$

for the vapour pressure depression (in ΔT units):

$$\lim_{c \to 0} \frac{\Delta T_p}{c} = \frac{R T^2}{\rho_1 \Delta H_v} \cdot \frac{M_1}{\bar{M}_n}. \tag{2.70}$$

Similar expression are valid for solidification point depression:

$$\lim_{c \to 0} \frac{\Delta T_f}{c} = - \frac{R T^2}{\rho_1 \Delta H_f} \cdot \frac{M_1}{\bar{M}_n},$$
(2.71)

and for the boiling point elevation:

$$\lim_{c \to 0} \frac{\Delta T_b}{c} = \frac{R T^2}{\rho_1 \Delta H_b} \cdot \frac{M_1}{\bar{M}_n}.$$
(2.72)

The vapour pressure relations determining solidification point depression and boiling point elevation are shown in Fig. 2.11. It can be seen that the original vapour pressure curve (a) intersects the line of atmospheric pressure at point A, while the vapour pressure curve of the solution (b) reaches the atmospheric pressure only at point B. Solidification begins at the intersection of the vapour pressure curves at the sublimation curve (c). This happens at point C in the case of the solvent and at point D in the solution. The difference between curves (a) and (b) is the vapour pressure depression which is used in vapour pressure osmometry—to be treated later—to determine the number average molecular mass.

In order to consider the evaluation possibilities and the measurable quantitative changes let us take a closer look at the property changes caused by a polymer with $\bar{M}_n = 10^4$ and 10^6 respectively at $c = 0.02$ g/cm³, the solvent being benzene (Margerison and East 1967). Calculations are based on the following set of data:

$p_1^0 = 13.265$ kPa (99.5 mmHg) at 25°C

$v_1 = 89.4$ cm³/mole at 25°C

$\dfrac{R T_f^2}{\Delta H_f} = 65.2$ K at 101.325 kPa (this is 1013 mbar = 1 atm),

$\dfrac{R T_b^2}{H_b} = 34.6$ K at the same pressure,

$R T = 2.48 \times 10^3$ J/mole at 25°C.

The results are summarized in Table 2.7. It can be seen that the $\bar{M}_n = 10^4$ value can be measured with all of the methods, but to determine the 10^6 molecular mass only osmotic pressure measurements are applicable, here the pressure difference 49.6 Pa (496 dyne/cm²) corresponds to about 6 mm difference in level of the benzene, and it can be measured reproducibly.

Table 2.7

Colligative property changes calculated by extrapolation

ΔQ	$M_n = 10,000$	$M_n = 1,000,000$
		$c = 2 \cdot 10^{-2}$ g/ml
Vapour pressure depression at 25 °C	2.37 Pa	0.0237 Pa
Freezing point depression at 1 atm	0.0116 K	0.000116 K
Boiling point elevation at 1 atm	0.0062 K	0.000062 K
Osmotic pressure at 25 °C	4.96 kPa	49.6 Pa ($\sim$6 mm benzene)

After Billmeyer 1984.

Osmotic pressure is expressed in Pa according to the SI system of units, but usage of mbar units is also allowed. Earlier these quantities have been expressed in atm or dyne/cm^2 units. In experimental work differences in level are detected in mm.

Let us denote the height of the solution level by h solution (in mm). This can also be expressed in mm of mercury (Hg):

$$h_{\text{mm Hg}} = h_{\text{solution}} \frac{\rho_{\text{solvent}}}{\rho_{\text{Hg}}} \tag{2.73}$$

where h_{Hg} is osmotic pressure expressed in mm of mercury (Torr). More exactly ρ_{solution} should be used instead of ρ_{solvent} but the difference is usually negligible.

The pressure can be calculated in Pascals using the following formula:

$$\pi(\text{Pa}) = \frac{h_{\text{solution}}(\text{mm}) \, \rho_{\text{solvent}} \cdot 101.325}{13.595 \cdot 760} . \tag{2.74}$$

The pressure in atmospheres can be calculated from the following expression (taking into account that 1 atm = 760 Torr = 101.325 Pa):

$$\pi(\text{atm}) = \frac{h_{\text{solution}}(\text{mm}) \, \rho_{\text{solvent}}}{\rho_{\text{Hg}} \cdot 760} = \frac{h_{\text{solution}} \cdot \rho_{\text{solvent}}}{10.330} . \tag{2.75}$$

(The density of mercury is 13.595 g/cm^3.)

Inter-relations between the osmotic pressures expressed in different units can be given as:

$$1 \text{ atm} = 760 \text{ mm Hg} = 76 \times 13.595 \times 980.66 = 1.013 \times 10^6$$

$$\text{dyne/cm}^2 = 1.013 \text{ bar} = 101.325 \text{ Pa} \tag{2.76}$$

$$1 \text{ microbar} = 0.1 \text{ Pa} \tag{2.77}$$

The principle of osmotic pressure determination is outlined in Fig. 2.12. Cell A contains the solvent, B is the membrane, cell C contains the solution, D denotes the two capillary tubes, π is the difference in level, which is the result of the osmotic pressure difference between cells A and C.

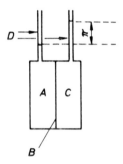

Fig. 2.12. Schematic representation of osmotic pressure measurements. (After Margerison and East 1967.)

Let us demonstrate the method by a practical example. When the osmotic pressure of a polystyrene solution is measured in toluene solvent at 25 °C ($\rho_{\text{toluene}} = 0.8618 \text{ g/cm}^3$, g $= 981 \text{ cm/s}^2$) data summarized in Table 2.8 are

Table 2.8

Result of osmotic pressure measurements on polystyrene

$c \text{ g/cm}^3 \cdot 1{,}000 = \text{kg/m}^3$	2.56	3.80	5.38	7.80	8.68
π cm toluene	0.325	0.545	0.893	1.578	1.856
π Pa	27.5	46.1	75.5	133.4	156.9
$\dfrac{\pi}{c} \text{ m}^2 \cdot \text{s}^{-2} \cdot 10$	10.7	12.1	14.0	17.1	18.1

After Margerison and East 1967.

obtained. Taking into account the first two terms of Eq. (2.49):

$$\frac{\pi}{c} = \frac{RT}{M_n} + Bc. \tag{2.78}$$

Experimental pressure data are shown in Fig. 2.10. The limiting value of osmotic pressure is:

$$\lim_{c \to 0}\left(\frac{\pi}{c}\right) = 0.77 \times 10^5 \, \text{cm}^2 \cdot \text{s}^{-2} = 7.7 \, \text{m}^2 \cdot \text{s}^{-2}.$$

From this the number average molecular mass in SI units is:

$$\bar{M}_n = \frac{RT}{\lim\left(\dfrac{\pi}{c}\right)}\left[\frac{\text{J} \cdot \text{mole}^{-1} \cdot \text{K}^{-1} \cdot \text{K}}{\text{m}^2 \cdot \text{s}^{-2}}\right] = \frac{8.314 \cdot 298.2}{7.7} = 322{,}000 \, \text{g/mole}$$

(the dimensions of the Joule being $\text{kg} \cdot \text{m}^2 \cdot \text{s}^{-2}$). The value of the second virial coefficient can be determined from the slope of the $\pi/c - c$ plot as:

$$B = \frac{12\text{m}^2 \cdot \text{s}^{-2}}{10\text{kg} \cdot \text{m}^{-3}} = 1.2\text{m}^5 \cdot \text{s}^{-2} \cdot \text{kg}^{-1}.$$

From this one can calculate the contribution of the second term to the osmotic pressure. Applying Eq. (2.78) where $c = 8.68 \, \text{kg/m}^3$ concentration:

$$18.1 = 7.7 + 1.2 \times 8.68 = 7.7 + 10.42,$$

that is, the contribution of the second term to the osmotic pressure is far larger than that of the first, "ideal" term.

It has to be noted, that this solution with a concentration close to 1% and exhibiting an excess osmotic pressure nearly 1.5 times higher than the ideal value, is usually very dilute; out of one million molecules only three are polymer.

As virial equations can be written in several different ways, the value and dimensions of the virial coefficients always depend on the actual form of the expression, for example:

$$\frac{\pi}{c} = \frac{RT}{\bar{M}_n} + Bc + Cc^2 + \ldots, \tag{2.79}$$

$$\frac{\pi}{c} = RT\left(\frac{1}{\bar{M}_n} + A_2c + A_3c^2 + \ldots\right), \tag{2.80}$$

$$\frac{\pi}{c} = \left(\frac{\pi}{c}\right)_{c \to 0}\{1 + \Gamma_2c + \Gamma_3c^2 + \ldots\}. \tag{2.81}$$

To sum up, the number average molecular mass of most soluble polymers can be determined by the methods described above. Deviation from ideal behaviour can be characterized by the second virial coefficient. This is influenced by various factors, such as:

    1. the chemical composition of the given polymer–solvent system;
    2. molecular mass and molecular mass distribution of the polymer;
    3. measuring temperature;
    4. stereoregularity of the polymer in question.

The effect of these factors on the second virial coefficients are summarized in Tables 2.9–2.12. Table 2.9 shows the effect of the solvents on the second virial coefficient of poly(methyl methacrylate) solutions. Table 2.10

Table 2.9

Dependence of the second virial coefficient on the type of solvent

Solvent	B, $m^5 \cdot s^{-2} \cdot kg^{-1}$
Chloroform	1.41
Toluene	0.82
Acetone	0.57
Diethyl ketone	0.24
m-Xylene	0.02

    Poly(methyl methacrylate) ($\bar{M}_n = 1.28 \times 10^5$ g/mole = 128 kg/mole) at 27 °C.
    After Margerison and East 1967.

Table 2.10

Dependence of the second virial coefficient on the molecular mass of the polymer

$\bar{M}_m$		B, $m^5 \cdot s^{-2} \cdot kg^{-1}$
g/mole	kg/mole	
20,600	20.6	1.35
39,500	39.5	1.19
80,500	80.5	1.14
130,000	130	0.99
350,000	350	0.77

    Poly(α-methyl styrene) in toluene at 30 °C.
    After Margerison and East 1967.

Table 2.11

Dependence of the second virial coefficient on temperature

In toluene (good solvent)		In cyclohexanone (poor solvent)	
°C	$B,$ $m^5 \cdot s^{-2} \cdot kg^{-1}$	°C	$B,$ $m^5 \cdot s^{-2} \cdot kg^{-1}$
22.0	0.77	27.0	−0.092
		31.0	−0.026
67.0	0.70	35.0	0
		39.0	+0.016

Polystyrene $\bar{M}_m = 1,610,000$.
After Margerison and East 1967.

Table 2.12

Dependence of the second virial coefficient on tacticity

Configuration	$B,$ $m^5 \cdot s^{-2} \cdot kg^{-1}$
Mainly isotactic	1.18
Mainly atactic	0.99

Polystyrene $\bar{M}_n = 100,000$ in toluene at
30 °C.

shows the variation of this quantity for poly(α-methyl styrene)–toluene
solutions as a function of average molecular mass. Table 2.11. summarizes the
variation of the second virial coefficient as a function of temperature. These
data refer to polystyrene, the solvents are toluene and cyclohexanone. The
effect of stereoregularity is demonstrated in Table 2.12, where the effect of
tacticity on the second virial coefficient is shown for polystyrene samples of
equal molecular mass.

These Tables show clearly that the second virial coefficients depend on so
many factors that their description in terms of basic molecular parameters is
not possible.

The virial expression of osmotic pressure is in many respects very similar
to the virial expansion of the equation of state for real gases:

$$Pv = nR\,T + \frac{n^2\beta'}{v} + \frac{n^3\gamma'}{v^2} + \dots \tag{2.82}$$

or after rearrangement:

$$\frac{P}{c} = \frac{RT}{\bar{M}} + \beta' c + \gamma' c^2 + \ldots \qquad (2.83)$$

where n is the number of moles of the gas, for m grams of a gas with molecular mass, M:

$$n = m/M = v \cdot c/M \left[\frac{\text{ml} \cdot \text{g/ml}}{\text{g/mole}} \right] = [\text{mole}] \qquad (2.84)$$

and c is a weight concentration or gas density term. The similarity between Eq. (2.82) and (2.83) and Eqs (2.78), (2.79) is obvious. At higher temperature β' is positive and by decreasing the temperature it reduces to zero. This is known as the Boyle temperature. Below this temperature β' gradually becomes negative, until liquefaction begins due to molecular aggregation.

Polymer–solvent systems behave similarly. At higher temperatures B is positive, with decreasing temperatures it approximates to zero, and the temperature where it reaches zero is the Θ-temperature. With lower temperatures B decreases further and becomes negative. Below the Θ-temperature molecular aggregation results in precipitation. At the Θ-temperature the macromolecular solution behaves ideally at a wide range of concentrations, the same way as real gases behave ideally at the Boyle temperature at a wide range of pressures.

The effect of Θ-temperature on the second virial coefficient has already been treated in terms of the Flory–Kriegbaum theory.

At this temperature attractive and repulsive forces acting on the segments are in equilibrium, the molecular conformation approximating the unperturbed state, and osmotic pressure can be given by the following simple relationship:

$$\pi = \frac{RT}{\bar{M}_n} c \qquad (2.85)$$

independent of concentration.

Above the Θ-temperature the polymer molecule is expanded in comparison to the unperturbed dimensions, and the second virial coefficient is large and positive. In this case the solvent is described as a "good" solvent. In contrast, poor solvents are those in which molecular dimensions approximate the unperturbed state, so that the second virial coefficient is close to zero.

In Table 2.13 the Θ-temperatures of some polymer–solvent systems are summarized. These values are extrapolated to infinite molecular mass.

Table 2.13

Θ-temperatures of some polymer–solvent pairs, extrapolated to infinite molecular mass

Polymer	Solvent	Θ °C
Polystyrene	Cyclohexane	34.4
Polystyrene	Toluene	~ -100.0
Poly(α-methyl styrene)	Cyclohexane	38.0
Polyisobutylene	Benzene	24.0
Poly(methyl methacrylate)	50 : 50 methyl-ethyl ketone	25.0
Poly(n-butyl methacrylate)	Isopropanol	21.0
cis-1,4-Polybutadiene	Isobutyl acetate	20.5
cis-1,4-Polyisoprene	Methyl-n-propyl ketone	14.5
trans-1,4-Polyisoprene	n-propyl acetate	60.0
Poly(vinyl acetate)	Ethyl-n-butyl-ketone	29.0
Poly(vinyl chloride)	Benzyl alcohol	155.4
Poly(dimethyl siloxane)	Methyl-ethyl-ketone	20.0

After Margerison and East 1967.

2.3.2. Osmometers

There is a wide variety of osmometer constructions, but most of them trace back to three basic types:
— Schulz-type;
— Fuoss–Mead type;
— Zimm–Myerson type osmometers.

The Schulz-type osmometer is the simplest. The membrane is mounted on a perforated stainless steel plate, the space above the membrane is filled with the solution and finally the osmometer is immersed in the solvent (Fig. 2.13).

The principle of measurement is demonstrated quite well by this type, but it is not often used in practice, since the method of measurement is static, and it takes more than one day. About 5 cm^3 of solution, and 500 cm^3 of solvent are needed.

The Fuoss–Mead osmometer consists of two stainless steel plates with special concentric grooves inside. These ensure the fixture of the membrane and a large contact area with the solution. A side view of the osmometer is

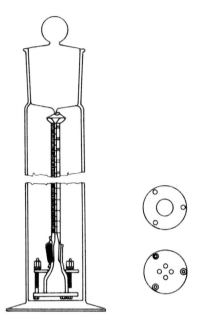

Fig. 2.13. Schulz-osmometer.

shown in Fig. 2.14, whereas the end-plate which holds the membrane is shown in Fig. 2.15. Width and depth of the grooves are both 2 mm, and the distance between them is 3 mm. The solvent and the solution move in a 3 mm vertical groove within the concentric ones.

The osmometer is filled via two sidecapillaries. Valves 1 and 2 are closed, and valve 3 is opened, the left hand side is filled with the solution, the right hand side being simultaneously filled with the solvent.

Care is taken to fill the two sides at a similar speed to avoid entrapped air. When the cells and capillaries are filled, valves 1 and 2 are opened for a moment to remove any air remaining in them.

By the correct manipulation of the valves the level in the solvent and solution capillaries can be set to any desired value. Since the heat capacity of the cell is large, thermal equilibrium can be readily maintained, the whole instrument is frequently placed into a heat-insulated box.

Osmotic pressure can be measured relatively easily by the so-called half-sum method. Some preliminary experiments are necessary to establish the approximate value of the equilibrium osmotic pressure, the exact value is then measured dynamically. To do so a $(\Delta h + x)$ cm difference in levels in produced on one side, where Δh is the estimated value of the equilibrium

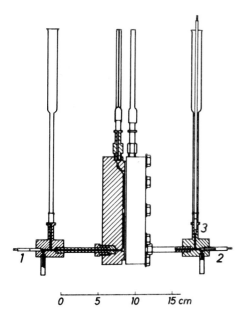

Fig. 2.14. Fuoss–Mead osmometer. (After Fuoss and Mead 1943.)

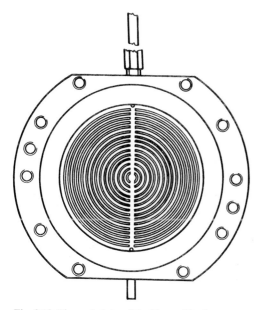

Fig. 2.15. The end-plate of the Fuoss–Mead osmometer. (After Fuoss and Mead 1943.)

difference is levels, and x is about 2 cm. The meniscus is read every half minute for about 8–10 minutes. Then the mesurement is repeated with a
$(\Delta h - x)$ initial value and exactly half minute time intervals.

Half of the sum (the arithmetic average) of the ascending and descending
levels is plotted versus time. This so-called half-sum gives the osmotic
pressure of the solution. This way the result can be obtained within a
relatively short time. This type of measurement is shown in Fig. 2.16.

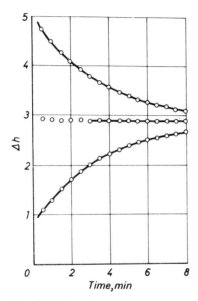

Fig. 2.16. Osmotic pressure measurements for poly(vinyl chloride) in methyl amyl ketone
according to the dynamic half-sum method. (After Fuoss and Mead 1943.)

When using the osmometer with a low-permeability membrane, the time
must be increased, in such cases thermostating is necessary. Several modified
versions of this method are known. Their common advantage is that the
membrane is in a well-defined, stretched position. For this type of
measurement about 10–12 cm^3 of solvent and solution are needed.

Zimm–Myerson osmometers are simpler than the Fuoss–Mead type.
One basic example is shown in Fig. 2.17. The osmometer consists of an open,
cylindrical cell of about 2 cm diameter (7) and two capillaries soldered to it
with diameters of 0.5 mm and 2 mm respectively. The thinner tube (6) is for
measurement, the thicker one (5) is for filling the system. The open ends of the
glass cell are exactly planparallel and polished. The two membranes are fixed

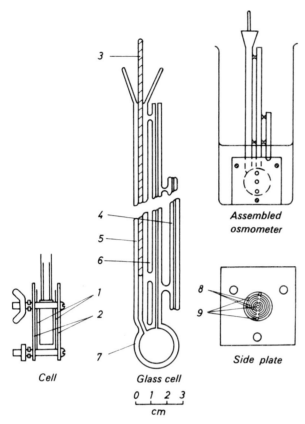

3

4

5

6

1

2

7

Cell

Glass cell

Assembled
osmometer

8

9

Side plate

0 1 2 3
cm

Fig. 2.17. Zimm–Myerson osmometer. (After Zimm and Myerson 1946.)

to the ends of the cells by means of two perforated steel or copper plates (2).
Holes in these perforated side-plates (8) ensure the solvent-membrane
contact, while the rest of the metal surface (9) serves to fix the membrane
mechanically.

Within the capillary of 2 mm in diameter a metal rod is placed (3) which
closes the hole after filling with solution. The diameter of this rod is chosen so
that movement of the fluid in the capillary is still possible. By moving the bar
the level of the measuring capillary can be changed and the half-sum method
can be applied. Measurement is performed by immersing the filled osmometer
in a solvent bath, the temperature of the solvent being regulated within
0.01 °C. The solvent level is read from external capillary (4) with high precision. For this type of measurement about 3 cm^3 of solution and 200 cm^3
of solvent are needed.

The Helfritz-type osmometer is also widely used. It combines the membrane mounting of the Fuoss–Mead osmometer (the membrane being stretched between metal plates) with the simple capillary system of the Schulz-osmometer. This kind of osmometer is shown in Figs 2.18 and 2.19.

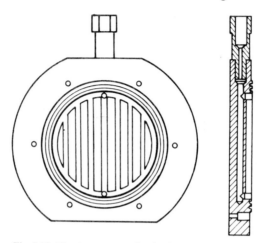

Fig. 2.18. Membrane mounting in the Helfritz osmometer.

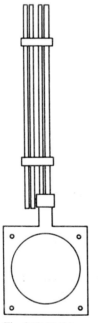

Fig. 2.19. Helfritz osmometer.

Fig. 2.18 demonstrates that part of the osmometer which serves for the mounting of the membrane, while Fig. 2.19 illustrates the osmometer itself with the measuring and reference capillaries.

The Helfritz-osmometer modified by Nagy and Varga (Fig. 2.20) can be used for dynamic (half-sum) measurements, since the capillary level can be set to any level. For this type of measurement 4 cm^3 of solution and 300 cm^3 of solvent are needed.

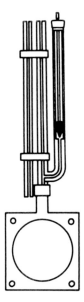

Fig. 2.20. Modified Helfritz osmometer.

Automatic osmometers using membranes are also available, where solution portioning, level-setting and osmotic pressure monitoring is automated. Such equipment works in a molecular mass range of 10^4–10^6, in both aqueous and non-aqueous solvents over a wide range of temperatures (5–130°C). Temperature is kept constant within 10^{-4}°C.

2.3.3. Membranes used for osmotic pressure determination

In the case of aqueous solutions mainly cellulose nitrate membranes are used, but for organic solutions poly(vinyl alcohol) or cellophane membranes are normal.

The quality of the membrane influences the accuracy of the measurement to a high degree. If the pores are too narrow, equilibrium is reached very slowly, whereas if the pores are too large the low molecular mass components can penetrate the membrane, so that the average molecular mass is overestimated. If the sample to be measured is highly polydisperse, fractionation is advisable.

Table 2.14 summarizes the pore sizes and upper limits of measurable molecular mass for membranes offered by the Membranfilter (Göttingen) Company.

Table 2.14

Pore size and measurable molecular mass range of various kinds of osmotic membranes

Grade	Coarse	Medium	Fine	Finer	Finest
Pore size, μm	200–80	80–50	50–10	10–5	<5
$\bar{M}_n$ max.	10^9	10^8	10^7	10^5	10^4

Selection of the correct membrane is a crucial step in the measurement. In order to get good osmometry results the macromolecules must not penetrate the membrane. This condition cannot be fulfilled for polymers with a wide scale of molecular mass distribution. To demonstrate how unrealistic results can be obtained in unfractionated samples the reduced osmotic pressure of a cellulose nitrate sample is plotted versus concentration in Fig. 2.21, with six different membranes. Dashed lines denote a sample of lower molecular mass, the continuous lines refer to a higher molecular mass sample. Deviations are notable in the case of the low molecular mass sample.

By fractionating sample B and removing about 30% low molecular mass fraction the osmotic pressure measurement in the same membranes gives unambiguous results (Fig. 2.22).

Low molecular mass materials profoundly influence the results of osmotic pressure measurements since this method yields number average molecular mass, and the weight factors of the individual molecules are equal. Thus, for example, a 1% component with a molecular mass of 500 reduces the average molecular mass of a second component with a molecular mass of 200,000 to 40,000.

Staverman (1951, 1952, 1957) studied the diffusion of low molecular mass polystyrene samples through some of the membranes detailed in Table 2.14.

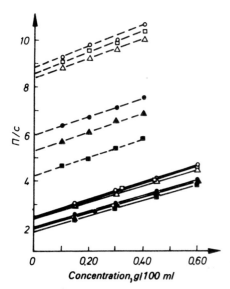

Fig. 2.21. Reduced osmotic pressure (π/c) plotted versus concentration. Data obtained from unfractionated nitrocellulose samples in n-butyl acetate. (After Alvang and Samuelson 1957.)
––– Sample A; ——— Sample B; ○ "Ultracella aller feinst", □ cellophane, △ ultracella feinst, ● denitrated collodium 1, ▲ denitrated collodium 2, ■ denitrated collodium 3.

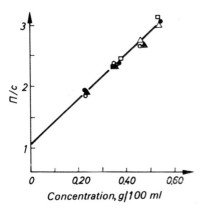

Fig. 2.22. Reduced osmotic pressure (π/c) plotted versus concentration. Fractionated nitrocellulose. Sample B of Fig. 2.21 after fractionation in n-butyl acetate. Notation as in Fig. 2.21. (After Alvang and Samuelson 1957.)

He established that an osmotic pressure value approximating equilibrium can be obtained within a few minutes, but the true equilibrium is not reached even after several hours. The results are shown in Figs 2.23 and 2.24 and in Table 2.15. From Fig. 2.23 one can see that the equilibrium is still not reached even after 1,300 minutes. The polymer molecules penetrate the membranes slowly and since they appear on the solvent-side, the concentration of the polymer solution decreases. This decrease of concentration is shown in Fig. 2.24 as a function of time.

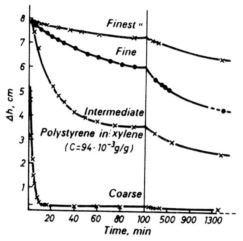

Fig. 2.23. Level difference of polystyrene solution as a function of time with different types of membrane. (After Staverman 1957.)

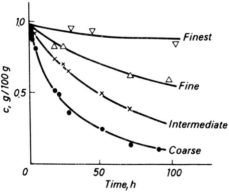

Fig. 2.24. Concentration depression of polystyrene caused by diffusion using differerent types of membrane. (After Staverman 1957.)

Table 2.15

Results of simultaneous diffusion and osmotic
pressure measurement (polystyrene in
chloroform)

Membrane	Time, h	Permeation, %	Apparent $\bar{M}_n \times 10^{-3}$
Coarse	25	62	504
	18	40	434
	2	30	406
Medium	149	34	232
	24	25	50
Fine	168	26	42
	24	21	33
Finest	144	16	31
	24	12	23

After Staverman 1957.

As a result of this diffusion process, the concentration of the solution in
question can change considerably which leads to inaccurate results. An
example of this is shown in Table 2.15 where an apparent molecular mass of
500,000 is obtained for a polysterene sample with $\bar{M}_m = 24{,}000$ and $\bar{M}_n$
$= 10{,}000$ if the measurement is performed with a coarse grade membrane.
Only about 38% of the original sample remains in the solution, the rest
penetrates into the solvent.

In conclusion, osmotic pressure measurements require extreme care and
caution, but the measurements can be performed with relatively simple
equipment.

2.3.4. Vapour-pressure osmometry

For the determination of number average molecular mass of relatively low
molecular mass molecules (below 20,000) vapour-pressure osmometers
(or electro-osmometers) are frequently used. The principle of measurement
differs from the previously described methods, since there is no use of
semipermeable membranes. The molecular mass is calculated directly from
vapour pressure depression. The vapour pressure depression phenomenon
has already been shown in Fig. 2.11 and it can be explained in the following
way. If a droplet of the solution gets into the saturated vapour of the solvent,

the solvent condenses on the droplet until the vapour pressure of the solvent and solution become equal. During condensation the solvent droplet heats up due to the heat of condensation. In the ideal case:

$$\Delta T_p = \frac{R T^2 M_1}{\rho_1 \Delta H_v} \cdot \frac{c}{\bar{M}_n} \tag{2.86}$$

so that the temperature increase of the droplet is inversely proportional to the number average molecular mass of the dissolved material, and proportional to its concentration. ΔH_v means the heat of evaporation at the temperature of the measurement in J/g units.

The temperature difference is measured between one thermistor in the vapour phase and a second, also in the vapour phase which has pure solvent dripped onto it continuously. The result is a few μ V voltage difference which is readily measurable.

In practice a series of measurements must be performed with different concentrations, so that temperature difference due to vapour pressure depression can be expressed as:

$$\lim_{c \to 0} \left(\frac{\Delta T_p}{c} \right) = \frac{R T^2 M_1}{\rho_1 \Delta H_v} \cdot \frac{1}{\bar{M}_n} \cdot \tag{2.87}$$

The temperature difference in electro-osmometers appears as a voltage difference, and the equation can be written in the following form:

$$\left(\frac{\Delta V}{c} \right)_{\lim c \to 0} = \frac{K}{\bar{M}_n} \tag{2.88}$$

where all constants of Eq. (2.87) are absorbed into K, and it contains also the proportionality constant between voltage and temperature and an empirical factor which takes into account the deviation from ideal conditions in the apparatus.

Temperature differences in vapour-pressure osmometry usually rank in the order of 0.1 °C, whereas measurable voltage differences are in the order of microvolts.

The values of virial coefficients can also be determined by vapour-pressure osmometry, according to Eqs (2.79), (2.80) or (2.81), similarly to classical osmotic pressure measurements. As one can see from Eq. (2.88), in vapour-pressure osmometric measurements the value of the empirical constant, K is needed. To determine the virial coefficients, however, no calibration is required if the following form of the virial equation is used:

$$\frac{\Delta V}{c} = \left(\frac{\Delta V}{c}\right)_{c\to 0} (1 + \Gamma_2 c + \Gamma_3 c^2 + \ldots) \tag{2.89}$$

where the meaning of Γ_2, Γ_3 is identical with those in Eq. (2.81), that is, they are the second and third virial coefficients respectively. The dimensions of the second virial coefficient are m^3/kg in SI units, and cm^3/g in the cgs system of units.

It is necessary to discuss in some detail the problem of concentration. So far by concentration, c, the g/cm^3 concentration has been intended. This type of concentration depends on temperature, since the volume of the solvent varies with temperature.

A temperature independent concentration value is obtained if the amount of dissolved material is related to the mass of the solvent, that is, if the concentration is given in g/kg units. This is especially useful in vapour-pressure osmometry where the temperature of the measurement can be any value ranging between the boiling and freezing points.

The c_m mass concentration in g/kg units can be calculated from the earlier $c\,(g/cm^3)$ concentration by the following relation:

$$c_m(g/kg) = \frac{c(g/cm^3) \cdot 1000}{\rho(g/cm^3)} \tag{2.90}$$

where ρ is the density of the solvent.

Two other notions, frequently used in connection with solutions are molarity and molality.

Molarity is an expression for volume-concentration; a solution exhibits unit molarity if it contains 1 gram-mole dissolved material in $1\ dm^3 = 1{,}000$ cm^3 solution $(kmole/m^3)$.

Molality is a mass concentration unit. If the molality is unity, the solution contains 1 gram-mole dissolved material in 1 kg solvent. Molality is independent of temperature $(mole/kg)$.

The schematic view of a vapour-pressure osmometer is shown in Fig. 2.25. Solutions of different concentration and the solvent, can be injected onto the thermistors by means of syringes. At the bottom of the system there is a cup filled with pure solvent. After reaching the desired temperature, and the vaporization equilibrium, the reference point of the system can be obtained if both thermistors are covered by a solvent droplet. Then the solution of the least concentration is dropped onto one of the thermistors, and the voltage difference between the two thermistors is measured. After performing a series of measurements for each concentration value, the concentration de-

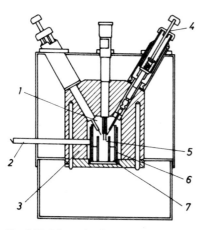

Fig. 2.25. Schematic of a vapour-pressure osmometer *1* measuring chamber, *2* mirror, *3* heating block, *4* syringe, *5* thermistor, *6* solvent cup, *7* filter paper.

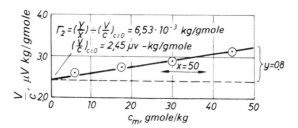

Fig. 2.26. Results of an osmotic pressure measurement.

pendence of the voltage is plotted as V/c_m versus c_m. The intercept yields $(V/c_m)_{c_m \to 0}$.

If the measurement is performed with a material of known molecular mass, the calibration constant can be calculated as:

$$K = M_{\text{standard}} \left(\frac{V}{c_m} \right)_{c_m \to 0}. \tag{2.91}$$

If an unknown material is studied, and K is known from earlier measurements, the number average molecular mass of the unknown sample is obtained from:

$$\bar{M}_n = \frac{K}{\left(\dfrac{V}{c_m} \right)_{c_m \to 0}}. \tag{2.92}$$

For calibration very pure compounds with known molecular mass should be used, for example, benzil ($M = 210.2$), squalene ($M = 422.8$) or pentaerythritol tetrastearate (PETS) ($M = 1202$).

The results of a measurement are shown in Fig. 2.26. As one can see on the ordinate, concentration is given in g/kg units. The low concentration limit, $(V/c_m)_{c_m \to 0}$, extrapolated from four points is 2.45 (μVkg/g). Taking into account the first two terms of Eq (2.89):

$$\Gamma_2 = \frac{\Delta \dfrac{V}{c_m}}{\Delta c_m \left(\dfrac{V}{c_m} \right)_{c_m \to 0}} = \frac{0.8}{50 \cdot 2.45} = 6.53 \cdot 10^{-3} \left[\frac{kg}{g} \right],$$

that is, 6.53 (g/g). As the density of toluene at the measuring temperature (65°C) is equal to 0.86 (g/cm^3), and the concentration is expressed in g/cm^3 units, then:

$$\Gamma_2 = \frac{6.53 \, [g/g]}{0.86 \, [g/cm^3]} = 7.6 \, [cm^3/g].$$

Data processing is more precise if the method of least squares is used to fit the measured points.

2.4. MOLECULAR MASS DETERMINATION FROM LIGHT SCATTERING MEASUREMENTS

From light scattering data the mass average molecular mass, the radius of gyration, describing the coil structure of the molecule, and the second virial coefficient, characteristic of the solvent–polymer interaction, can be obtained.

The basic phenomenon is well known in everyday life; when entering a dusty room the path of a beam of light is quite clear, since the light is scattered from the dust particles. This kind of scattering should be distinguished from reflection as in this case (similarly to the mirror) incident and reflection angles are bound to certain values.

2.4.1. Scattering in gases

2.4.1.1. Scattering by a single particle. The Rayleigh equation

First let us investigate the interaction of a single molecule in vacuo with a plane-polarized, monochromatic light beam. Let us assume that the molecule is spherical, and so the position of the molecule with respect to the light beam causes no special problem. It is also assumed that the diameter of the particle is much smaller than the wave length of the light as shown in Fig. 2.27. The arrow shows the direction of the light.

It is known that light (which in an electromagnetic radiation) has electric and magnetic fields oscillating perpendicular to each other.

Both fields move at a speed of c, and the direction of the light is perpendicular to that of field oscillations. The position of the fields at a given instant is shown in Fig. 2.28. In the Figure the light moves along the X-axis, the polarization plane is the XY plane, and electrical (E) and magnetic (H) fields oscillate along the Z and Y axes respectively. At a given instant and at point 0 both fields are equal to zero, along the X-axis both fields change, maximum values, E_0 and H_0 are found at a distance of $\lambda/4$ from the origin. At point 0,

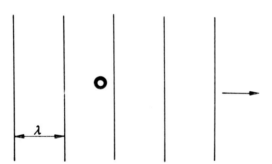

Fig. 2.27. Planar wave fronts distant from the source of radiation. (After Margerison and East 1967.)

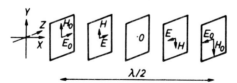

Fig. 2.28. Vectorial representation of the electrical (E) and magnetic (H) fields in the electromagnetic radiation. (After Margerison and East 1967.)

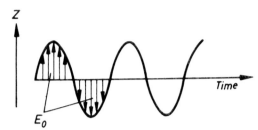

Fig. 2.29. Time dependence of the electrical field in a given point.

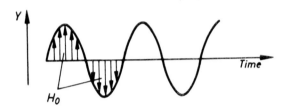

Fig. 2.30. Time dependence of the magnetic field at a given point. (After Margerison and East 1967.)

both fields vary as a function of time. This function is shown in Fig. 2.29 for the electric field, where the lengths of the arrows indicate the magnitude of the field, and their direction shows the acceleration of a free positive charge in the field.

Simultaneously the magnetic field changes along the Y-axis (Fig. 2.30). Here again the lengths of the arrows are proportional to the magnitude of the magnetic field, and their direction gives that in which the field would accelerate a free N-monopole.

The oscillation of the electrical field perturbs the outer electrons of the atoms in the molecule, and an oscillating induced dipole is produced. Its magnitude, p, at any given time, t, is proportional to the inducing field, E, the proportionality constant is the polarizability of the molecule, α, so that:

$$p = \alpha E, \tag{2.93}$$

where:

$$E = E_0 \cos \frac{2\pi ct}{\lambda} = E_0 \cos 2\pi \nu t, \tag{2.94}$$

and:

$$\nu = c/\lambda, \tag{2.95}$$

the frequency of light.

This oscillating dipole moment, p, produces secondary electromagnetic radiation, consequently energy, or more exactly light, is scattered by the molecule when illuminated with monochromatic light. The radiated energy, I_s, is proportional to the square average of the second derivative of the induced dipole moment with respect to time:

$$I_s = \frac{2}{3c^3} \overline{\left(\frac{d^2p}{dt^2}\right)^2}$$

(2.96)

where c is the velocity of light.

The intensity of the incident light, I_0, the amount of energy passing through unit surface, is proportional to the square of the amplitude of the electric field, E_0:

$$I_0 = \frac{c}{8\pi} E_0^2.$$

(2.97)

The first and second derivatives of the induced dipole moment can be calculated from Eqs (2.93) and (2.94) as:

$$\frac{dp}{dt} = -2\pi v\alpha E_0 \sin 2\pi vt,$$

(2.98)

$$\frac{d^2p}{dt^2} = -4\pi^2 v^2 \alpha E_0 \cos 2\pi vt.$$

(2.99)

The square of the second derivative is:

$$\left(\frac{d^2p}{dt^2}\right)^2 = 16\pi^4 v^4 \alpha^2 E_0^2 \cos^2 2\pi vt.$$

(2.100)

The average is determined as follows, taking into account that:

$$\overline{\cos^2 2\pi vt} = 0.5$$

(2.101)

and so

$$\overline{\left|\frac{d^2p}{dt^2}\right|^2} = 8\pi^4 v^4 \alpha^2 E_0^2.$$

(2.102)

Inserting Eq. (2.102) into Eq. (2.96) the scattered intensity for a single isolated particle is given by:

$$I_s = \frac{2}{3c^3} 8\pi^4 v^4 \alpha^2 E_0^2$$

(2.103)

or by introducing the wavelength of light:

$$\frac{v}{c} = \frac{1}{\lambda_0} \tag{2.104}$$

where λ_0 is the wavelength in vacuo, then:

$$I_s = \frac{16}{3} \pi^4 \frac{c}{\lambda_0^4} \alpha^2 E_0^2. \tag{2.105}$$

Inserting the intensity of the incident light, according to Eq. (2.97) the famous Rayleigh equation results:

$$I_s = \frac{16}{3} \pi^4 \frac{c}{\lambda_0^4} \alpha^2 \frac{8\pi}{c} I_0, \tag{2.106}$$

or

$$I_s = \frac{8}{3} \pi I_0 \left(\frac{2\pi}{\lambda_0}\right)^4 \alpha^2. \tag{2.107}$$

The scattered intensity is proportional to the intensity of the primary beam, and inversely proportional to the fourth power of the wavelength.

2.4.1.2. Scattering by multiple particles

Energy scattered by individual particles remains far below the detection limit. In practice, scattering properties of individual particles are always inferred from measurements on a large number of particles. Scattered intensities originating from individual centres can be modified by interference, since interfering waves can amplify or cancel each other. If the scattering centres are distributed randomly, and the distance between them is large enough, the interference effect can be neglected. In such cases intensity is simply the sum of those scattered by the individual particles, so that, if there are N statistically distributed Rayleigh-particles present in unit volume (1 cm^3), the total emitted energy will be:

$$I = NI_s = \frac{8}{3} \pi N I_0 \left(\frac{2\pi}{\lambda_0}\right)^4 \alpha^2. \tag{2.108}$$

Combining all constants into a single proportionality coefficient:

$$I = \tau I_0 \tag{2.109}$$

where τ is the turbidity coefficient, or simply turbidity, which is the total intensity scattered from unit volume of the scattering medium. The value can be given as:

$$\tau = \frac{8}{3}\pi N \left(\frac{2\pi}{\lambda_0}\right)^4 \alpha^2 = \frac{128\,\pi^5}{3\,\lambda_0^4} N\alpha^2. \tag{2.110}$$

2.4.1.3. Modification of the Rayleigh equation

The Rayleigh equation (as given by Eqs (2.106) and (2.107)) contains the polarizability of the individual molecules. In the case of dilute gases this is related to the dielectric permittivity of the medium:

$$\alpha = \frac{\varepsilon-1}{4\pi}\frac{1}{N}. \tag{2.111}$$

According to the Maxwell relationship, on the other hand, relative permittivity is equal to the square of the refractive index of the medium:

$$\varepsilon = n^2, \tag{2.112}$$

and so finally the polarizability can be given by:

$$\alpha = \frac{n^2-1}{4\pi}\frac{1}{N}. \tag{2.113}$$

Inserting this into Eq. (2.108), turbidity can be expressed as a function of the refractive index of the medium, n:

$$\tau = \frac{8}{3}\pi N \left(\frac{2\pi}{\lambda_0}\right)^4 \frac{(n^2-1)^2}{16\pi^2}\frac{1}{N^2}, \tag{2.114}$$

or

$$\tau = \frac{8}{3}\pi^3 \frac{(n^2-1)^2}{N\lambda_0^4}. \tag{2.115}$$

Some authors use the following form:

$$\tau = \frac{32}{3}\pi^3 \frac{(n-1)^2}{N\lambda_0^4} \tag{2.116}$$

where it is taken into account that if $n \approx 1$, then:

$$n^2-1 = (n+1)(n-1) \approx 2(n-1). \tag{2.117}$$

Eqs (2.115) and (2.116) represent the Rayleigh equation for turbidity. Using this equation the number of molecules in unit volume can be measured, so that the Avogadro number, or the Loschmidt number (N_L, the number of molecules in 1 cm^3 of an ideal gas in standard state) can be determined.

2.4.2. Scattering by particles in dilute solutions

Scattering phenomena considered so far are the result of scattering from particles surrounded by a vacuum. Solutions differ from gases in that the scattering particles themselves are surrounded by scattering and refracting solvent molecules. Equations derived for gases have to be modified as far as the calculation of polarizabilities, α, are concerned.

Let us assume that we are interested in the scattering of the dissolved molecule only, that is, not in the total induced dipole moment, but only in the difference between the dipole moments of the solution and the solvent. Let us also assume that the total induced dipole moment of the solution, p is the weighted sum of those of the components, p_1 and p_2, the weight factors being the mole fraction of the solvent, Y_1 and the dissolved material, Y_2 respectively:

$$p = Y_1 p_1 + Y_2 p_2 \tag{2.118}$$

where

$$Y_1 = \frac{N_1}{N_1 + N_2},$$

and

$$Y_2 = \frac{N_2}{N_1 + N_2},$$

N_1 and N_2 denote the number of molecules in unit volume of solvent and solute respectively.

The dipole moments of the solution, solvent and solute can be calculated from Eqs (2.93) and (2.113) as follows:

$$p = \frac{n^2 - 1}{4\pi} \frac{1}{N_1 + N_2} E, \tag{2.119}$$

$$p_1 = \frac{n_1^2 - 1}{4\pi} \frac{1}{N_1} E, \tag{2.120}$$

$$p_2 = \alpha_2 E. \tag{2.121}$$

Inserting all these into Eq. (2.118) gives:

$$\frac{n^2 - 1}{4\pi} \frac{1}{N_1 + N_2} E = \frac{N_1}{N_1 + N_2} \frac{n_1^2 - 1}{4\pi} \frac{1}{N_1} E + \frac{N_2}{N_1 + N_2} \alpha_2 E,$$

$$n^2 - 1 = n_1^2 - 1 + 4\pi N_2 \alpha_2,$$

where the polarizability of the solute is equal to:

$$\alpha_2 = \frac{n^2 - n_1^2}{4\pi N_2}. \tag{2.122}$$

For dilute solutions the expression can be further simplified, since the difference between n and n_1 is very small and:

$$n^2 - n_1^2 = (n + n_1)(n - n_1) \approx 2n_1(n - n_1). \tag{2.123}$$

Introducing the c (g/cm^3) concentration, the molecular mass of the solute, M and taking into account that the mass of a single molecule is equal to M/N_{Av}, furthermore that of N_2 molecules is $N_2 \cdot M/N_{Av}$, the following expression results for N_2:

$$N_2 = N_{Av} \frac{c}{M} \tag{2.124}$$

where the polarizability of the solute becomes:

$$\alpha_2 = \frac{2n_1 M}{4\pi N_{Av}} \frac{n - n_1}{c}. \tag{2.125}$$

The $(n - n_1)/c$ quotient can be replaced by dn/dc, the differential refractive index increment, which is nearly constant for dilute solutions. Inserting N_2 instead of N and α_2 instead of α into Eq. (2.110) the following result is obtained for the turbidity of a dilute solution:

$$\tau = \frac{8}{3} \pi N_2 \left(\frac{2\pi}{\lambda_0}\right)^4 \alpha_2^2, \tag{2.126}$$

or by using Eqs (2.124) and (2.125):

$$\tau = \frac{8}{3} \pi N_{Av} \frac{c}{M} \left(\frac{2\pi}{\lambda_0}\right)^4 \frac{4n_1^2 M^2}{16\pi^2 N_{Av}^2} \left(\frac{dn}{dc}\right)^2, \tag{2.127}$$

$$\tau = \frac{32}{3} \pi^3 \frac{n_1^2}{N_{Av} \lambda_0^4} \left(\frac{dn}{dc}\right)^2 cM. \tag{2.128}$$

The turbidity of the pure solvent (i.e. the intensity of light scattered by the solvent) has to be substracted from the experimentally measured values when calculating the turbidity of the dissolved particles.

2.4.2.1. Angular dependence of the scattered intensity

Scattering caused by dissolved molecules can be measured, in principle, in two different ways. One is the measurement of the intensity of the incident and leaving light beam, the second is the measurement of the scattered intensity. In practice, the second method is always used, since the intensity change of the primary radiation is hardly measurable. Besides the angular dependence of the scattered intensity provides important information about the size of the macromolecules in the solution.

Let us imagine a spherical surface around the scattering particle or volume element. Intensity measured on this surface will be denoted as I. As mentioned in connection with Figs 2.28 and 2.29 it is assumed that the polarized incident radiation oscillates along the Z-axis. The induced dipole situated at the origin of the coordinate system of Fig. 2.31 also oscillates along

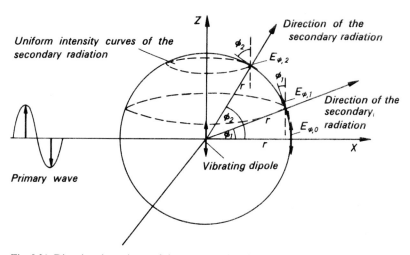

Fig. 2.31. Direction dependence of the scattered light intensity. (After Springer 1970.)

the Z-direction, and the amplitude is the function of molecular polarizability. Electrical vectors of both primary and secondary radiation are perpendicular to the direction of light. This direction is the X-axis in the case of the primary radiation whereas the secondary radiation moves from the origin in all directions. Fig. 2.31 shows the electrical vectors of the secondary radiation at the intersection of the spherical surface and the X–Z plane. Two scattering directions are shown, the angles between the original direction (X-axis) and the scattered ones are Φ_1 and Φ_2 respectively. One can see from the Figure that:

$$\cos \Phi = \frac{E_\Phi}{E_{\Phi,0}},\qquad (2.129)$$

that is, the intensity of the scattered electrical vectors are proportional to the cosine of the scattering angle:

$$E_\Phi = E_{\Phi,0} \cos \Phi. \qquad (2.130)$$

Since according to Eq. (2.97) the intensity is proportional to the square of the electrical vector, then the scattered intensity can therefore be given as:

$$I_\Phi = I_{\Phi,0} \cos^2 \Phi, \qquad (2.131)$$

where $I_{\Phi,0}$ is the scattered intensity measured perpendicular to the direction of the electrical vector of the primary radiation, and to the direction of polarization. (This is equal to the original direction.) In order to calculate the total scattered intensity all scattered intensities, I_Φ, have to be calculated along the spherical surface:

$$I = \int_F I_\Phi dF_\Phi. \qquad (2.132)$$

For the sake of simplicity Fig. 2.31 shows the X–Z plane only. The intensity of secondary radiation originating from the oscillating dipole is uniform along the circles shown by the dashed lines in Fig. 2.31. So the spherical surface integral can be calculated by summing infinitesimal rings, along which the intensity is uniform (Fig. 2.32). The area of such a ring (dashed area in Fig. 2.32) can be given as:

$$dF = 2r(\cos \Phi)\pi r d\Phi = 2\pi r^2 \cos \Phi d\Phi. \qquad (2.133)$$

Taking into account Eqs (2.131) and (2.133) the integral in Eq. (2.132) can be expressed as:

$$I = \int_{-\pi/2}^{\pi/2} I_{\Phi,0} \cos^2 \Phi 2\pi r^2 \cos \Phi d\Phi = 2\pi r^2 I_{\Phi,0} \int_{-\pi/2}^{\pi/2} \cos^3 \Phi d\Phi, \qquad (2.134)$$

and its solution:

$$I = 2\pi r^2 I_{\Phi,0} \left[\sin \Phi - \frac{1}{3} \sin^3 \Phi \right]_{-\pi/2}^{\pi/2} = \frac{8}{3}\pi r^2 I_{\Phi,0}. \qquad (2.135)$$

This is the result for a single particle, or an elementary scattering volume, the total scattered intensity on a spherical surface with radius r around the scattering centre.

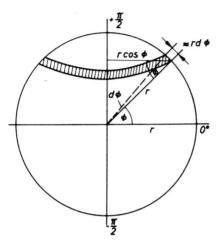

Fig. 2.32. Calculation of the volume integral of a spherically symmetric function with respect to polar coordinates. (After Springer 1970.)

In practice, three types of incident radiation, with different states of polarization, are important. These are the vertically or horizontally polarized, and nonpolarized light. Scattered intensity is measured in the $X–Y$ plane. Besides the scattering angle, Φ, it is advantageous to introduce Θ, the angle of observation. Let us take a close look at the dependence of scattered intensity on the angle of observation in the three different states of primary polarization.

2.4.2.2. Scattering of vertically polarized incident light

The incident light and the excited particles oscillate along the Z-axis. Scattering phenomena are shown in Fig. 2.33. It can be seen that in the $X–Y$ plane, for all values of the Θ angle $\Phi = 0°$. According to Eq. (2.131) for all angles of observation, Θ, since $\cos \Phi = 1$:

$$I_{\Theta, \mathrm{v}} = I_{0°, \mathrm{v}} \tag{2.136}$$

where the index V indicates vertical polarization. This means that the scattered intensity in the $X–Y$ plane is uniform and independent of the angle of observation.

Θ, the angle of observation, or, according to a different nomenclature, the scattering angle, is introduced at this point. Later on we shall see that in reflection experiments Θ denotes the angle between the incident light beam

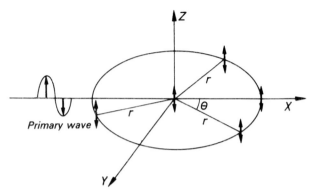

Fig. 2.33. Scattering diagram of a primary radiation beam polarized in the X–Z plane. (After Springer 1970.)

and the reflecting plane. In such cases the scattering angle is equal to 2Θ. Therefore, when reading the literature it is always advisable to clarify what is meant by Θ.

2.4.2.3. Scattering of horizontally polarized incident light

In this case both the incident light and the generated particles oscillate along the Y-axis. In the X–Y plane (the plane of observation) for all values of Θ, $\Phi = \Theta$, (that is, the angle of observation is equal to the angle of deviation from the plane of oscillation), as can be seen from Fig. 2.34. So $\cos \Phi = \cos \Theta$ and according to Eq. (2.131):

$$I_{\Theta,H} = I_{0^\circ,H} \cos^2 \Theta \tag{2.137}$$

where subscript H indicates horizontal polarization. In the case of a horizontally polarized incident light beam, the scattered intensity in the X–Y plane depends on the scattering angle, and is proportional to $\cos^2 \Theta$.

Angular dependence of the scattered intensity can be demonstrated by the scattering surface. This surface can be obtained if, from the scattering centre, vectors proportional to the scattered intensity are drawn in all directions. The ends of these vectors produce the scattering surface.

Variation of the electrical vectors of the secondary radiation, induced by the incident light, has already been shown in Fig. 2.31 in the X–Z plane. Points of equal intensity can be produced by rotation around the Z-axis. When plotting the intensity, which is proportional to $\cos^2 \Phi$ a toroidal body, similar to a doughnut is obtained. Its shape is thinner in the central region, with a hole in the centre (Fig. 2.35).

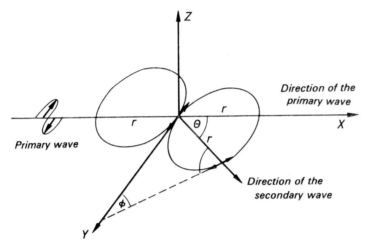

Fig. 2.34. Scattering diagram of a primary radiation beam polarized in the Y–Z plane. (After Springer 1970.)

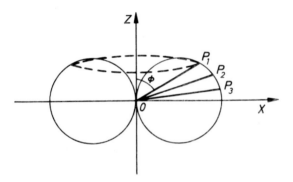

Fig. 2.35. Scattering envelope of the scattered radiation polarized in the X–Z plane, toroidal form. (After Margerison and East 1967.)

As the length of a line drawn from the centre to the surface of the toroid is proportional to the scattered intensity in the given direction, it can be seen from Fig. 2.35 that the value of the scattered intensity is constant in the X–Y plane, it shows great variation in the X–Z plane, maximum in the X-direction, and zero in the Z-direction.

If the plane of polarization is changed from the Z-axis to the Y-axis, the scattered intensity greatly varies in the X–Y plane and will be constant in the X–Z plane (Fig. 2.36).

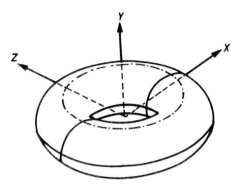

Fig. 2.36. Scattering envelope of the scattered radiation polarized in the X–Y plane as viewed from the Y-axis.

2.4.2.4. Scattering of non-polarized incident light

Non-polarized light can be regarded as a superposition of two, perpendicularly polarized light beams of equal intensity. Using Eqs (2.136) and (2.137) and denoting the non-polarized light with index U, the following equation results:

$$I_{\Phi,U} = I_{\Phi,V} + I_{\Phi,H} = I_{0^\circ,V} + I_{0^\circ,H}\cos^2\Theta. \tag{2.138}$$

Since

$$I_{0^\circ,V} = I_{0^\circ,H} = \frac{1}{2}I_{0^\circ,U}, \tag{2.139}$$

then the scattered intensity induced by a non-polarized incident light, as a function of the scattering angle can be described by the following relation:

$$I_{\Phi,U} = I_{0^\circ,U}\frac{1+\cos^2\Theta}{2}. \tag{2.140}$$

The scattering surface of the unpolarized radiation is shown in Fig. 2.37. It can be seen from Eq. (2.140) and Fig. 2.37, that if the angle of observation is equal to 90°, even in the case of non-polarized incident light only vertically polarized scattered light is observable. It is also clear that the intensity of the forward and backward scattered light is symmetrical with respect to the 90° value.

The angular dependence of the scattered radiation is summarized in Fig. 2.38 where the plane of observation (the X–Y plane) is viewed from above (from the positive side of the Z-axis) and the intensity of the scattered light is shown as the function of the angle of observation.

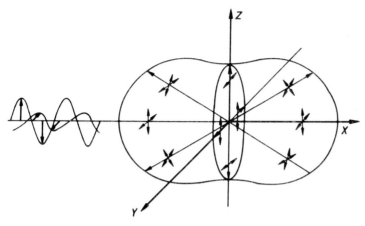

Fig. 2.37. Scattering envelope of unpolarized primary radiation. (After Springer 1970.)

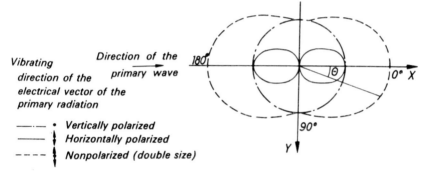

Fig. 2.38. Direction dependence of scattered radiation from small molecules. (After Springer 1970.)

2.4.2.5. Definition of the reduced scattered intensity

Introduction of this quantity is advantageous in the sense that it is independent of the initial intensity of the primary beam and of the distance between the scattering centre and the place of observation.

The value of turbidity can be expressed using Eqs (2.109), (2.128) and (2.135) as:

$$\tau = \frac{I}{I_0} = \frac{8\pi r^2 I_{\Phi,0}}{3I_0} = \frac{32}{3}\pi^3\frac{n_1^2}{N_{Av}\lambda_0^4}\left(\frac{dn}{dc}\right)^2 cM, \qquad (2.141)$$

or

$$\frac{r^2 I_{\Phi,0}}{I_0} = \frac{4\pi^2 n_1^2}{N_{Av} \lambda_0^4} \left(\frac{dn}{dc}\right)^2 cM. \tag{2.142}$$

By definition, the reduced scattered intensity, R_Θ, is described by the following expression:

$$R_\Theta \equiv \frac{r^2 I_\Theta}{I_0}, \tag{2.143}$$

where I_Θ is the scattered intensity at a distance r from the scattering volume unit, Θ the angle of observation, and I_0 the intensity of the primary radiation. Using Eqs (2.136), (2.137) and (2.140) which give the Φ and Θ dependence of the scattered radiation for differently polarized incident lights, the following relations can be derived for the reduced scattered intensities:

$$R_{\Theta,V} = \frac{4\pi^2 n_1^2}{N_{Av} \lambda_0^4} \left(\frac{dn}{dc}\right)^2 cM = K_V^* cM \tag{2.144}$$

for the vertically polarized incident light, where:

$$K_V^* = \frac{4\pi^2 n_1^2}{N_{Av} \lambda_0^4} \left(\frac{dn}{dc}\right)^2 \tag{2.145}$$

and

$$R_{\Theta,H} = K_V^* (\cos^2 \Theta) cM \tag{2.146}$$

for the horizontally polarized incident light, and

$$R_{\Theta,U} = K_V^* \frac{1 + \cos^2 \Theta}{2} cM \tag{2.147}$$

for the non-polarized primary radiation.

2.4.2.6. Concentration dependence of the scattered intensity

According to Eqs (2.144) to (2.147) the R_Θ/c expression is independent of the concentration of the sample. This condition is fulfilled for ideal solutions only. Einstein and Smoluchowsky explained the scattered intensity increment of solutions with respect to the pure solvent by the temporary and spatial concentration fluctuations in the volume elements. They derived an equation which contains the $\partial \Delta \mu / \partial c$ term determining the concentration fluctuation. As we have seen, in the case of dilute solution this derivative can be approximated by a power series, and so the equation will contain the virial coefficients characteristic of the solute–solvent interaction:

$$\tau = \frac{32}{3}\pi^3 \frac{n_1^2}{N_{Av}\lambda_0^4}\left(\frac{\partial n}{\partial c}\right)_T^2 \times \frac{c}{\left(\frac{1}{M} + 2A_2c + 3A_3c^2 + \dots\right)}. \tag{2.148}$$

This equation differs from Eq. (2.128) only in terms containing the virial coefficients. A_2 and A_3 coefficients are identical with those appearing in Eq. (2.80). The coefficients related to A_2 and A_3 are the result of differentiation with respect to c, since the derivation (not shown in detail here) contains the $\partial\pi/\partial c$ differential quotient. The two different expressions for the refractive index increment are essentially identical. Combining Eqs (2.126) and (2.148) and inserting them into Eq. (2.143) the following equations can be obtained for the different states of polarization of the primary beam:

$$R_\Theta = K^* \frac{c}{\frac{1}{M} + 2A_2c + 3A_3c^2 + \dots} \tag{2.149}$$

or

$$\frac{K^*c}{R_\Theta} = \frac{1}{M} + 2A_2c + 3A_3c^2 + \dots \tag{2.150}$$

where

$$K^* = K_V^* = \frac{4\pi^2 n_1^2}{N_{Av}\lambda_0^4}\left(\frac{\partial n}{\partial c}\right)_T^2 \tag{2.151}$$

for vertically polarized, and

$$K^* = K_H^* = K_V^* \cos^2\Theta \tag{2.152}$$

for horizontally polarized, and

$$K^* = K_U^* = K_V^* \frac{1 + \cos^2\Theta}{2} \tag{2.153}$$

for non-polarized incident radiation.

The series expansion in Eq. (2.150) is practically negligible after the second term as higher order coefficients are usually very small.

If K^*c/R_Θ is plotted versus concentration the curve intersects the ordinate at $1/M$, and the slope gives the second (osmotic) virial coefficient.

Other authors have used equations similar to, but formally non-identical with Eq. (2.150). As was mentioned during the introduction of virial expansion Eqs (2.79) to (2.81) the values and dimensions of virial coefficients used for the description of the concentration dependence of π/c, are determined by the actual form of the equation (i.e. on the terms in front of the brackets).

The relationship between the second virial coefficients used in Eqs (2.79) to (2.81) is as follows:

$$\Gamma_2 = \frac{BM}{RT} = A_2 M \tag{2.154}$$

and so the dimensions of Γ_2 are:

$$[\Gamma_2] = \left[\frac{cm^5 \cdot s^{-2} \cdot g^{-1} \cdot gmole^{-1}}{g \cdot cm \cdot cm \cdot s^{-2} \cdot mole^{-1} \; K^{-1} \cdot K} \right] = [cm^3 \cdot g^{-1}], \tag{2.155}$$

where the dimensions of B are taken from Eq. (2.78).

Dimensions of A_2 are:

$$[A_2] = \left[\frac{cm^3 \cdot g^{-1}}{mole^{-1}} \right] = [cm^3 \cdot g^{-1} \cdot mole]. \tag{2.156}$$

Inserting Eq. (2.128) into the expression in Eq. (2.124) gives:

$$\frac{K^*c}{R_\Theta} = \frac{1}{M}(1 + 2\Gamma_2 c + 3\Gamma_3 c^2 + \ldots) \tag{2.157}$$

This relation is often quoted as the basic expression for molecular mass determination from light scattering measurements.

2.4.2.7. The effect of polymolecularity on light scattering

Up to this point it has been assumed that only a single type of molecule with a molecular mass of M is present. If the polymer consists of more components, with different molecular masses, the scattered intensity at the Θ angle is the sum of the individual scattered intensities in the volume element.

Rearranging Eq. (2.157) and noting Eq. (2.124), then:

$$\frac{R_\Theta}{K^*(1 + 2\Gamma_2 c + 3\Gamma_3 c^2 + \ldots)} = cM = \frac{N_2 M^2}{N_{Av}}. \tag{2.158}$$

If there are many different species present denoted by subscript i in the solution, then these are summed, such that:

$$\frac{R_\Theta}{K^*(1 + 2\Gamma_2 c + 3\Gamma_3 c^2 + \ldots)} = \frac{\Sigma N_i M_i^2}{N_{Av}}. \tag{2.159}$$

Since

$$\Sigma N_i M_i^2 = \bar{M}_m \Sigma N_i M_i = \bar{M}_m c N_{Av}, \tag{2.160}$$

then the scattered intensity of the polydisperse system gives the result:

$$\frac{R_\Theta}{K^*(1+2\Gamma_2 c+3\Gamma_3 c^2+\ldots)} = \bar{M}_m c, \tag{2.161}$$

so that this method yields the mass average molecular mass.

Light scattering measurements are very important methods for mass average molecular mass determination, at present, the only absolute one for this purpose. This kind of average is regarded as the most characteristic for the description of molecular behaviour.

It is worth mentioning here that a further expression for the mass average molecular mass is also widespread:

$$\bar{M}_m = \frac{\Sigma N_i M_i^2}{\Sigma N_i M_i} = \frac{\Sigma c_i M_i}{\Sigma c_i}. \tag{2.162}$$

This is identical with the earlier definition, since:

$$c = \frac{\Sigma N_i M_i}{N_{Av}}. \tag{2.163}$$

2.4.2.8. Depolarization of the scattered light

It has been assumed so far that the scattering particles are isotropic, that is, that the direction of the secondary, induced radiation is parallel with that of the primary one. In the case of randomly distributed anisotropic molecules, however, the scattered light partially depolarizes. The degree of depolarization is characterized by the intensity ratio of the horizontal and vertical scattered light components observed at $\Theta = 90°$, and its value is denoted by Δ where:

$$\Delta = \frac{\text{intensity of the horizontal}}{\text{intensity of the vertical component}} \atop {\text{of the scattered light}} \tag{2.164}$$

As shown earlier in the case of isotropic molecules and non-polarized primary radiation, the scattered light observed at $\Theta = 90°$ contains a vertically polarized component only; where $\Delta_U = 0$. If there is a horizontal component as well, then its intensity in relation to the vertical components gives molecular anisotropy according to Eq. (2.139). To perform such a measurement polarization filters have to be applied at $\Theta = 90°$ to resolve the H_U (horizontal) and V_U (vertical) components of the scattered intensity induced by the unpolarized primary radiation.

Polarized primary beams are frequently used for the determination of the degree of depolarization, and the Krishnan relationship is utilized:

$$H_{U, 90°} = U_{H, 90°}; \qquad V_{U, 90°} = U_{V, 90°} \qquad (2.165)$$

which states that the unpolarized scattered intensity of a horizontally polarized light beam is equal to the horizontal component of the scattered intensity originating from a non-polarized incident light beam at $\Theta = 90°$. Similarly, the unpolarized scattered intensity of a vertically polarized light beam is equal to the vertical component of the scattered intensity caused by a non-polarized incident radiation at the same angle of observation.

The degree of polarization can therefore be given alternatively as:

$$\Delta_U = \frac{H_{U, 90°}}{V_{U, 90°}} = \frac{U_{H, 90°}}{U_{V, 90°}}. \qquad (2.166)$$

According to this relationship, the degree of polarization can be determined by measuring the total scattered intensity of a horizontally polarized light beam and dividing it by the total scattered intensity caused by a vertically polarized light beam. When we are interested in depolarization caused by the dissolved material only, the scattered intensities caused by the pure solvent have to be substracted from the scattered intensities of the solution. So the degree of polarization caused by the solute can be expressed by:

$$\Delta_U = \frac{I_{H, \text{solution}} - I_{H, \text{solvent}}}{I_{V, \text{solution}} - I_{V, \text{solvent}}}. \qquad (2.167)$$

If the scattered light is depolarized, the K^* value obtained in the equations derived for isotropic molecules Eqs (2.151), (2.152) and (2.153) has to be multiplied by a correction factor, f. Its value according to Cabannes is equal to:

$$f = \frac{6 + 6\Delta_U}{6 - 7\Delta_U}. \qquad (2.168)$$

It follows that instead of K^* used in Eqs (2.149), (2.150) and (2.157)–(2.160) the following product should be used:

$$K = K^* f. \qquad (2.169)$$

Summarizing the results, the mass average molecular mass of the solute, and the second virial coefficient characteristic of the solvent–solute interaction can be determined by the following relations, depending on the state of polarization of the primary radiation:

$$\frac{Kc}{R_\Theta} = \frac{1}{\bar{M}_{\mathrm{m}}} + 2A_2 c + 3A_3 c^2 + \ldots \tag{2.170}$$

where

$$K = K_{\mathrm{V}} = \frac{4\pi^2 n_1^2}{N_{\mathrm{Av}} \lambda_0^4} \left(\frac{\partial n}{\partial c}\right)_{\mathrm{T}}^2 f \tag{2.171}$$

for a vertically polarized primary beam

$$K = K_{\mathrm{H}} = K_{\mathrm{V}} \cos^2 \Theta \tag{2.172}$$

for a horizontally polarized primary radiation and

$$K = K_{\mathrm{U}} = K_{\mathrm{V}} \frac{1 + \cos^2 \Theta}{2} \tag{2.173}$$

for unpolarized primary radiation.

The analogue of Eq. (2.157), which is principally similar to Eq. (2.170), but differs in its precise form and in the numerical values of the virial coefficients, can also be written for anisotropic molecules using the Cabannes correction factor, so that:

$$\frac{Kc}{R_\Theta} = \frac{1}{\bar{M}_{\mathrm{m}}} (1 + 2\Gamma_2 c + 3\Gamma_3 c^2 + \ldots). \tag{2.174}$$

2.4.3. Scattering in macromolecular solutions

The dimensions of molecules treated so far have been much smaller than the wavelength of the radiation (smaller than $\lambda/20$). For such molecules the generating electrical field can be regarded as homogeneous, and radiation of the dipoles can be described as that of a point dipole.

The situation is more complicated with the macromolecules, where the diameter of the molecular coil is comparable to the wavelength of the radiation, in other words, it exceeds the $\lambda/20$ value. Electrons in the disparate portions of the molecule become generated in different phases. Secondary waves originating from different parts of the macromolecules are coherent, as they are induced by the same primary radiation. They can therefore produce

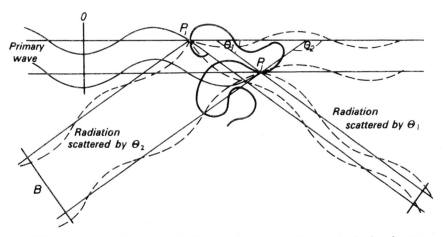

Fig. 2.39. Intensity loss of scattered radiation as a consequence of intramolecular interference. (After Springer 1970.)

interference phenomena, which, depending on the angle of observation, leads to more or less complete extinction. The situation is outlined in Fig. 2.39. The scattering molecule can be divided into small volume elements, from which we will be concerned with P_i and P_j. Plane 0 is perpendicular to the direction of the primary radiation, where the phase of the radiation is a fixed, constant value. Phase difference of the primary radiation at scattering elements P_i and P_j depends on the angle of observation. Along the direction of the primary radiation there is no path-difference between the secondary radiations originating from the two scattering elements. With Θ, the angle of observation, the path-difference also increases, whereas the scattered intensity decreases, owing to interference. This increase of path-difference with increasing angles of observation is noted by the increasing path-differences between the 0 plane and the A, B planes situated perpendicular to the scattered light beams. Partial extinction of the scattered light, due to interference phenomena, leads to an asymmetry of the scattering diagram; macromolecular materials exhibit higher scattered intensity at angles of observation lower than 90°, rather than above this value. Such a distorted scattering surface is shown in Fig. 2.40.

This kind of smooth scattering contour is observed, however, only if the size of the dissolved molecule does not exceed half of the wavelength. In the case of larger molecules, besides asymmetry, maxima and minima also appear. Since at $\Theta = 0°$ there is no decrease in the intensity of the scattered

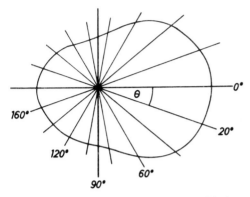

Fig. 2.40. Scattering diagram of molecules with size exceeding $\lambda/20$. (After Springer 1970.)

light, then for macromolecules the intensity of the scattered light is measured at different angles and this function is extrapolated to zero degree where formulae found for the Rayleigh-range are valid.

2.4.3.1. Dependence of the scattering curve on molecular size and shape

Chain-like macromolecules usually form random coils in solution. As has been mentioned in the introduction, in connection with the description of molecular conformations, that two different parameters are used to characterize such coils: the end-to-end distance and the radius of gyration.

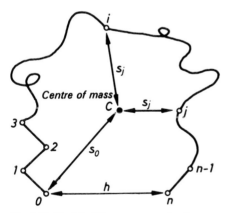

Fig. 2.41. Model of linear macromolecule.

A model of the end-to-end distance, h, and of the radius of gyration, or scattering mass radius, s is shown in Fig. 2.41. The definition of the radius of gyration, s, for a single molecule is:

$$s^2 \equiv \frac{1}{n+1} \sum_{i=0}^{n} s_i^2.$$ (2.175)

If more molecules are present, $\bar{H} = (\overline{h^2})^{0.5}$ represents the average end-to-end distance, and $\bar{S} = (\overline{s^2})^{0.5}$ the average radius of gyration.

2.4.3.2. Angular dependence of the scattering function

Interference effects in macromolecular solutions can be characterized by the $P(\Theta)$ scattering function. Since at $0°$ there is no interference, then $P(\Theta)$ is equal to 1, while at higher Θ values owing to interference effects, $P(\Theta) < 1$.

The $P(\Theta)$ scattering function is dependent upon the angle of observation, the wavelength used, and the shape of the dissolved macromolecule. According to Debye the scattering function can be written in the following form:

$$P(\Theta) = 1 - a_1 \sin^2 \Theta/2 + a_2 \sin^4 \Theta/2 - \ldots$$ (2.176)

where:

$$a_1 = \frac{16}{3} \pi^2 \frac{(\bar{S})^2}{\lambda^2}.$$ (2.177)

$\bar{S}$ is the average radius of gyration, and λ is the wavelength in the solution. For dilute solutions the following approximate equality is valid:

$$\lambda \approx \frac{\lambda_0}{n_1}.$$ (2.178)

At small angles the reciprocal scattering function can be approximated as:

$$1/(1-x) \approx 1+x$$ (2.179)

and

$$\frac{1}{P(\Theta)} = 1 + \frac{16\pi^2(\bar{S})^2}{3\lambda^2} \sin^2 \Theta/2 + \ldots.$$ (2.180)

Using this formula the average radius of gyration can be calculated from the scattering function measured at low angles.

The average radius of gyration can be determined for various molecular shapes, and so for a sphere with radius r:

$$(\bar{S})^2 = \frac{3}{5}r^2 \tag{2.181}$$

for an ellipsoid with half-axes a and b

$$(\bar{S})^2 = \frac{a^2 + b^2}{4} \tag{2.182}$$

and for an ideal random coil

$$(\bar{S})^2 = \frac{1}{6}(\bar{H})^2. \tag{2.183}$$

If the shape of the dissolved molecule is known, the scattering function gives information about the size of the molecule, or the average end-to-end distance. Various tables are available which detail the parameters of molecules with other, more complicated shapes.

2.4.3.3. Measurement of the scattering function

The scattering function can be determined from the angular dependence of the scattered intensity. The asymmetry coefficient, z is characteristic of the scattering function, and can be obtained from observed scattered intensities at $\Theta = 45°$ and $\Theta = 135°$:

$$z = R_{45}/R_{135}. \tag{2.184}$$

As mentioned earlier the scattering function at $\Theta = 0°$ is equal to unity, and so:

$$R_\Theta = R_{0°}P(\Theta). \tag{2.185}$$

The dependence of the asymmetry coefficient on the shape parameters (as the molecular radius in the case of spherical molecules or end-to-end distance in the case of random coils) has been determined for various regular shape models. These data are available from Tables, consequently the average dimensions of molecules with known shape can be determined from the measured asymmetry coefficient. Its practical application will be discussed later.

According to Eq. (2.185), the scattered radiation R_Θ observed at angle Θ should be divided by $P(\Theta)$, the scattering function, to obtain the interference-free scattered radiation without extinction at $\Theta = 0°$. Zimm has elaborated a method to determine the main parameters of such a system.

2.4.4. Determination of molecular mass, size and shape

According to Zimm the scattered intensity of macromolecular solutions exhibiting intramolecular interference phenomena, and with asymmetry coefficients higher than unity, can be described by the following equation (instead of Eq. (2.170)):

$$\frac{Kc}{R_\theta} = \frac{1}{P(\Theta)}\left[\frac{1}{\bar{M}_m} + 2A_2c + 3A_3c^2 + \dots\right] \tag{2.186}$$

or approximately

$$\frac{Kc}{R_\theta} = \frac{1}{P(\Theta)\bar{M}_m} + 2A_2c + \dots. \tag{2.187}$$

Two methods are used to process the measured data: one is the asymmetry measurement introduced by Debye, the other is the extrapolation method introduced by Zimm.

2.4.4.1. Asymmetry method (Debye method)

R_{45}, R_{90} and R_{135} are determined in uncontaminated polymeric solutions. The asymmetry coefficient, z from Eq. (2.184), and Kc/R_{90} are determined at different concentrations.

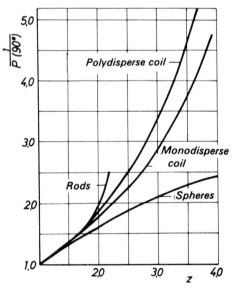

Fig. 2.42. Dependence of the reciprocal scattering function on the asymmetry coefficient.

The limiting value of the asymmetry coefficient at $c=0$ is determined by extrapolation, where the $1/(z-1)$ value is plotted, and a straight line is often obtained. A practical example is shown in Fig. 2.46. According to Eq. (2.177) the scattering function can be calculated for various molecular models. The reciprocal scattering function at $90°$ for several molecular models is plotted versus the asymmetry coefficient in Fig. 2.42. This way the reciprocal scattering function $(1/P(90°))$ can be determined from the asymmetry coefficient.

This method does not allow the determination of molecular shape, it has therefore to be assumed or obtained from other measurements. If, for example, it is known that the molecule is a random coil, the scattering function, the end-to-end distance, and average radius of gyration can be determined. An example of such a calculation will be considered later. In this method $P(90°)$ has to be calculated from z.

According to Eq. (2.187), by extrapolating to $c=0$:

$$\lim_{c \to 0} \frac{Kc}{R_{90}} = \frac{1}{P(90)\bar{M}_m}, \tag{2.188}$$

the mass average molecular mass can be determined from $P(90°)$ and Kc/R_{90}.

2.4.4.2. The Zimm extrapolation method

This method yields the mass average molecular mass independent of the molecular shape, but requires measurements at a wider range of angles. From Eqs (2.180) and (2.187):

$$\frac{Kc}{R_\Theta} = \frac{1}{\bar{M}_m}\left[1 + \frac{16}{3}\pi^2 \frac{(\bar{S})^2}{\lambda^2}\sin^2\frac{\Theta}{2} + \ldots\right] + 2A_2 c + \ldots. \tag{2.189}$$

By neglecting higher order terms:

$$\frac{Kc}{R_\Theta} = \frac{1}{\bar{M}_m} + \frac{1}{\bar{M}_m}\frac{16}{3}\pi^2 \frac{(\bar{S})^2}{\lambda^2}\sin^2\frac{\Theta}{2} + 2A_2 c. \tag{2.190}$$

At a constant Θ value Kc/R_Θ is a linear function of c, whereas at a constant c value Kc/R_Θ varies linearly with $\sin^2 \Theta/2$. Plotting these functions and extrapolating them to $c=0$ and $\Theta=0°$ the following limiting values can be obtained:

$$\lim_{c \to 0} \frac{Kc}{R_\Theta} = \frac{1}{\bar{M}_m} + \frac{1}{\bar{M}_m}\frac{16}{3}\pi^2 \frac{(\bar{S})^2}{\lambda^2}\sin^2\frac{\Theta}{2} \tag{2.191}$$

and

$$\lim_{\Theta \to 0} \frac{Kc}{R_\Theta} = \frac{1}{\bar{M}_\mathrm{m}} + 2A_2 c. \tag{2.192}$$

If these extrapolated values are plotted versus $\sin^2 \Theta/2$ and c for Eqs (2.191) and (2.192) respectively, the intercepts give the reciprocal mass average molecular mass, whereas the slopes give the radius of gyration and the second virial coefficient respectively.

These extrapolations are performed by means of a single diagram according to the method introduced by Zimm, where Kc/R_Θ values are plotted against $\sin^2 \Theta/2 + bc$. b is an arbitrarily chosen constant to expand the abscissa scale. Fig. 2.43 shows a Zimm plot of this kind. Along the dashed lines

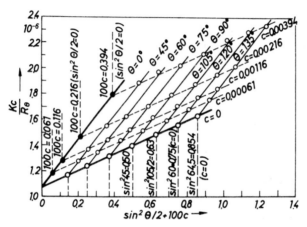

Fig. 2.43. Zimm-plot for polystyrene. $M = 940,000$ solvent methylethyl ketone.

Kc/R_Θ values are plotted against $\sin^2 \Theta/2 + bc$ at constant concentration values. Extrapolating these lines to bc abscissa values Kc/R_Θ values measured at 0 degree are obtained at the particular concentrations (full points). The intercept of the straight line connecting these full points gives $1/\bar{M}_\mathrm{m}$.

Points interconnected by continuous lines show the Kc/R_Θ values at a particular angle as a function of concentration. Extrapolating these lines to the $\sin^2 \Theta/2$ abscissa values (shown by empty points in the Figure) Kc/R_Θ values relevant to zero concentration ($bc = 0$) are obtained as a function of the angle of observation. The ordinate intercept of the straight line drawn through these points also gives $1/\bar{M}_\mathrm{m}$.

Slopes of the straight lines used for extrapolation can be used to obtain the second virial coefficient (points extrapolated to $\Theta = 0°$), and the average radius of gyration (points extrapolated to $c=0$), in accordance with the following relations:

$$A_2 = \frac{1}{2} \lim_{c \to 0} \left\{ \frac{d}{dc} \left(\lim_{\Theta \to 0} \frac{Kc}{R_\Theta} \right) \right\},$$

(2.193)

and

$$(\bar{S})^2 = \bar{M}_m \frac{3}{16} \frac{\lambda^2}{\pi^2} \lim_{\Theta \to 0} \left\{ \frac{d}{d \sin^2 \Theta/2} \left(\lim_{c \to 0} \frac{Kc}{R_\Theta} \right) \right\}.$$

(2.194)

In the case of polydisperse systems, because of reasons not discussed here, the so-called z-average defined by Eq. (2.7) can be obtained from the radius of gyration.

Up to this point the radius of gyration has been determined from Eq. (2.189). No information is gained, however, about molecular shape. To solve this problem the sample has to be fractionated, and from the Zimm plot of the fractionated samples Kc/R_{45}, Kc/R_{90} and Kc/R_{135} must be determined at infinite dilution. Using these values the asymmetry coefficient can be determined according to Eq. (2.186):

$$\lim_{c \to 0} \frac{Kc/R_{135}}{Kc/R_{45}} = \lim_{c \to 0} \frac{R_{45}}{R_{135}} = z.$$

(2.195)

Using Eq. (185), and knowing the average molecular mass the experimental value of the reciprocal scattering function can also be determined:

$$\bar{M}_m \lim_{c \to 0} \frac{Kc}{R_{90}} = \frac{1}{P(90)}.$$

(2.196)

Finally, a careful study of this experimental $1/P(90)$ function plotted versus z fixes information about the shape of the dissolved particles, using theoretical predictions shown in Fig. 2.43.

2.4.4.3. Measurement of the scattered intensity

The absolute value of the reduced scattered intensity can be defined as the intensity of the scattered light scattered by 1 cm^3 of the scattering medium, at a radius of 1 cm from the scattering centre, measured at Θ degrees, provided that the intensity of the primary beam is 1 candela (cd). Since these conditions are never fulfilled in practice, commercial equipment gives relative rather than absolute values. Absolute scattered intensities for standard materials, such as benzene, are available in the literature. Using carefully purified standards, relative scattered intensity is measured at a fixed angle, usually at $\Theta = 90°$.

Using this experimental value and that found in the literature for absolute
scattered intensity, a correction factor can be calculated for the equipment,
and absolute intensities may then be determined for any particular apparatus.
By simple proportions:

$$\frac{R_{90,\text{sample}}}{R_{90,\text{standard}}} = \frac{G_{90,\text{sample}}}{G_{90,\text{standard}}}$$

or

$$R_{90,\text{sample}} = \frac{R_{90,\text{standard}}}{G_{90,\text{standard}}} G_{90,\text{standard}} \cdot \qquad (2.197)$$

2.4.4.4. Correction for the scattering volume

The calibration procedure described above is only valid for a single angle of
observation, $\Theta = 90°$. At other angles it must be taken into account that the
scattering volume, V_Θ is a function of the angle. Fig. 2.44 shows the scattering
volume at angles other than $\Theta = 90°$. It is obvious that:

$$V_{90} = V_\Theta \cdot \sin \Theta. \qquad (2.198)$$

and absolute scattered intensity at any particular angle of observation can be
obtained if, in addition to the equipment constant, the sine of the angle of
observation is used as a multiplying factor:

$$R_{\Theta,\text{sample}} = \frac{R_{90,\text{standard}}}{G_{90,\text{standard}}} G_{\Theta,\text{sample}} \sin \Theta. \qquad (2.199)$$

Sometimes the refractive index of the medium around the measuring cell
is very different from that of the solution. In this case the scattered light is
refracted in the surrounding medium, the observed scattering volume is
further modified, and further correction factors have to be used. If the liquid
surrounding the measuring cell has a refractive index close to that of the
solution and the wall of the cell is surrounded by air, whose refractive index is

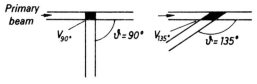

Fig. 2.44. Dependence of the effective scattering volume on the scattering angle. (After
Margerison and East 1967.)

close to unity, the refractive index correction is independent of the shape of the cell, and its value is equal to n^2, where n is the refractive index of the solution. In this case this value appears as a further correction factor in Eq. (2.199).

2.4.5. Equipment

The scheme of a light scattering instrument is shown in Fig. 2.45. Scattered intensity is measured by a photomultiplier, set to various angles of observation. Raw experimental data are the current galvanometric readings

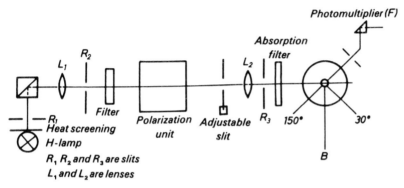

Fig. 2.45. Schematic of light scattering apparatus.

which are proportional to the intensity detected by the photomultiplier. From these galvanometric readings, G_θ, the reduced scattered intensity, R_θ, can be calculated using Eq. (2.199).

2.4.6. A practical example of the asymmetry method

Asymmetry parameters measured in a benzene–polystyrene solution are given in Table 2.16. The z-value relevant to infinite dilution is determined by plotting $1/(z-1)$ versus c (Fig. 2.46). This relationship is usually linear, and is therefore more suitable for extrapolation than z, the asymmetry coefficient itself. This is shown well in Fig. 2.46 where both functions are plotted. The limiting value is equal to $z_{c=0} = 1.641$. $1/P(90)$, the reciprocal value of the

Table 2.16

Dependence of the asymmetry coefficient, on concentration
(Polystyrene sample)

$c \cdot 10^3$ g $\cdot$ cm^{-3}	2.00	1.50	1.00	0.75	0.50
z	1.277	1.326	1.385	1.428	1.494

After Margerison and East 1967.

scattering function at $\Theta = 90°$, can be calculated, but instruments are
normally provided with tabular information assuming various molecular
shapes. Such data have been shown plotted in Fig. 2.42.

If there is no other information available concerning the form of the
dissolved molecule, random coils can be assumed, and an approximate value
of $1/P(90)$ can be taken from Fig. 2.42. Fig. 2.47 shows an alternative diagram
for random coils, from which the scattering function can be determined with
higher precision. In this case $1/P(90)$ is equal to 1.46. This value will be used in
molecular mass determination.

From $1/P(90)$, the average radius of gyration, $\bar{S}$ can be calculated using
Eq. (2.180). This relation is shown in Fig. 2.48. In our case the radius of
gyration, $\bar{S}$ is equal to 48 nm. The average end-to-end distance calculated for
ideal random coils using Eq. (2.183) is also plotted in the same Figure. In our
example this value is equal to 116 nm. Molecular mass of the sample can be
obtained using the asymmetry method if Kc/R_{90} values are plotted versus
concentration, and the low concentration limit is determined by extra-

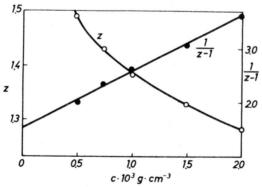

Fig. 2.46. Determination of the asymmetry coefficient. (After Margerison and East 1967.)

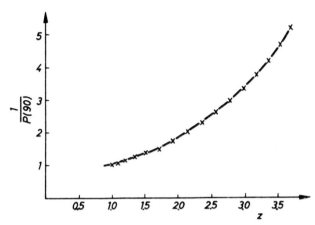

Fig. 2.47. Scattering variation of a random coil macromolecule with the asymmetry coefficient.

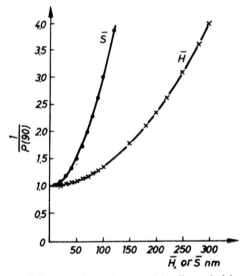

Fig. 2.48. Dependence of the inertial radius and of the average end-to-end distance on $1/P$ (90°).

polation. Kc/R_{90} values obtained in this example are shown in Table 2.17. Plotting these values versus c, a straight line is obtained with an intercept at $c = 0$ equal to 2.04×10^{-6} (mole $\cdot$ g^{-1}) (Fig. 2.49). Such that:

$$2.04 \cdot 10^{-6} \text{ mole} \cdot \text{g}^{-1} = \frac{1}{\bar{M}_m P(90°)} .$$

And from this relation $\bar{M}_m = 720{,}000$ (g $\cdot$ mole^{-1}).

Table 2.17

Dependence of Kc/R_{90} on concentration

$c \cdot 10^3 \text{ g} \cdot \text{cm}^{-3}$	2.00	1.50	1.00	0.75	0.50
$\dfrac{Kc}{R_{90}} \cdot 10^6 \text{ mole} \cdot \text{g}^{-1}$	3.72	3.27	2.85	2.64	2.51

After Margerison and East 1967.

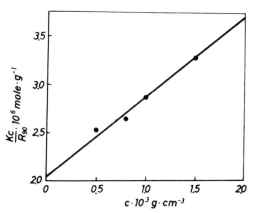

Fig. 2.49. Kc/R_{90} plotted versus concentration. (After Margerison and East 1967.)

2.4.7. A practical example of the Zimm method

Scattering data for a polystyrene–benzene solution measured with an unpolarized primary beam of wavelength 4.561×10^{-5} cm are summarized in Table 2.18. Data refer to galvanometric readings. Absolute scattered intensities can be calculated using Eq. (2.170). The Rayleigh factor for benzene is $R_{90} = 16.3 \times 10^{-6}$ cm. The G_{90} value is equal to 100. (This value can be set using a glass standard.) The equipment constant is expressed as $k = 16 \cdot 3 \times 10^{-6}/100 = 16.3 \times 10^{-8}$. Using this value, scattered intensities calculated from galvanometric readings are given in Table 2.19.

In order to obtain the molecular mass using Eq. (2.161) we should like to plot Kc/R versus the angle of observation. To do so Table 2.20 has been calculated from data given in Table 2.19. Material constants used in the calculation are given with the Table.

The curves in Fig. 2.50 have been drawn using the data from Table 2.20, so that the intercepts can be determined at different concentrations. Plotting

Table 2.18

G_Θ galvanometric readings

c, g·cm^{-3}	Θ										
	30°	37.5°	45°	60°	75°	90°	105°	120°	135°	142.5°	150°
$2.00 \cdot 10^{-3}$	1,542	1,153	917	607	461	408	440	540	755	936	1,235
$1.50 \cdot 10^{-3}$	1,383	1,045	820	550	413	363	384	475	660	820	1,080
$1.00 \cdot 10^{-3}$	1,158	872	682	455	343	301	319	392	540	660	380
$0.75 \cdot 10^{-3}$	998	750	590	396	297	263	275	339	464	565	755
$0.50 \cdot 10^{-3}$	803	607	477	319	241	214	224	275	376	454	607
Pure benzene	282	213	170	128	105	100	105	127	170	207	285

Non-polarized, monochromatic light, wavelength is 546.2 nm.
After Margerison and East 1967.

Table 2.19

$R_\Theta \times 10^6$ cm values

c, g·cm^{-3}	Θ										
	30°	37.5°	45°	60°	75°	90°	105°	120°	135°	142.5°	150°
$2.00 \cdot 10^{-3}$	102.7	92.3	86.1	67.6	56.0	50.2	52.7	58.3	67.4	72.3	77.4
$1.50 \cdot 10^{-3}$	89.7	82.6	74.9	59.6	48.5	42.9	43.9	49.1	56.5	60.8	64.8
$1.00 \cdot 10^{-3}$	71.4	65.4	59.0	46.2	37.5	32.8	33.7	37.4	42.6	45.0	48.5
$0.75 \cdot 10^{-3}$	58.4	53.3	48.4	37.8	30.2	26.6	26.8	29.9	33.9	35.5	38.3
$0.50 \cdot 10^{-3}$	42.5	39.1	35.4	27.0	21.4	18.6	18.7	20.9	23.7	24.5	26.2

$R_\Theta = 16.3 \times 10^{-8} \times G_\Theta \sin \text{cm}^{-1}$

After Margerison and East 1967.

these intercepts versus concentration (Fig. 2.51) allows the intercept relevant to zero concentration and 0° angle of observation to be obtained. This intercept is the reciprocal value of the average molecular mass to be determined, $1/\bar{M}_m$. The intercept is 1.37×10^{-6} (mole·g^{-1}), and so the mass average molecular mass of the sample is 730,000 (g·mole^{-1}).

The slope of the straight line in Fig. 2.51 is equal to $2A_2 = 8.76 \times 10^4$ (mole·g^{-2}·cm^3), and so:

$$A_2 = 4.38 \times 10^{-4} \text{ (mole·g}^{-2}\text{·cm}^3)$$

or according to Eq. (2.128):

$$\Gamma_2 = A_2 M = 3.20 \times 10^2 \text{ (cm}^3\text{·g}^{-1}).$$

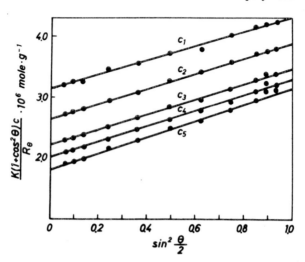

Fig. 2.50. $K(1+\cos^2\Theta)\, c/R_\Theta$ versus $\sin^2\Theta/2$. Data are taken from a polystyrene solution in benzene. Temperature is 25°C. $c_1 = 2.0 \times 10^{-3}$; $c_2 = 1.5 \times 10^{-3}$; $c_3 = 1.0 \times 10^{-3}$; $c_4 = 0.75 \times 10^{-3}$; $c_5 = 0.5 \times 10^{-3}$, all concentrations given in g/cm^{-3} units. (After Margerison and East 1967.)

Table 2.20

$$\frac{K(1+\cos^2\Theta)c}{2R_\Theta}\, 10^6 \text{ mole} \cdot \text{g}^{-1} \text{ values}$$

c, g·cm^{-3}	Θ										
	30°	37.5°	45°	60°	75°	90°	105°	120°	135°	142.5°	150°
$2.00 \cdot 10^{-3}$	3.18	3.26	3.25	3.45	3.56	3.72	3.78	4.01	4.16	4.21	4.22
$1.50 \cdot 10^{-3}$	2.73	2.76	2.81	2.94	3.08	3.27	3.40	3.57	3.72	3.75	3.78
$1.00 \cdot 10^{-3}$	2.29	2.33	2.37	2.53	2.66	2.85	2.96	3.12	3.29	3.38	3.37
$0.75 \cdot 10^{-3}$	2.10	2.14	2.17	2.32	2.47	2.64	2.79	2.93	3.10	3.21	3.20
$0.50 \cdot 10^{-3}$	1.92	1.95	1.98	2.16	2.33	2.51	2.66	2.79	2.96	3.11	3.12

$K = 9.34 \times 10^{-8}$ cm$^2 \cdot$ g$^{-2} \cdot$ mole, where
$K = 4\pi^2 n_1^2, (dn/dc)^2 / \lambda_0^4 N_{Av}$,
using the following data: $n_1 = 1.5014$,
$\lambda_0 = 5.461 \times 10^{-5}$ cm, $dn/dc = 0.106$ cm$^3 \cdot$ g^{-1}
$N_{Av} = 6.023 \times 10^{23}$ mole^{-1}.

After Margerison and East 1967.

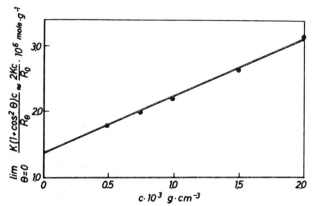

Fig. 2.51. Dependence of Kc/R_θ on the concentration. Data as in Fig. 2.50. (After Margerison and East 1967.)

As mentioned in connection with Fig. 2.43, both extrapolations can be performed using a single diagram, the Zimm plot. Here Kc/R_θ is plotted versus $\sin^2 \Theta/2 + kc$ where k is an arbitrary constant, usually 100 or 1,000. This value is used to shift the lines belonging to a single concentration value (Fig. 2.52). Consequently, two sets of straight lines are obtained, one of them refers to constant angles of observation and different concentrations, whereas the other to constant concentrations and varying angles of observation. Determinations of the average molecular mass and the second virial

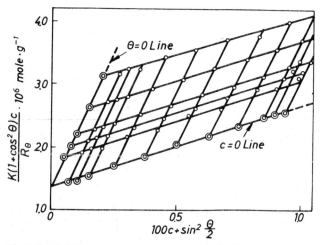

Fig. 2.52. The Zimm plot. Data are the same as in Figs 2.50 and 2.51. Circles are extrapolated values. (After Margerison and East 1967.)

coefficient has been shown above. Now let us calculate the radius of gyration. The slope of the line belonging to $c=0$, according to Eq. (2.191), is equal to:

$$\frac{16\pi^2(\bar{S})^2}{3\lambda^2\,\bar{M}_m},$$

the numerical value from the Zimm plot is 1.29×10^{-6} (mole $\cdot$ g^{-1}).

From this follows:

$$(S)^2 = \frac{1.29 \cdot 10^{-6} \cdot 730{,}000 \cdot 3}{16\pi^2}\left[\frac{546.1}{1.5014}\right]^2 [\text{nm}^2]$$

$$\bar{S}^2 = 2369, \quad \bar{S} = 48\,[\text{nm}].$$

The average end-to-end distance, assuming a random coil configuration and using Eq. (2.183) is $\bar{H} = 120$ (nm).

To conclude this section on light scattering measurements it should be noted that the applicability of this method to polymer solutions depends on the fulfillment of two important conditions:

— The polymer molecule must be isotropic.
— There must be a finite, but not excessive refractivity difference between the solvent and the solution.

The first condition is met by random coil polymers. For disk- or rodlike polymers correction due to depolarization must be used at $\Theta = 90°$. The second condition limits the number of solvents. The pivotal parameter in this respect is dn/dc, the refractivity increment, which has a squared relationship with the R values measured. In the special case where dn/dc is equal to zero, that is, if the refractive indices of the polymer and the solvent are equal, there is no excess scattering in comparison to the pure solvent. For example, in case of PVC dissolved in trichloro-benzene.

If, however, refractivity increment is large, higher than, say, 0.2 cm$^3 \cdot$ g^{-1}, scattered intensity is not a linear function of $(dn/dc)^2$, and the equations given are not valid for such a case. Within these limits refractivity increment must be measured with high precision, as a two per cent error causes one of four per cent in the molecular mass.

As a practical example let us consider the polystyrene system treated previously. Here $dn/dc = 0.106$ cm$^3 \cdot$ g^{-1}, so that if 1 g of polystyrene is dissolved in 100 cm^3 solution, then:

$$\Delta n = 0.106 \times 0.01\ \text{cm}^3 \cdot \text{g}^{-1} \cdot \text{g} \cdot \text{cm}^{-3} = 0.00106.$$

If concentrations are in the range used for light scattering measurements, the accuracy of the refractive index determination must be at least 0.00002 in order to get less than four per cent error in the molecular mass. Evidently, for this purpose a special instrument is needed, which will not be dealt with in detail here. Since the scattered intensity is proportional to λ^{-4} (see Eq. (2.105)) the shortest possible wavelength should be used with a very narrow spectral distribution. In fact, modern instruments always use lasers as light sources.

Finally the problem of solvent purification should be mentioned.

Solvents used for light scattering measurements are usually purified by centrifuge and filtration. The use of dust- and contamination-free laboratory vessels, and the manipulation of solvents and solutions in them requires great care and routine. Even the slightest impurity leads to high errors, and techniques must be rigorous.

2.5. VISCOSITY MEASUREMENTS IN DILUTE SOLUTIONS

Measurement of the viscosity of dilute polymer solutions is one of the most frequently used, and easily available methods, for the characterization of the average molecular mass of polymers. Being a relative method, the result is obtained using empirical relations determined experimentally, but its accuracy is frequently good enough to solve many problems.

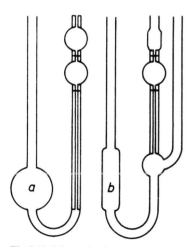

Fig. 2.53. Schematic of the Ostwald *a* and Ubbelohde *b* viscometers. (After Margerison and East 1967.)

Viscosity of solvent and solutions are usually determined in capillary viscometers. Two widely used types are the Ostwald and Ubbelohde viscometers shown in Fig. 2.53. Viscosity is measured by determining the flow time of the solvent or the solution between two marks on the glass capillary. Flow is monitored by the movement of the liquid meniscus. Viscosity is calculated from the Poiseuille formula:

$$v = \frac{\pi P r^4}{8 \eta l} \tag{2.200}$$

where v is the volumetric flow rate, P the pressure difference, the driving force of the flow, r the radius of the capillary, η the viscosity of the fluid medium, and l the length of the capillary. This equation is valid for the laminar flow of Newtonian liquids.

Eq. (2.200) can be simplified by introducing the viscosity ratio, earlier referred to as relative viscosity:

$$\frac{\eta_{\text{solution}}}{\eta_{\text{solvent}}} = \frac{\eta_2}{\eta_1} \tag{2.201}$$

where it is assumed that both measurements are performed at the same temperature. If the volume of the liquid between the two marks is equal to V, and the outflow times for the solvent and the solution are t_1 and t_2 respectively, then:

$$\frac{V}{t_1} = \frac{\pi P_1 r^4}{8 \eta_1 l}, \tag{2.202}$$

and

$$\frac{V}{t_2} = \frac{\pi P_2 r^4}{8 \eta_2 l}, \tag{2.203}$$

so that

$$\frac{\eta_2}{\eta_1} = \frac{t_2 P_2}{t_1 P_1}. \tag{2.204}$$

If the height of the liquid column, h, is constant, P is proportional to the density of the liquid, and so:

$$\frac{\eta_2}{\eta_1} = \frac{t_2 h \rho_2}{t_1 h \rho_1} = \frac{t_2 \rho_2}{t_1 \rho_1}. \tag{2.205}$$

In the case of dilute solutions the two density values are practically equal, so to a fair approximation:

$$\frac{\eta_2}{\eta_1} = \frac{t_2}{t_1}.$$

(2.206)

Concentrations and viscosities are usually chosen so that the outflow time should be greater than 100 sec, whereas the optimum value of viscosity ratio is 1.1–1.5. The concentration dependence of the viscosity ratio is shown in Fig. 2.54 for a polystyrene–toluene solution.

The viscosity ratio–concentration function can be described by a power series:

$$\frac{\eta_2}{\eta_1} = 1 + [\eta]c + kc^2 + \ldots$$

(2.207)

where $[\eta]$ and k are constants. This equation shows that at zero concentration the viscosity ratio is equal to unity, and the concentration dependence at low concentrations can be readily described by a quadratic function. Further terms of the power series become important only at higher concentrations.

In the case of relatively dilute solutions Eq. (2.207) can be rearranged to determine the characteristic constants:

$$\frac{\eta_2 - \eta_1}{\eta_1 c} = [\eta] + kc.$$

(2.208)

Plotting the left hand side of Eq. (2.208) versus the concentration results in a straight line with a slope k and intercept $[\eta]$.

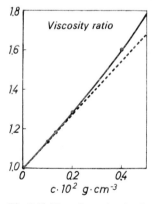

Fig. 2.54. Viscosity ratio of polystyrene/toluene solution as a function of concentration. (After Margerison and East 1967.)

This is shown in Fig. 2.55 using the data of the previous Figure. From the Figure it follows that:

$$[\eta] = 129.2(\text{cm}^3 \cdot \text{g}^{-1})$$

$$k = 5.65 \times 10(\text{cm}^6 \cdot \text{g}^{-2}).$$

The quantity $(\eta_2 - \eta_1)/\eta_1$ is known as the specific viscosity, its relationship with the directly measurable viscosity ratio is shown in Table 2.22. The left hand side of Eq. (2.208) is called the viscosity number. Its limiting value, $[\eta]$ is the so called limiting viscosity number, sometimes referred to as the intrinsic viscosity. In the case of a given polymer–solvent system both k and $[\eta]$ are

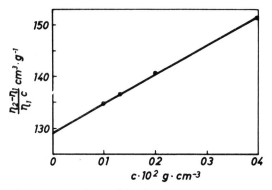

Fig. 2.55. Dependence of the viscosity number on concentration. (After Margerison and East 1967.)

Table 2.21

Limiting viscosities of PS samples in toluene at 30 °C and the k values

$\bar{M}_n$	$[\eta]$ cm$^3 \cdot$ g^{-1}	$k \cdot 10^{-3}$ cm$^6 \cdot$ g^{-2}
76.000	38.2	0.45
135.000	59.2	1.17
163.000	69.6	1.58
336.000	105.4	3.90
440.000	129.2	5.65
556.000	165.0	8.40
850.000	221.0	15.0

After Margerison and East 1967.

functions of molecular mass. Table 2.21 shows limiting viscosity numbers and k values for polystyrene/toluene solutions measured at 30°C, as a function of the number average molecular mass. k values can also be expressed by a term containing the square of the limiting viscosity number:

$$k = k \cdot [\eta]^2. \qquad (2.209)$$

Eq. (2.208) can be re-written using this notation in the form:

$$\frac{\eta_2 - \eta_1}{\eta_1 c} = [\eta] + k'[\eta]^2 \qquad (2.210)$$

which is known as the Huggins equation, k' being the Huggins constant. Its value is approximately 0.38 in good solvents, and lies between 0.5 and 0.8 in Θ-solvents, and between 1 and 1.3 in poor solvents. Previously used, and currently accepted nomenclature of the various quantities occurring in Eq. (2.210) is summarized in Table 2.22.

Table 2.22

Symbols and notations used in viscometry

Present	Earlier	Formula
Notation		
Viscosity ratio	Relative viscosity	$\eta_r = \dfrac{\eta}{\eta_0} \approx \dfrac{t}{t_0}$
Specific viscosity	Specific viscosity	$\eta_{sp} = \eta^{-1} = \dfrac{\eta - \eta_0}{\eta_0} \approx \dfrac{t - t_0}{t_0}$
Viscosity number	Reduced viscosity	$\eta_{red} = \dfrac{\eta_{sp}}{c} = \dfrac{\eta - \eta_0}{\eta_0 c}$
Logarithmic viscosity number	Inherent viscosity	$\eta_{inh} = \dfrac{\ln \eta_r}{c}$
Limiting viscosity number (LVN)	Intrinsic viscosity	$[\eta] = \left(\dfrac{\eta_{sp}}{c}\right)_{c \to 0} = \left[\dfrac{(\ln \eta_r)}{c}\right]_{c \to 0}$

2.5.1. Characterization of rigid molecules by the limiting viscosity number

According to the calculations of Einstein the viscosity ratio of the polymeric solution can be described with the following relationship if the molecules behave as rigid spheres:

$$\frac{\eta_2}{\eta_1} = 1 + 2.5\,\Phi \tag{2.211}$$

where Φ is the volume fraction of the solute. If mass concentration $c\ [\text{g/cm}^3]$ is used, then:

$$\Phi = c/\rho_2 \tag{2.212}$$

where ρ_2 is the density of the dissolved particles. The Einstein equation can also be written in the form:

$$[\eta] = 2.5/\rho_2. \tag{2.213}$$

The viscosity of such systems also depends on concentration.

Using the Huggins relations (Eq. (2.210)) this concentration dependence can be described by:

$$\frac{\eta_2 - \eta_1}{\eta_1 c} = [\eta] + k'[\eta]^2 c. \tag{2.210}$$

According to Simha:

$$[\eta] = 2.5/\rho_2$$

and

$$k' = 12.6/[\eta]^2 \rho_2^2.$$

From the limiting viscosity number the hydrodynamically effective (apparent) volume and density of the dissolved molecules can be determined using:

$$\rho_2 = \frac{m}{V_e} \tag{2.214}$$

where V_e is the effective volume, and m the mass of the dissolved particles.

True and apparent volumes differ if the molecules do not behave as rigid spheres, for example, if solvation occurs, which increases the effective (hydrodynamically active) volume.

For rigid rodlike molecules there is a relationship between the limiting viscosity of the dilute solution and the molecular mass of the dissolved molecules, which can be determined in the following way.

Let us assume that Eqs (2.208) and (2.209) are applicable, such that:

$$[\eta] = 2.5/\rho_2,$$

where

$$1/\rho_2 = \frac{V_e}{m}.$$

The hydrodynamically effective density is calculated for rodlike molecules from an effective volume which corresponds to the volume of a sphere produced by the rotation of rodlike molecules:

$$V_e = \frac{L^3 \pi}{6}, \tag{2.215}$$

where L is the length of particles. m, the mass of the molecule is proportional to the length of the molecule:

$$m = L m_0, \tag{2.216}$$

where m_0 is the mass of unit length of the molecule. From this it follows that:

$$\frac{1}{\rho_2} = \frac{V_e}{m} = \frac{L^3 \pi}{6 L m_0} = \frac{L^2 \pi}{6 m_0} = \frac{m^2 \pi}{6 m_0^3}. \tag{2.217}$$

If the dissolved particles are molecules, m means the molecular mass, so that, in the case of rigid rodlike molecules, the limiting viscosity number should be proportional to the square of the molecular mass, in accordance with experimental observations.

2.5.2. Characterization of random coil molecules by the limiting viscosity number

Coiled molecules can also be described by the Einstein formula, albeit the effective density of the coil, ρ_{eff} is very low. In the case of dense coils its value is around 10^{-2}, but here it has to be taken into account that the coil contains immobilized, hydrodynamically inactive solvent, therefore this type of molecule is known as an "impermeable coil".

The degree of coiling can be relatively low, in which case solvent molecules may flow freely among the macromolecular segments. Such molecules are referred to as "freely permeable coils", with effective densities between 10^{-3} and 10^{-4}.

In practice the state of dissolved macromolecules is between these limiting cases. Apparent density can be calculated from the ratio of molecular mass and mole volume:

$$\rho_{eff} = \frac{M}{N_{Av} V_e} \tag{2.218}$$

where N_{AV} is the Avogadro number, V_e the apparent or effective volume of the coil.

Coil density varies with molecular mass as:

$$\rho_{eff} = kM^{-a}. \tag{2.219}$$

Applying the Einstein relation to coils:

$$[\eta] = 2.5/\rho_{spherical\ equivalent} \tag{2.220}$$

where $\rho_{spherical\ equivalent}$ is the density of that sphere viscosity which would be equal to that of the molecular coil. This quantity is in the same order of magnitude as ρ_{eff}, and so it is expected that the limiting viscosity number of coiled molecules should lie between 10^2 and 10^4 cm^3/g, which is confirmed by experiments. If it is assumed that $\rho_{spherical\ equivalent}$ varies with the molecular mass similarly to ρ_{eff}, the effective coil density, according to Eq. (2.219), then it can be written that the limiting viscosity number:

$$[\eta] = KM^a. \tag{2.221}$$

This is one of the most important equations in viscometry, and has been derived by Staudinger, Mark, Houwink, Kuhn and others. In the literature it is usually referred to as the SMH equation from the initials of the first three scientists, or as the Mark–Houwink equation.

As shown in the introduction of the excluded volume concept, the ratio of real and ideal end-to-end distances, α, the Flory expansion parameter, characterizes the deviation from the ideal random coil state, that is, the degree of perturbation, so that:

$$\bar{H} = \alpha\bar{H}_0. \tag{2.18}$$

Variation of average end-to-end distance can be described in another way, and this modified form will be used in further discussion of the SMH equation. This alternative form describes the perturbation of the end-to-end distances by way of an exponent:

$$(\bar{H})^2 = C(\bar{H}_0)^2 P^{1+\varepsilon} \tag{2.222}$$

where P is the degree of polymerization, and C a constant.

Comparing Eq. (2.222) with Eq. (2.18) it follows that $\alpha^2 \approx P^\varepsilon$, and so:

$$\varepsilon = \frac{d\ln \alpha^2}{d\ln P} = \frac{P}{\alpha^2} \frac{d\alpha^2}{dP}. \tag{2.223}$$

The exponent, a, in Eq. (2.221) depends on the degree of coiling, and is related to the parameter ε, so that:

$$a = \frac{3\varepsilon + 1}{2}. \tag{2.224}$$

In the case of impermeable coils $a = 0.5$ and $\varepsilon = 0$, whereas for freely permeable coils $a = 1$ and $\varepsilon = 0.3$. For real molecules the value of a usually ranges between 0.6 and 0.9. For rigid rods $a = 2$ and $\varepsilon = 1$. In the case of dense spheres $a = 0$, where the limiting viscosity number is independent of M. In the case of densified impermeable coils $a < 0.5$. Such values are found, for example, in branched molecules.

The limiting viscosity of coiled molecules highly depends on the solvent. In good solvents its value changes only slightly with temperature. In poor solvents, however, it shows a strong temperature dependence, the coils expand, and the limiting viscosity number increases.

At the Θ-temperature, or in the Θ-state, the Flory expansion parameter appearing in Eq. (2.18) is equal to unity. This means that perturbation forces are cancelled by attractive forces arising between the segments. The ideal solvent and temperature are denoted as the Θ-solvent and Θ-temperature respectively. In this case ε is equal to zero, and $a = 0.5$.

2.5.3. Molecular mass determination from the limiting viscosity number

Eq. (2.221) is used mainly to determine the average molecular mass from the measured limiting viscosity number. K and a values are taken from Tables. These Tables are based on measurements with samples of narrow molecular mass distribution (molecular mass standards). Plotting the limiting viscosity numbers versus molecular mass (determined by an independent method) the slope and intercept of the fitted straight line gives a and K respectively. Such experimental points are shown in Fig. 2.56.

It is important to perform this calibration with samples of narrow molecular mass distribution, since in this case:

$$\bar{M}_n \approx \bar{M}_v \approx \bar{M}_m \tag{2.225}$$

where $\bar{M}_v$ denotes the viscosity average molecular mass, which can be given by (Schaefgen 1948):

$$\bar{M}_v = \left\{ \frac{\Sigma N_i M_i^{a+1}}{\Sigma N_i M_i} \right\}^{1/a}. \tag{2.226}$$

This expression is equal to the mass average if $a=1$. Viscosity average molecular mass lies between the mass average and the number average,

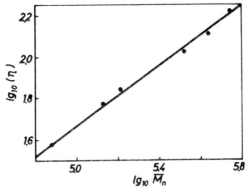

Fig. 2.56. Dependence of limiting viscosity number on log M_n (After Margerison and East 1967.)

usually closer to the mass average. If the distribution is not too broad (if $\bar{M}_m/\bar{M}_n < 2$) its value is only a few per cent lower than $\bar{M}_m$, and can be taken as practically equal to the mass average.

Experimentally determined K and a constants in the literature are numerous, and frequently differ, even for the same system. Since limiting viscosity measurements are fairly reproducible, it is advisable to give the K and a parameters used for data processing. K and a parameters for some polymer/solvent system are given in Table 2.23.

As mentioned earlier, under Θ-conditions $a=0.5$. It would therefore be expected that coil densities of molecules with equal molecular mass would be uniform, and the K constants for various systems under these conditions should be similar. This expectation is confirmed by experiments. K values under Θ-conditions are usually between 0.08 and 0.1. However, Eq. (2.221) is valid for linear polymers only.

Table 2.23

Kuhn–Mark–Houwink constants of some polymer/solvent pairs

Polymer	Solvent	T, °C	$\dfrac{K \cdot 10^2}{cm^3 \cdot g^{-1}}$	a	Method
Polystyrene	Toluene	25	1.00	0.73	Light scattering
Polystyrene	Toluene	30	1.05	0.73	Osmometry
Polystyrene	Benzene	25	1.03	0.74	Osmometry
Poly(methyl methacrylate)	Toluene	25	0.71	0.73	Light scattering
Poly(methyl methacrylate)	Benzene	25	0.57	0.76	Light scattering
Polyethylene	Trichloro benzene	135	4.60	0.73	Light scattering
Polypropylene	Trichloro benzene	135	1.37	0.75	Light scattering
Polyamide 6	*m*-cresol	25	32.00	0.62	End group method
Polyisobutylene	Toluene	25	4.08	0.62	Light scattering
Polyisobutylene	Benzene	24	8.30	0.5	Osmometry
Polyisobutylene	Benzene	60	2.60	0.66	Osmometry
cis-1,4-polyisoprene	Toluene	25	5.00	0.67	Osmometry
Natural rubber	Toluene	25	5.00	0.67	Light scattering
Polyvinyl chloride	Cyclohexanone	25	0.015	0.62	Light scattering
Poly(acrylonitrile)	Dimethyl formamide	25	2.33	0.75	Light scattering
Poly(vinyl alcohol)	Water	25	3.00	0.5	Sedimentation
Poly(vinyl acetate)	Acetone	25	1.02	0.72	Light scattering
Poly(vinyl pyrrolidone)	Water	25	5.65	0.55	Light scattering
Polybutadiene	Cyclohexane	20	3.60	0.7	Light scattering
Poly(dimethyl siloxane)	Toluene	25	0.74	0.72	Light scattering
Cellulose	Couxam	20	0.099	1.0	Calculation

2.5.4. Determination of coil dimensions from the limiting viscosity number

According to the studies of Debye (1948), Kirkwood (1948) and Flory (1949, 1950, 1951) the limiting viscosity number of polymeric solutions are proportional to the ratio of the effective hydrodynamic volume and the molecular mass. Effective volume is proportional to the third power of the linear dimension of the randomly coiled molecular chain. If the average end-to-end distance $\bar{H}$ is chosen as a linear dimension, then:

$$[\eta] = \Phi \frac{(\bar{H})^3}{M} \tag{2.227}$$

where Φ is a universal constant, its value being 2.84×10^{23}.

If instead of $\bar{H}$, $\alpha\bar{H}_0$ is substituted, as in Eq. (2.18), and it is taken into account that:

$$\frac{(\bar{H}_0)^2}{M} = \frac{h_0^2}{M} \qquad (2.16)$$

which is characteristic of the chain structure of the polymer (independent of molecular mass and environment), it can be written that:

$$[\eta] = \Phi\left(\frac{(\bar{H}_0)^2}{M}\right)^{3/2} M^{1/2}\alpha^3 = K M^{1/2}\alpha^3 \qquad (2.228)$$

where $K = [(\bar{H}_0)^2/M]^{3/2}$ is constant for a given polymer independent of the molecular mass and of the solvent.

At the Θ-temperature, α, the Flory expansion coefficient, is equal to 1, and so Eq. (2.228) is equal to the special form of Eq. (2.226) written for the Θ-temperature:

$$[\eta]_\Theta = K M^{1/2}. \qquad (2.229)$$

This is known as the Flory–Fox equation.

The ratio of the limiting viscosity numbers measured under the Θ-condition and under some other ones, allows α to be calculated by dividing Eq. (2.228) by Eq. (2.229), so that:

$$\frac{[\eta]}{[\eta]_\Theta} = \alpha^3. \qquad (2.230)$$

The linear expansion coefficient can therefore be determined from limiting viscosity measurements. If $\bar{H}_0$, the perturbation free dimension is known, then the actual end-to-end distance, $\bar{H}$, can be calculated using Eq. (2.18).

Molecular mass distribution in polymers. General characterization of the distribution curve

As has been mentioned, polydisperse systems can be characterized by different molecular mass averages. Characterization by molecular mass averages is not, however, sufficient. It is clear that two samples with the same mass average molecular mass (100,000), where one consists of molecules with molecular masses of 50,000 and 150,000, the other consisting of molecules with molecular masses of 50 and 10,000,000, will exhibit quite different properties. The latter is said to have a wide distribution.

Polymer scientists have consistently wished to describe their samples by molecular mass distribution instead of averages, but experimental techniques have made routine molecular mass distribution measurements possible only in relatively recent history. This has been achieved due to the rapid development of gel permeation chromatography, during the 1960's.

$\bar{M}_n$ and $\bar{M}_m$ values can be calculated from the distribution curves or measured by the methods discussed previously. The ratio of these two averages characterizes the spread of the distribution. $P = \bar{M}_m/\bar{M}_n$ is a measure of polydispersity. Its value is 1 in the case of monodisperse systems, and higher than unity for polydisperse ones. Typical values lie between 3 and 10. In the case of polypropylene, for example, a necessary condition for good fibre-formation is a relatively narrow molecular mass distribution, where the P value should not exceed 4.

An accurate distribution can be determined by fractionation, and a subsequent determination of molecular mass and relative amounts of the fractions. Molecular mass fractionation was first based on the molecular mass dependence of solubility. To describe this phenomenon let us first investigate the principle of fractionation, and the phase separation phenomenon.

3.1. PHASE SEPARATION IN POLYMER SOLUTIONS

Consider first the equilibrium of two liquid phases, both containing amorphous polymer and a solvent or solvents. If the temperature of a polymeric solution is decreased, the solvent becomes poorer in a thermodynamic sense. Eventually a temperature is reached below which the polymer and the solvent are no longer miscible at any ratio. In this case phase separation occurs. It results as well if a poorer solvent is added to the system. In this case an upper critical solution temperature (UCST) is obtained. Such polymer solutions are endothermic.

It may occur that phase separation is the result of increasing temperature, this is known as the lower critical solution temperature (LCST). Such polymer solutions are exothermic.

3.1.1. Primary polymer–solvent systems

The necessary condition for equilibrium in a two-component system is the equality of partial molar free enthalpies of the components in the two phases.

In other words, the first and second derivative of ΔG_1 as given by Eq. (2.36) with respect to v_2 must be zero:

$$\Delta \bar{G}_1 = kT\left[\ln[1-v_2] + \left[1 - \frac{1}{x}\right]v_2 + \chi_1 v_2^2 \right],$$ (2.36)

so that:

$$\left[\frac{\partial G}{\partial v_2}\right]_{\chi_1 x} = 0,$$

and

$$\left[\frac{\partial^2 G}{\partial v_2^2}\right]_{\chi_1 x} = 0.$$ (3.1)

From this condition the critical concentration where phase separation occurs can be calculated:

$$v_{2,\,crit} = \frac{1}{1+x^{1/2}} \approx \frac{1}{x^{1/2}}.$$ (3.2)

This is a relatively low volume fraction, and for typical polymers (with $x \approx 10^4$) is approximately 0.01.

The critical value of the interaction parameter is:

$$\chi_{1,\,crit} = \frac{(1+x^{1/2})}{2x} \approx \frac{1}{2} + \frac{1}{x^{1/2}}, \tag{3.3}$$

that is, it is only slightly higher than 1/2, dependent upon the molecular mass, and tends to 1/2 if the molecular mass approaches infinity.

The phase separation temperature is given by:

$$\frac{1}{T_{crit}} = \frac{1}{\Theta}\left[1 + \frac{1}{\psi_1}\left(\frac{1}{x^{1/2}} + \frac{1}{2x}\right)\right] \approx \frac{1}{\Theta}\left(1 + \frac{C}{M^{1/2}}\right) \tag{3.4}$$

where C is constant for a given polymer–solvent system.

This $1/T_{crit}$ quantity is a linear function of the reciprocal square root of the molecular mass. It follows from the equation that the Θ-temperature is the critical solution temperature at infinite molecular mass.

There is a qualitative agreement between this theory and experiment. The phase separation temperature is plotted versus concentration in Fig. 3.1. Higher molecular mass fractions precipitate at higher temperatures (Schultz 1952).

T_{crit} and $v_{2,\,crit}$ values correspond to the maxima of these curves. In the neighbourhood of the critical point the polymer concentration in the dilute phase is very low. The phenomenon of the coexistence of two liquid phases, one the dilute solution, the other the nearly pure solvent, is known as coacervation.

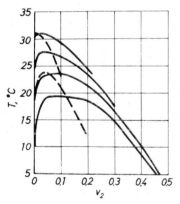

Fig. 3.1. Variation of precipitation temperature with concentration of the polymer component. Polystyrene in diisobutyl ketone. Continuous lines are experimental, whereas dashed lines are theoretical curves. Molecular mass of the polymer is higher for the upper curves.

T_{crit} is plotted versus molecular mass in Fig. 3.2, the points lie on a straight line, and intercepts agree with values obtained from osmotic pressure measurements (under the condition $A_2 = 0$ in Eq. (2.80)) within the limits of experimental error. The intercept is T_{crit} extrapolated to infinite molecular mass, and is equal to Θ. One of the best ways to determine the Θ-temperature is precipitation measurement on a series of fractionated samples.

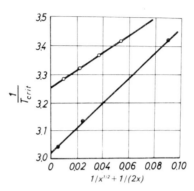

Fig. 3.2. Dependence of the critical temperature on molecular mass. ○ *PS* in cyclohexane, ● polyisobutylene in diisobutylene ketone.

3.1.2. Ternary systems

Most ternary systems contain polymer, solvent and a precipitating agent (non-solvent). Phase relations are usually demonstrated by triangular diagrams (Fig. 3.3).

The binoidal curve, where the two phases are in equilibrium, is a function of the molecular mass. The critical point related to infinite molecular mass corresponds to the Θ-temperature of the binary system. A ternary system can be characterized by three $\chi_{i,j}$ interaction parameters, related to each of the component-pairs. By a correct combination of $\chi_{i,j}$ values the polymer can be dissolved in solvent mixtures consisting of non-solvent components, where neither of the individual components dissolves the polymer. A well known example is cellulose acetate which is soluble in a mixture of ethanol and chloroform, but not in the component solvents.

In contrast, two different polymers in a single solvent frequently form two separate phases, both containing only a single type of polymer. As an example the natural rubber and polystyrene solution in benzene can be mentioned. This general incompatibility can be explained in terms of low mixing entropy (see Eq. (2.29)), the number of mixing molecules is too low.

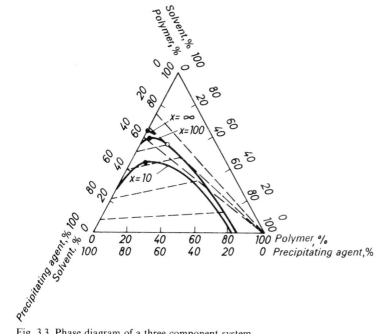

Fig. 3.3. Phase diagram of a three component system.

3.1.3. Higher component systems

The theory of phase separation in polydisperse polymeric solutions is based on the simplification that the $\chi_{i,j}$ interaction parameters for all j values between 0 and x are uniform and equal to $\chi_{i,j}$, only the x value changes for the various components, x being the number of segments, as defined earlier.

The theory gives a full description in terms of polymer segment distribution and equilibrium conditions, but the main point is that the efficiency of separation into two phases can be given by a simple equation, known as the Bronsted–Schulz formula:

$$\ln \frac{v'_x}{v_x} = \sigma x \tag{3.5}$$

where v'_x and v_x are the concentrations of the samples containing x segments in the precipitating (more concentrated) and dilute phases respectively, whereas σ is obtained from:

$$\sigma = v_2 \left[1 - \frac{1}{\bar{x}_n} \right] - v'_2 \left[1 - \frac{1}{\bar{x}_n} \right] + \chi_1 \left[[1 - v_2]^2 - [1 - v'_2]^2 \right] \tag{3.6}$$

where v'_2 and v_2 are total polymer concentrations in the two phases, and $\bar{x}_n$ the number average of x. A number of conclusions can be drawn from the theory:

1. All components of the polymer sample are always present in both phases. All components exhibit a better solubility in the precipitating phase, $v'_x > v_x$ is valid for all x values.

2. Since the σ-value depends on the relative concentration of all components (i.e. on the molecular mass distribution), then the result of the fractionation cannot be given in advance, at least only if the molecular mass distribution is known in detail.

3. The more concentrated phase will be richer in longer chains, the phase separation efficiency, $e^{\sigma x}$ grows exponentially with x.

4. Volume ratio should be large, so that most of the low molecular mass components remain in solution because of the large volume.

Flory gave a theoretical expression for the efficiency of fractionation. Accordingly, if V' and V are the volumes of the precipitated and dilute phase respectively, and f_x is the fraction of dissolved molecules with length x, then f_x is given by:

$$f_x = \frac{Vv_x}{Vv_x + V'v'_x} = \frac{1}{1 + Rv'_x/v_x} = \frac{1}{1 + Re^{\sigma x}} \tag{3.7}$$

where $R = V'/V$. The fraction of the same molecules in the precipitated (concentrated) phase is equal to:

$$f'_x = \frac{Re^{\sigma x}}{1 + Re^{\sigma x}}. \tag{3.8}$$

If $R \ll 1$ the majority of shorter chains remain in the dilute phase because of its large volume.

Because of the higher distribution coefficient, v'_x/v_x, longer chains appear in a higher concentration in the precipitated phase, in spite of its smaller volume.

Figures 3.4 and 3.5 give examples of the fractionation. Fig. 3.4 shows the fractionation of a sample with a given molecular mass distribution at various R values. The σ value has been chosen arbitrarily, so that for all R values $f = 0.5$ if $x = 2,000$.

The upper curve shows the molecular mass distribution curve, the lower three curves represent the composition of the dilute phase at $R = 0.1$, 0.01 and 0.001 respectively. The difference between the original and composition curve of the dilute phase gives the composition of the precipitated phase.

The efficiency of fractionation can be improved by decreasing R, but it will be noted from Fig. 3.4 that the molecular mass distribution of the precipitated phase can never be sharp.

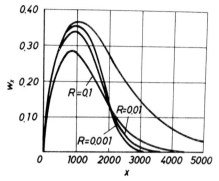

Fig. 3.4. Results for fractionations performed at different volume ratios. (After Flory 1953.)

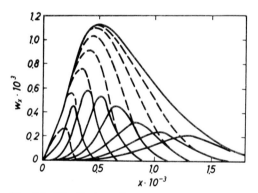

Fig. 3.5. Distribution of 8 fractions of a polymer characterized by the envelope curve. (Calculations of Flory 1953.)

In practice low R values have to be maintained, that is, one has to work with dilute solutions and small precipitates.

Flory (1953) calculated the results of fractionation. Fig. 3.5 shows the molecular mass distribution of a polymer sample decomposed into eight fractions and distribution curves of fractions at $R = 0.001$. It can be seen that the individual fractions still cover a relatively large molecular mass range. The dashed lines show the distribution of molecular chains remaining in the dilute phase.

3.2. FRACTIONATION METHODS FOR POLYMERS

The two main methods of fractionation are preparative and analytical.

In the case of preparative fractionation, fractions are collected and studied subsequently. The mass of the individual fractions is usually quite low, normally less than one gram.

In the case of analytical fractionation the fractions themselves are usually not collected.

If fractionation is to be performed with respect to molecular mass, it is important that the polymer in the polymer-rich phase should be in the amorphous rather than crystalline state; and other parameters disturbing fractionation, such as branching or tacticity, must be kept under control.

Fractionation itself does not give information about the molecular mass distribution of the sample, it must be combined with a simultaneous or subsequent molecular mass determination.

3.2.1. Preparative fractionation methods.

3.2.1.1. Preparative fractionation by precipitation

(i) Fractionation using a precipitating agent. For fractionation from a solvent–nonsolvent (precipitating agent) mixture, at the temperature chosen for fractionation, the precipitating agent is gradually added to the solution until turbidity appears. To ensure equilibrium conditions the solution is heated until homogeneity (transparency) is reached, it is then cooled back slowly to the desired temperature, where the system is strictly thermostated. The precipitated phase is allowed to settle and removed. Subsequently, further precipitating agent is added and the procedure repeated.

To achieve a greater efficiency the fractionation procedure is repeated several times. Efficiency is improved if the initial precipitation is performed from a very dilute solution, or if the initial fractions are re-precipitated from more dilute solutions, and the dissolved part is added to the solution before precipitating the next fraction.

Solvent and precipitating agent must be chosen with care to allow precipitation over a relatively wide solvent/nonsolvent composition range, however, the precipitation must be complete, before the nonsolvent/solvent ratio becomes too high.

Stability and volatility of the liquid components must also be taken into account, and the gel-phase formed by precipitation must be sufficiently swollen to allow ease of handling.

(ii) Precipitative fractionation by the triangle method. The scheme of this method is shown in Fig. 3.6. Initially sufficient precipitating agent is added to precipitate about one half of the dissolved polymer. The gel is dissolved, so that two solutions are obtained, noted as *a* and *b*. The procedure is repeated with these fractions, about half of the dissolved fractions being precipitated.

From solution *a* gel *c* and solution *d* are obtained, the result of solution *b* being gel *e* and solution *f*.

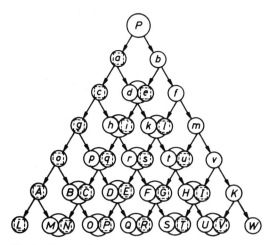

Fig. 3.6. Schematic of the triangular cross-fractionation method.

Solution *d* and gel *e* is brought together to give solution *d/e*; and gel *c* is dissolved, and so finally three solutions are obtained. Further precipitating agent is added to all three solutions to precipitate approximately half of the dissolved polymer. From solution *c*, gel *g* and solution *h* result, solution *d/e* gives gel *i* and solution *k*; and from solution *f* we obtained gel *l* and solution *m*. The procedure is continued until sufficient fractions are obtained. The final number of precipitated and dissolved fractions is equal, and these can be precipitated alone or together with the corresponding gel fractions. The distribution of the fractions is fairly sharp, since subsequent fractions exhibit more narrow distributions. The method is fast, as it makes possible parallel processing of the solutions, and high solution volumes can be avoided.

(iii) Other precipitation methods. Solvent evaporation methods involve the evaporation of large volumes of the solvent, where the solvent has to be more volatile than the precipitating agent.

Precipitation by cooling is simple in the sense that a single liquid is needed, and the volume is constant, but many polymers cannot be precipitated completely by cooling alone.

3.2.1.2. Preparative fractionation by selective dissolution

(i) Extraction. In these methods contact is maintained between the polymer and a partially dissolving liquid. Components with lower molecular mass are dissolved first, and after removing this fraction, the procedure is repeated with a better solvent. This is known as direct extraction.

In the case of the coacervation method the polymer is precipitated from solution and the swollen polymer gel extracted.

In the film extraction method the polymer is precipitated in the form of a very thin film onto a supporting foil to ensure rapid equilibration.

(ii) Fractional solution. Good results can be obtained by gradient solution methods. Here the polymer is deposited onto a dispersed support medium packed into a column, and successive elution carried out with a series of solvents of increasing solvent power. This method can be made continuous,

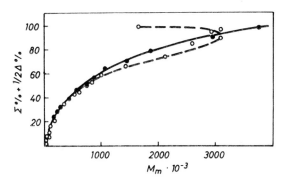

Fig. 3.7. Integral distribution curve of polypropylene. Successful fractionation, ● Unsuccessful fractionation, ○. (After Shyluk 1962.)

but the means chosen for initial precipitation is important, since the efficiency of fractionation can be substantially improved if precipitation is performed selectively, that is, if the lighter molecules are precipitated later. Efficiency of fractionation can be improved by, for example, slow cooling of the solution.

As an example the integrated distribution curves of a polypropylene sample are shown in Fig. 3.7, where the solutions have been cooled from 126°C to room temperature in 1 and 6 hours respectively. The solution cooled down in 1 hour, represented by empty circles, shows backlash, while the solution cooled down within 6 hours, the filled circles, shows no irregularity, the polypropylene molecules having precipitated in increasing order of molecular mass. The precipitating support medium was sand. One particular approach is attributed to Shyluk, and is described below for the fractionation of isotactic polypropylene. The apparatus used is shown in Fig. 3.8. The column is filled with fine meshed sand and is heated to 126°C by refluxing butyl-acetate, the reflux is not shown in the Figure.

A polypropylene solution of 1–2% in p-xylene solvent is added to the system, and the upper half of the column is soaked with the sample. This can be achieved by careful selection of concentration and viscosity. The column is cooled slowly, at a rate of 0.2°C/min, to room temperature. The cooling rate is maintained by evacuating the heat exchanger and regulating the heating power. On reaching room temperature sand is spread on the top of the column, glass wool is placed over it, and the column covered by a perforated disk, shown in the Figure, to prevent displacement of the support bed by the eluating fluid.

The atactic part of the sample is removed by washing through with 200 ml p-xylene at room temperature. This fraction is precipitated by methanol

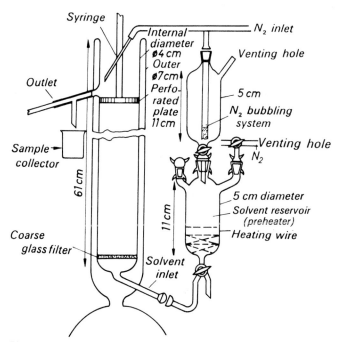

Fig. 3.8. Schematic of a column-elution fractionation system. (After Shyluk 1962.)

after the partial distillation of p-xylene. The rest of the extraction is performed at elevated temperatures. Eluating fluids containing 0.2% of phenyl-β-naphtylamine are purged with nitrogen gas immediately before use. At the top of the column a protective nitrogen atmosphere is maintained above the sand layer covered by the polymer, the gas being pumped through a syringe.

The p-xylene is removed by flushing with 250 ml of nonsolvent (2-buthyloxy-methanol containing 10% ethylene glycol), the column is then heated to 156°C, the temperature of fractionation, by refluxing cyclohexanone in the outer jacket.

The solvent is a high boiling point petrol fraction, its boiling point range falls between 109 and 114°C, at 50 mm Hg vacuum. Combinations of the solvent and nonsolvent are purged with preheated nitrogen gas, the purged liquid is then heated to 156°C in the preheater and pumped through the column at a rate of about 5 ml/min.

Fractions are collected and precipitated in flasks containing 35 ml of methanol and a few grams of dry ice. Usually 15–20 fractions are collected. The final fraction is eluted with pure solvent and collected in acetone, since high boiling point hydrocarbons do not mix with methanol.

To increase the particle size of the concentrate the precipitated polymer is heated in a water bath for an hour. The solvent is decanted, and the precipitated polymer is repeatedly washed through with methanol to remove antioxidants; the polymer is heated, cooled and decanted in each washing cycle. The fractions are dried at 65°C in a vacuum oven overnight, they are then cooled to room temperature and weighed.

The molecular mass of the fractions is determined by specific viscosity measurements, at 135°C in decaline, at a concentration of 0.1%.

The results are shown in Fig. 3.7. If precipitation is performed with due care, the results of fractionation are quite good, and the molecular mass difference between the first and last fractions covers several orders of magnitude. The method is reproducible, selective and relatively rapid. The fractionation itself takes about 6 hours, although the whole procedure requires 3–4 days. The procedure can be performed by skilled technicians.

(iii) The Baker–Williams method. Baker and Williams introduced a chromatographic method for polymer fractionation in 1956. The schematic diagram of their method is shown in Fig. 3.9.

The sample to be studied is precipitated onto glass beads by evaporating the solvent, and the beads are placed on the top of the column.

During the fractionation procedure the extracting fluid, flowing from above the beads, gradually enriches in good solvent. It can therefore dissolve fractions with higher and higher molecular mass. Moving downstream the lower molecular mass components of the solution move faster, since the higher molecular mass fractions precipitate. At the same time, the extracting fluid contains more and more solvent and dissolves primarily the low molecular mass components from the precipitated fractions.

Fractionation with respect to molecular mass along the column is improved by a gradual decrease in temperature along the column, but good results can also be obtained in the absence of a temperature gradient.

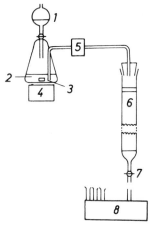

Fig. 3.9. Schematic of the Baker–Williams chromatographic fractionating system *1* solvent, *2* mixing flask to produce solvent gradient, *3* mixing motor, *4* magnetic mixer, *5* valve, *6* fractionating column with support material, *7* valve, *8* sample collector.

Theories proposed for the mechanism of fractionation using this method are still not conclusive. Since repeated precipitation is assumed, the method is frequently referred to as precipitation chromatography, and has been applied to several polymers with great success. The column is made of glass or metal, its length is between 35 and 90 cm. Heating is provided by coils around the column, or by other suitable heating elements, and regulated by a temperature programmer. Kriegbaum and Kurz applied resistor coils to the outer surface of the glass column to produce an almost linear temperature gradient.

Support media usually consist of glass beads with diameters in the range of 40 and 70μm. This medium has to be carefully purified before repeated usage. Purifications achieved by washing through the column with hot concentrated HCl solution several times until the yellow colour of the washing disappears. The column is then washed through with hot nitric acid, and finally with water and a volatile solvent. As alternatives to glass beads cellite or crosslinked polystyrene resin are frequently used. It is important to avoid porous support media, since when polymer molecules are allowed to penetrate the beads, gel permeation is achieved (to be discussed later). With porous media, components of higher molecular mass leave the column faster, whereas lighter molecules permeate, and are retained longer. The two effects are almost opposite in purpose, so that in the case of porous supporting media there is no fractionation or its efficiency is relatively low.

A well-established form of solvent metering is the logarithmic gradient suggested by Alm. This is achieved in a mixing flask and an attached metering funnel. The mixing flask contains the nonsolvent to be diluted by the solvent from the metering funnel at a rate proportional to the flow rate of the nonsolvent. Fractions can be collected manually or automatically at regular time intervals, or as a function of eluent volume or weight.

From the point of view of column performance the means used for polymer precipitation is of great importance. Schneider et al. have shown that the polymer is frequently not precipitated onto the beads of the support medium, but rather, polymer particles are formed between them. Uniform precipitation can be achieved by slow precipitation from thermodynamically poor solvents, or as indicated in Fig. 3.7, by extremely slow cooling.

(iv) Criteria for efficient fractionation. Success of fractionation is determined by three criteria:

1. The total mass of the fractions has to be equal to the initial mass of the polymer sample.

2. Because of possible degradation and cross-linking processes the fractionation procedure can be regarded as successful only if the mass average of the limiting viscosity numbers of the individual fractions is equal to the limiting viscosity number of the unfractionated sample:

$$[\eta] = \Sigma m_i [\eta_i] \tag{3.9}$$

This is the so called Philippoff rule, where m_i is the mass fraction of the i-th fraction, and $[\eta_i]$ is the corresponding limiting viscosity number.

3. Molecular masses must vary monotonically with the number of the fraction, backlash should occur only if there is a structural reason for it.

The main purpose for polymer fractionation is normally the study of molecular mass distribution, and to this end, molecular masses of the fractions have to be measured. Molecular mass determination is usually performed by viscometry because of the relative ease and availability of this approach. However, this method can lead to considerable errors, especially if branching is present.

Polydispersity of linear homologous polymers is described by the normalized molecular mass distribution function:

$$\sum_i f(M_i) = 1. \tag{3.10}$$

The integral molecular mass distribution function, $I(M_j)$, is the mass fraction of those molecules which exhibit molecular mass less than or equal to M_j:

$$I(M_j) = \sum_{i=1}^{j} f(M_i) \tag{3.11}$$

where $M_1 < M_2 < M_3 < \ldots < M_j < \ldots$. This function contains the cumulative or summed mass fractions.

As mentioned earlier the molecular mass distributions of the individual fractions overlap. If m_i denotes the mass, and $\bar{M}_i$ the average molecular mass of this fraction, it is usually assumed that:

$$I(\bar{M}_j) = \sum_{i=1}^{j-1} m_i + \frac{1}{2} m_j. \tag{3.12}$$

Using this assumption compensation is made for the overlapping due to the relatively wide distribution of the molecular mass of the individual fractions.

An integrated distribution curve obtained by this method is shown in Fig. 3.10. It should be noted here that curves shown in Fig. 3.7 have also been calculated using Eq. (3.12), but are expressed in percentage terms.

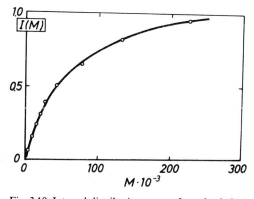

Fig. 3.10. Integral distribution curve of a polyethylene sample.

The molecular mass distribution of a polymer sample is perhaps described better by the differential distribution function, $w(M)$ defined as:

$$w(M) = \frac{dI(M)}{dM},$$ (3.13)

or

$$I(M) = \int_0^M w(M)\,dM.$$ (3.14)

This differential distribution curve can be obtained by analytical or numerical differentiation of the integrated distribution curve. The position of the peak, or peaks, and the spread of the distribution can be seen on the differential curve (Fig. 3.11). If fractionation or differential calculation has not been performed with due care the distribution function may show spurious multiple peaks. If the differential distribution curve exhibits a number of true

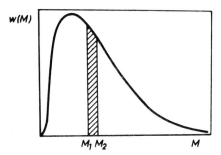

Fig. 3.11. Differential distribution curve.

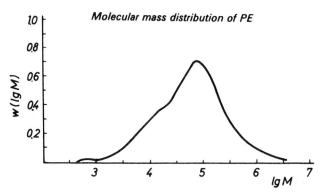

Fig. 3.12. Molecular mass distribution of a high pressure (low density) polyethylene sample.

peaks, that is, the sample contains more species with different molecular mass distributions, then the system is referred to as polydisperse or polymodal.

Differential distribution curves can be calculated, in principle, without integral distribution curves. In practice, when data processing is performed by computers, both types of distribution curve are calculated. Differences between samples can be demonstrated better if the abscissa scale is log M instead of M. This is indicated in Fig. 3.12 which shows the molecular mass distribution of a commercial PE sample.

3.2.2. Analytical methods of fractionation

Preparative fractionation and the construction of a distribution curve from these data is tedious, and the method itself has certain theoretical and practical limitations. Efforts have been made to gain information about molecular mass distribution in a less laborious way.

3.2.2.1. Summative fractionation

In the case of cellulose and several other polymers it is important to minimize the contact time between the polymer and its solvent. If fractionation is by the summative method, the contact time is relatively short.

The essence of this method is as follows. The polymer is dissolved in its solvent, and slowly, a relatively large amount of precipitating agent, about 1/3 by volume, is added, and some portion of the polymer precipitates. The mixture is centrifuged and the precipitated polymer discarded. The dissolved polymer is regenerated from a proportion of the solution, the mass per cent concentration of the solvent is determined and denoted as $F_{(p)}$. The molecular mass of the polymer regenerated from the solution is determined by some of the methods, for example, by viscometry, and denoted as $\bar{P}$.

The procedure is repeated with a new polymer sample, but this time poorer solvent is used as the precipitating agent in order to precipitate a higher proportion of the polymer. By changing the composition of the precipitating agent any number of $F_{(p)}$ and $\bar{P}$ values can be obtained.

Plotting $F_{(p)}$ versus $\bar{P}$ yields the summative distribution function.

In order to make the procedure simple, summative fractionation is performed at constant volume.

The advantage of this method is that all fractions are functions of the polymer and precipitating agent. They are not the product of an intermediate distribution disturbed by previous fractionation steps, as is the case with other methods.

Both integral and differential distribution functions can be calculated from the summative distribution function, but the results are not very accurate.

3.2.2.2. Turbidimetry

In turbidimetry, precipitating agent is added slowly to the polymer solution, whilst being continuously stirred, and the amount of precipitated polymer is measured by the decrease in transparency.

3.2.2.3. Gel permeation

Gel permeation is the most important form of analytical fractionation, and because of its importance, a separate, more detailed section is devoted to it.

3.2.3. Gel permeation chromatography as a rapid method for molecular mass distribution determination in polymers

The phenomenon of gel permeation has been systematically applied to routine fractionation of polymers since 1964 (Moore). It has been observed earlier that nonionic macromolecular solutions pumped through gels (e.g. through ion-exchanging resin columns), fractionate with respect to their molecular mass. Moore produced gels with well-defined, reproducible pore sizes and used them to determine molecular mass distribution functions.

Gel permeation chromatography has rapidly gained importance for molecular mass distribution determination since 1963. Whilst earlier methods required several weeks to perform such measurements, gel permeation chromatography (later simply GPC) yields the whole distribution function and the desired averages within a few hours. Furthermore, high pressure GPC can characterize polymers within fifteen minutes.

Separation with respect to molecular size in GPC is due to the selective permeation of polymer molecules into the gel. Low molecular mass fractions can penetrate the pores, high molecular mass fractions simply by-pass the gel particles. High molecular mass species leave the column first, but the passage of low molecular mass components is retarded. The reverse is true for other chromatographic methods.

The idealized process is outlined in Fig. 3.13. Full details of the mechanism of separation in the column is still not clear, this, however, did not influence the widespread application of equipment based on the principle. A schematic diagram of the instrument is shown in Fig. 3.14, and it works as follows.

The eluating medium is continuously circulated through the column. The sample to be studied is dissolved in the eluating medium, typically at concentration between 0.1 and 0.5%. The sample should be injected within a

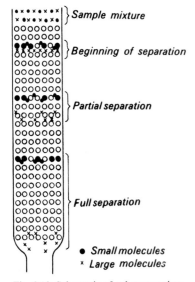

Fig. 3.13. Schematic of gel permeation.

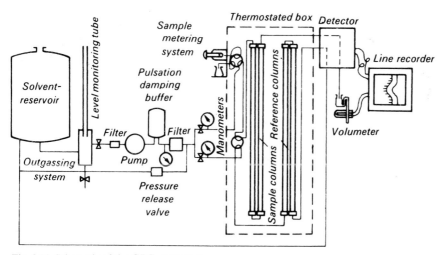

Fig. 3.14. Schematic of the GPC apparatus.

very short time in order to eliminate the effect of injection speed on separation. This injection point is the top of the column, and is achieved by use of a syringe, although commercial equipment frequently uses a calibrated valve. A schematic representation of this metering system is shown in Fig. 3.15.

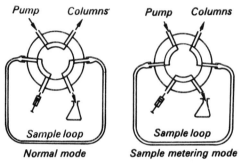

Fig. 3.15. Metering by a six-input valve.

3.2.3.1. Fractionation by gel chromatography

The principle of fractionation has already been described. Homogeneity of the column and of the solvent flow is very important. If part of the solvent could find a faster way through the column, for example, through fissures, it could distort the results of fractionation. The filling of GPC columns has to be performed with extreme care, usually with a special machine by the manufacturer. Homogeneous filling without cracks and fissures is inevitable for accurate measurements. The gel is packed into the column in a swollen state, so that if it dries and the pores are filled with air, the original state can only be restored by swelling and outgassing.

The efficiency of gel chromatography is characterized by a special theoretical plate number. In contrast to other branches of chromatography described here the theoretical plate number depends strongly on the sample to be studied, the notion cannot therefore be used in its original sense. This characteristic can be used, however, if different columns are compared with the same kind of sample.

3.2.3.2. Detection and collection of samples

Detection of polymer molecules leaving the column can be either continuous or intermittent. The advantage of the continuous method is that it can be automated, however, intermittent sample collection offers the possibility of studying the fractions later by independent methods. The methods may be combined: the solution leaving the column flows via a detector cell, and subsequently the fractions being collected.

It is worth mentioning that besides these analytical fractionation purposes GPC can also be used as a means of preparative fractionation, known as preparative gel permeation chromatography.

In the case of analytical GPC, a typical size can be as low as a few milligrams. The total eluated volume ranges between 25 and 100 ml, so the eluated solution contains less than one mg solute in 1 cm^3 of the solvent, which means that the detection method must be very sensitive. Suitable techniques include, on the one hand refractivity, or dielectric permittivity measurements, or spectroscopic methods such as ultraviolet, and infrared absorption spectroscopy, or perhaps fluorescence spectroscopy. If the molecules are labelled with radioactive nuclei, their concentration can be measured by activity measurements. Where fractions are collected, colourimetry, and if the solute itself is colourless dyeing colourimetry, or fluorescence measurements after the application of suitable reagents can be used for concentration determination.

Differential measurements are widespread, especially differential refractometry. The system is the analogue of a two-path spectrometer, a double stream of pure solvent and solution being compared. Automatic recording differential refractometers can be used for this purpose, the only condition for their application is that the difference in refractive index between the solvent and the sample to be studied should exceed 10^{-7}.

Fraction collection is usually performed by a syphon, with overflow after, for example, 5 cm^3 of eluent. Besides concentration detectors, automatic viscometers are also used, these can monitor continuously the viscosity of the fractionated solution. Another novel detection system is the small angle light scattering (SALS) detector cell which continuously measures the absolute molecular mass of the flowing solution, that is, absolute molecular mass can be assigned to every point of the chromatogram.

3.2.3.3. Molecular mass distribution determination from GPC measurements

Gel permeation chromatography is an ideal method for molecular mass distribution determination. The chromatogram contains data which relate elution volume to the relative concentration, to refractivity or other values proportional to concentration. In order to find out the relationship between elution volume and molecular size, the column has to be calibrated. Reference to the size of macromolecules usually means the length of the chain, but the GPC method fractionates the macromolecular system with respect to the hydrodynamic volume, it is therefore more correct to speak about calibration with respect to hydrodynamic radius. Calibration is performed using standard samples with narrow molecular mass distribution and known average molecular mass. By plotting the molecular mass of these standard samples versus the measured GPC elution volume, the relationship between elution volume and molecular mass for a polymer of given chemical composition can be established. If no molecular mass standard is available with the required chemical composition, then calibration can be performed with standards of different chemical composition using the hydrodynamic radii only. More accurate results are obtained if the polymer to be studied is fractionated by some methods (e.g. by gel permeation chromatography) and the molecular mass of these fractions is determined by other, independent methods. Once this relationship is known, we have our own standards to calibrate the GPC column.

Consequently, the relation between the elution volume, V_e and the molecular mass of polymer–solvent pairs having similar composition and

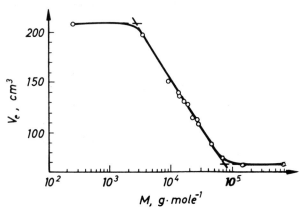

Fig. 3.16. V_e elution volume plotted versus molecular mass, M. Investigation of proteins in Sephadex G 75 gel. (After Andrews 1964.)

interactions is empirical. For a given column, however, the elution volume depends on the molecular mass in a limited molecular mass range; at too low or too high molecular mass no separation occurs. This effect is shown in Fig. 3.16, where the column can be used only in the 10^3–10^5 molecular mass range. These limiting values depend primarily on the composition of the gels, and are given by the manufacturer. Since the columns in the GPC can be changed, the correct column can be chosen for the given measurement range, assuring the required selectivity.

A so-called "universal" GPC calibration curve can also be costructed. As mentioned earlier, elution volumes of a homologous polymer series depend on the molecular mass, since the hydrodynamic volume of the macromolecules depends on the molecular mass in a regular way. Hydrodynamic volume is proportional to the $[\eta]M$ product.

If the $\log([\eta]M)$ values for different polymers are plotted versus their elution volumes, a calibration curve independent of the type of polymer is obtained. As an example Fig. 3.17 shows a "universal calibration curve" of this type. It is clear that various polymers exhibiting the random coil structure in solution give an identical relationship, that is, the calibration is universal as far as molecular volumes are concerned (Samay 1979).

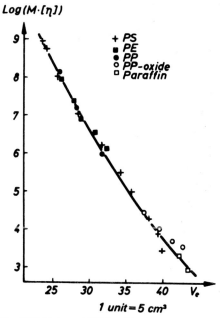

Fig. 3.17. Universal GPC calibration curve. (After Benoit 1966.)

The width of the chromatogram increases monotonically with the spread of the distribution function. The elution curve for a monodisperse sample is, however, still a Gaussian distribution function with a finite width, owing to the fact that sample injection is not perfectly instantaneous, and separation is the consequence of statistical processes. This effect is of course taken into account when calculating the molecular mass distribution function. It is assumed that the total standard deviation in the case of a monodisperse sample is the result of the molecular mass distribution on one hand, and a zone-broadening effect, on the other. The total broadening can be described by the following relationship:

$$\sigma_{total}^2 = \sigma_{molecular}^2 + \sigma_{instrumental}^2 \qquad (3.15)$$

where the instrumental broadening can be determined by the reversal of the flow direction after finishing the elution fractionation (Tung 1966).

The GPC method can be applied to polymers which dissolve at elevated temperatures only. An important application is the molecular mass determination of polyolefins, to be performed at 135 °C with trichlorobenzene as solvent. In this case, of course, the whole system, including the refractometer has to be thermostated to maintain this temperature.

As an example the molecular mass distribution curves of two polypropylene samples are shown in Fig. 3.18. Experience shows that only those polypropylene grades which can be characterized by a narrow molecular mass distribution can be used for fibre production. For this application there are also limitations on the mass average molecular mass to be used.

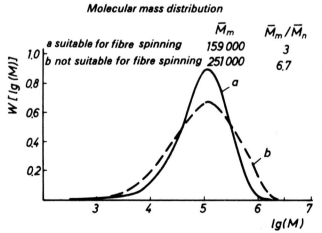

Fig. 3.18. Molecular mass distribution of polypropylene samples.

3.2.4. High pressure GPC

A modification to standard GPC equipment is that of high pressure GPC. The working pressure of such instruments is 400 bar, and the total time for the analysis is only some 7 to 15 min. Supporting media for these columns can withstand high pressures without the alteration of pore sizes. Owing to the high working pressure of these instruments, it has been necessary to develop special pumps and metering systems.

PART TWO

POLYMER MORPHOLOGY

CHAPTER 4

The two phase structure of polymers. Crystallinity

The arrangement of polymer molecules in the solid phase is studied by morphology, dealing mainly with the phenomenological description of structures found in the samples. Evolution of these structures can be monitored using kinetic methods.

In the subsequent sections the methods used to acquire morphological and kinetic information will also be considered.

Firstly, however, a short overview is given on the main differences between polymeric and low molecular crystals. Then, after a brief discussion of polymer morphology the crystallization kinetics of polymers are discussed, and finally the more important methods of investigation and their results are outlined.

4.1. DIFFERENCES IN THE BEHAVIOUR OF HIGH AND LOW MOLECULAR COMPOUNDS

There are a number of basic differences between the behaviour of high and low molecular materials as described below.

4.1.1. Long-range rubber elasticity

Under the action of external forces, low molecular materials deform with no deliberate tendency for recovery. High molecular substances, however, can regain their original shape even after 100% deformation. This phenomenon is known as rubber elasticity. On stretching, molecular chains form an extended configuration, which is made possible by rotations around the C–C single bonds (Fig. 4.1).

Since the configurational probability of extended chains is lower, after the removal of the external force the more favourable *a* position is regained. This phenomenon is characteristic mainly of rubber-like materials with a very low resistance to deformation.

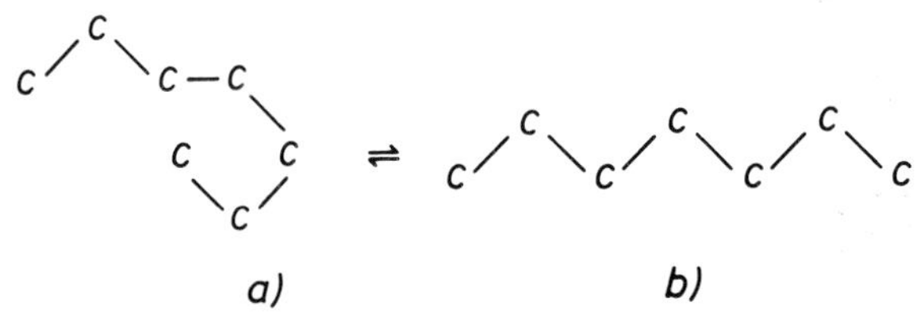

a)   b)

Fig. 4.1. Conformation *a* of the carbon–carbon chain can be transformed into the conformation *b* by extension and rotations around single C–C bonds. After the cessation of the external load the more probable *a* conformation recovers. (After Sharples 1966.)

As will be shown later the presence of crystalline areas increases this resistance by about two orders of magnitude, but there remains about 5% elastic deformation, which is enough, for example, in the textile industry, to avoid tearing.

This reinforcing effect of crystallinity can be demonstrated by comparing slightly cross-linked natural rubber with polyethylene, which has a more or less similar chemical structure. At room temperature rubber is amorphous, whereas polyethylene is crystalline, their moduli are 10^6 and 10^8 Pa respectively, a ratio of about 100. The main reason for this difference is the absence or presence of crystallinity (see Table 4.3).

4.1.2. Differences in crystallinity

When low molecular substances are cooled from the molten state, they either form a glassy solid, or crystallization occurs. Crystallization with a few exceptions means the formation of a more ordered, relatively denser structure in comparison to the melt. It is therefore accompanied by a considerable volume contraction, which is quite clearly indicated in volume–temperature diagrams (Fig. 4.2).

In the case of low molecular materials the glassy or crystalline state can be distinguished simply by inspection. If the product is crystalline, then this crystallinity is uniform throughout the whole sample.

When cooling high molecular substances from the molten state, similar phenomena are observed. By cooling polystyrene or poly(methyl methacrylate) melts glassy materials are formed. On cooling polyethylene, polypropylene or polyester melts a curve corresponding to a crystallization phenomenon is observed (Fig. 4.3). This indicates the possibility of crystallization in polymers.

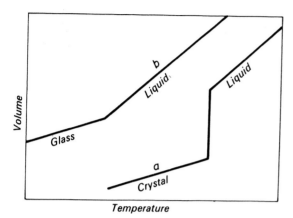

Fig. 4.2. Volume changes of a low molecular solid and of an amorphous glass as a function of temperature. *a* First order transition; *b* second order transition, or glass temperature. (After Sharples 1966.)

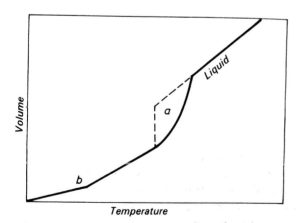

Fig. 4.3. Volume change of a partially crystalline polymer sample as a function of temperature. *a* First order transition; *b* second order transition or glass temperature. Continuous line is the melting, dashed line is the cooling (crystallization) curve. (After Sharples 1966.)

In the case of polystyrene or poly(methyl methacrylate) only a slight inflection is observable on the volume–temperature diagram. Such inflections are also observed in crystallizable polymers, but usually at well below the melting point. Such points are known as the glass temperature, or secondary transition temperature. This secondary transition point is the temperature at which the glass formation begins. Primary transition can be characterized by a continuous change of G, the free-enthalpy or the Gibbs free energy, as a function of temperature, while entropy S, and volume V changes abruptly. S and V are the first derivatives of G, consequently the first order transition can be defined as a process in which G changes continuously, but its first derivatives:

$$S = \left(\frac{\partial G}{\partial T}\right)_P$$

and

$$V = \left(\frac{\partial G}{\partial P}\right)_T \tag{4.1}$$

are discontinuous functions. Secondary transitions are processes in which both free enthalpy and its first derivatives, S and V, change continuously, whereas the second derivatives show an abrupt change (Guggenheim 1949), so that in this case:

$$\frac{\partial V}{\partial T} = \frac{\partial^2 G}{\partial P \, \partial T}$$

and

$$c_p = T \frac{\partial^2 G}{\partial T^2} \tag{4.2}$$

are discontinuous.

The secondary transition temperature is frequently referred to in the literature as glass temperature or brittle point. The latter term comes from mechanical methods of determination. Brittle points determined mechanically can differ from glass temperature values obtained by specific volume measurements by a few degrees (depending on the actual measuring conditions), nevertheless all these data refer to the same process.

All of the fibre-forming polymers belong to the crystallizable group, that is, various grades of polyolefins, polyesters and polyamides; furthermore, cellulose and protein derivatives also exhibit some degree of crystallinity.

Some rubbers also crystallize at low temperatures. Typical noncrystalline polymer materials are the highly crosslinked thermosetting resins, such as phenol–formaldehyde systems, and the atactic forms of polystyrene, atactic polypropylene and poly(methyl methacrylate). Physical order is the main structure-determining factor of crystallizable polymers, chemical composition being only of secondary importance.

The main differences between the crystallization of high- and low molecular materials can be summarized as follows:

(i) The crystallization of low molecular substances is total, that is, even if there are several crystalline nuclei present, the evolving structures are fully crystalline. Crystallization in polymers is different, the disordered structure characteristic of the liquid state can superimpose the crystallization process and remains in the evolved structural units. While in low molecular crystals the crystallites are divided by sharp interfaces, they are absent in polymers. The phenomenon of a glass temperature in the solidified sample suggests, besides crystallites, the presence of amorphous or disordered components (see Fig. 4.3). Since crystallizing polymers always contain a proportion of amorphous component, such polymers are frequently referred to as semi-crystalline. The amorphous content can vary depending on chemical composition and crystallization conditions, its value usually ranging between 20 and 80%.

(ii) An important characteristic of crystalline polymers is that their melting extends over a broad temperature interval sometimes as great as 100°C, in sharp contrast to low molecular crystals, where very narrow melting ranges are observed. In the case of a given crystal, melting point depression may result from the small size of crystallites; such a phenomenon

Table 4.1

Differences in the crystallization of low and high molecular materials

	Crystallinity	Melting	Hysteresis	Character of the transition
Low molecular mass	Total	Sharp melting point	None	Through equilibrium states
High molecular mass	Partial	Interval	Yes	Far from equilibrium states

occurs if their size is so small that the surface energy is comparable with that of crystallization. The wide melting point range in polymers may also be explained by this effect. It can be concluded from a broad melting range that the size of crystallites is very small, of the order of 10 nm.

(iii) The third basic difference is that in low molecular materials overcooling can be eliminated, in the case of polymers, however, there is always a hysteresis cycle between freezing and melting curves (See Fig. 4.3), melting through equilibrium states cannot be realized. The main differences are summarized in Table 4.1.

4.1.3. Coexistence of various levels of morphological unit

The crystallization processes of low molecular materials have been under study for a long time. As shown later, polymers exhibit additional problems compared with those observed in low molecular crystals. This results from the coexistence of morphological units differing in size by several orders of

Table 4.2

An overview of polymer structures with respect to their dimension and methods of investigation

	Atomic structure within the crystalline structure	Molecular structure	Supermolecular structure
Dimension	10^{-8}–10^{-7} cm	10^{-7}–10^{-5} cm	10^{-6}–10^{-1} cm
Level	X-ray level		Optical level
Structure	Lattice structure	Size, shape and order of crystallites, crystallinity, order of the chains, amorphous and mesomorphic structures, chain orientation	Single crystals, spherulites, hedrites, dendrites
Suitable methods of investig-ation	X-ray diffraction, electron diffraction, density measurements, infrared spectroscopy, NMR spectroscopy	Wide and small angle X-ray diffraction, electron microscopy, density measurement, infrared spectroscopy, birefringence measurements, NMR, ultra-sound wave propagation, DSC, DTA	Density measurements (dilatometry), optical microscopy, electron microscopy, wide and small angle X-ray diffraction, small angle light diffraction, DSC, DTA

magnitude, from visible to molecular size dimensions. Amorphous areas with elastic properties are always present, and mix with crystalline components at the molecular level, resulting in a very complex structure observed only in polymers. Making distinction between the various kinds of structural unit is not always an easy task. Nevertheless, in order to get a better insight, such structural units are frequently grouped together from a morphological point of view. The various structural features found in polymers are outlined in Table 4.2, and a more detailed treatment is given below in a morphological review which progresses from optical to molecular and atomic levels of detectability.

4.2. SURVEY OF POLYMER MORPHOLOGY

4.2.1. Crystallites

An optical microscopic investigation of polymers crystallized from the melt, such as polyethylene, shows the lack of well-defined crystal surfaces characteristic of low molecular crystals, such as benzoic acid.

The visual appearance of polymers does not indicate their crystalline nature. Even today it is not possible to manufacture a "polyhedron limited by regular planes," which is the classical definition of crystalline structures. All the same, the crystalline nature of certain polymers has been known for many years. The crystalline nature of cellulose was recognized by Scherrer and Zsigmondy, furthermore Scherrer et al. in 1920, and that of stretched rubber by Katz in 1925 from X-ray diffraction patterns. The presently accepted theory of polymer crystallinity and related structural theories, are the result of

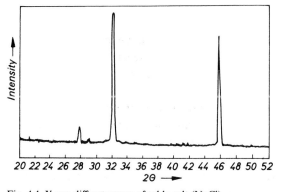

Fig. 4.4. X-ray diffractogram of table salt (NaCl).

much research and development, and knowledge in this field continues to grow. Here, it is sufficient to discuss crystallinity problems posed by stereospecific catalysts, and that of chain folding.

It has already been made clear that the degree of crystallinity of polymer chains can be quite different and that polymers possess non-crystalline areas

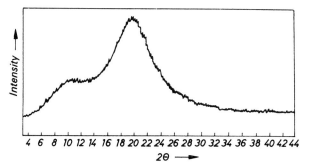

Fig. 4.5. X-ray diffractogram of polystyrene.

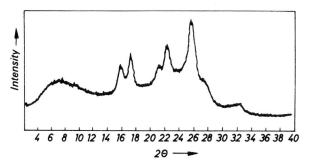

Fig. 4.6. X-ray diffractogram of a polyester.

within the crystal structure. These areas are frequently referred to as amorphous regions, albeit the adequacy of this nomenclature is disputable. In any event, much more experimental and theoretical work is required to clarify the true nature of these areas.

The crystallinity of polymers depends on their thermal and mechanical history, and its value influences processability and other properties. Consequently the determination of crystallinity, and its control is of great practical importance.

Recognition of polymer crystallinity is related to X-ray patterns, and the only unambiguous sign of crystallinity is the presence of X-ray interference.

Gases, liquids and solids, in their amorphous state, show diffuse interference patterns, while crystalline materials yield sharp interference lines. This is a generally accepted criterion for crystallinity. As examples, X-ray diffract-ograms of table salt, which is common crystalline material, polystyrene, an amorphous polymer, and of a polyester, typical of a crystalline polymer, are shown in Figs 4.4, 4.5 and 4.6 respectively. The fact that these crystals are invisible by optical microscopy shows that their dimensions rank below 10^{-6} m. Broadening of X-ray lines suggests dimensions in the $10^{-7} - 10^{-8}$ m range.

4.2.1.1. Micelle theory of crystallites

In order to fit in the existence of very small crystals with the known chain-structure of polymer molecules, early theories assumed that some molecules of similar size formed aggregates known as micelles. Accordingly, crystalline micelles are held together by secondary forces resulting in a macroscopic structure. Mark et al. have tried to prove these ideas by data obtained from the X-ray investigation of unit cells. It has since been shown, however, that the micellar concept is inadequate, as in some cases, for example, cellulose, the length of the macromolecule exceeds that of the micelle by two orders of magnitude. It is now clear that the size of crystallites is independent of the molecular mass, so long as the molecular mass exceeds a particular relatively low limit.

4.2.1.2. The fringed micelle theory of crystallites

Keeping to historical sequence, the next theory to be developed was the fringed micelle theory, stating that the molecular chain is allowed to participate in several crystalline regions. The structure of a polymer as depicted by this theory is shown in Fig. 4.7. This view, according to current knowledge, is somewhat oversimplified, nevertheless it is worth recording owing to its historical importance and descriptive nature.

The theory proposes that ordered structures are not separated as they are in low molecular materials. Under the action of mechanical forces or solvents low molecular crystalline materials separate into elementary crystallites, and mainly secondary bonds are disrupted. Such separation is impossible in the case of semicrystalline polymers, since the crystalline regions are coupled by primary forces through chain molecules. Thus, notions such as crystals or micelles are somewhat misleading ordered regions, typical of semicrystalline polymers, will therefore be referred to here as crystallites.

A further characteristic of the fringed micelle theory is that the amorphous phase is taken to be continuous (as shown in Fig. 4.7). The amorphous phase is regarded as the matrix with embedded crystallites. Recently, however, the study of polymer single crystals grown from very dilute solutions, has led to a reversed hypothesis that polymers should be

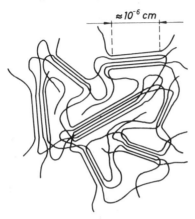

$\approx 10^{-6}$ cm

Fig. 4.7. Fringed micellar structure of semicrystalline polymers. Molecular chains pass through several crystalline and amorphous zones. (After Sharples 1966.)

regarded as defective single crystals, where the amorphous parts correspond to the defect regions. Accordingly, the amorphous parts are inclusions in a single crystal matrix. It is clear that these two theories are somewhat different descriptions of the same phenomenon, and usually the majority phase is regarded as being continuous. Many polymer properties can be explained by assuming a continuous amorphous matrix. Examples are the long-range recoverable deformation; the large, albeit limited swelling of polymers observable in good swelling media; furthermore, the high diffusivity of swelling liquids, and their effect on elastic moduli.

Use of an effective degrading agent which penetrates the amorphous regions, and is able to disrupt primary chemical bonds of amorphous chain segments without destroying the crystalline structure, allows the crystallites to be isolated and studied by electron microscopy. This technique has proven to be successful in the case of cellulose, where acidic–aqueous solutions specifically dissolve the amorphous regions.

Palmer and Cobbold (1964) studied the structure of synthetic polymers, such as polyethylene very successfully using suitable reagents. Isolated crystalline fragments were found to be similar to the original crystallites, but some important differences were evident owing to the removal of mechanical stresses exerted by amorphous segments on the original crystallite.

In the case of cellulose the size of these particles varies depending on the nature of the sample. A typical example would be 46 nm in length and 7.3 nm in width (Ranby 1951). These dimensions well accord with those estimated from the broadening of X-ray lines of the original material.

From electron diffraction patterns for the individual crystallites it follows that the molecular chains are ordered in parallel bundles along the longer axis, and it can be concluded from the 7.3 nm width that about 100–150 chain segments build up the crystallite. Such an agreement can be checked if the crystalline parts are separated by an appropriate solvent, and their molecular size determined by osmometry, light scattering or any independent method, and the results compared to the electron microscopic pattern. In Ranby's work the degree of polymerization turned out to be 94. The length of monomeric units is 0.515 nm, so that the predicted length of the molecule would be 48.4 nm, in good agreement with the experimental value (46 nm).

Similar investigations were made later on synthetic polymers, but for polyethylene, for example, Palmer found the fragments to be much larger than in the case of cellulose; their diameter being about 10^3 nm, and thickness 20–30 nm. There is a basic difference in polyethylene compared with the cellulose crystals, that is, the chains are oriented along the shorter axis direction, and the lamellar shape of the fragments is somewhat similar to the solution-grown monocrystals to be considered later.

One further feature of polyethylene crystallites is the presence of internal defects. Thus the amorphous content of the original sample consists of two components: the disordered, presumably continuous, chain matrix connecting the crystallites, and the internal defect phases within the crystallites.

4.2.1.3. Paracrystalline structure

Hosemann developed a theory in 1950 to explain the action of lattice distortions on X-ray diffractograms. According to his theory the crystalline lattice can deform under the effect of various external of internal forces, resulting in the so-called paracrystal. The theory explains the curious properties of polymer crystals in terms of lattice defects instead of fringed micelles and distinguishes two classes of defects, known as first and second order lattice distortions.

The defect is considered to be first order distortion if the defect sites are displaced along a regular lattice, whereas second order defects are accumulative, that is, defects spread over the deformed environment, and multiply. In polymers both types of defect are present, but second order defects are more typical.

Optical analogues and the corresponding diffractograms of these lattices are shown in Figs 4.8, 4.9 and 4.10. Fig. 4.8 shows the ideal lattice and the first and second order lattice distortions.

Paracrystalline structures in polymer can be characterized by the presence of second order lattice distortions. These distortions lead to the

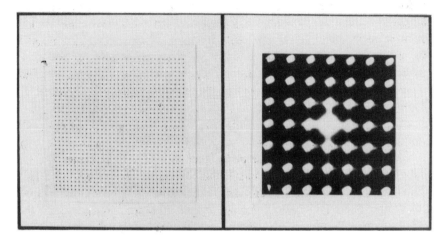

Fig. 4.8. Ideal lattice and its light diffraction pattern.

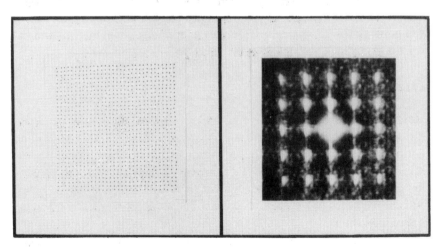

Fig. 4.9. First order lattice distortion and its light diffraction pattern.

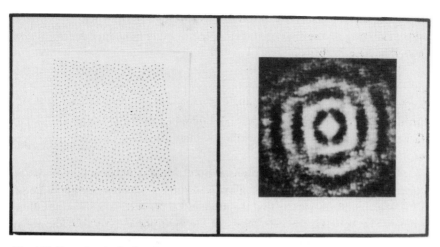

Fig. 4.10. Second order lattice distortion and its light diffraction pattern.

broadening of interference lines, similar to the broadening effect of decreasing crystallite size. The main difference between these broadening mechanisms is that, while the crystallite size has a similar effect on interference lines of different order, the line broadening caused by second order lattice distortions increases dramatically with the orders of interference.

Crystal defects can be very important in polymers, as with metals. For example irregularities formed by non-fitting methylene groups in poly-ethylene can be detected by the NMR study of oriented samples (Reneker 1962). Such defects can migrate within the crystal and disappear at the surface of crystallites or at chain foldings, or can be transformed into other defects at the chain ends. Similar processes occur during annealing when the crystalline structure becomes more perfect.

4.2.2. Amorphous zones

According to the simplified fringed micelle theory only two components are present: crystallites and amorphous areas. The latter are not independent structural units, but since they are complementary to crystalline areas and important in determining the physical properties of semicrystalline polymers, their characteristics will be discussed.

As shown in Figs 4.2 and 4.3 the amorphous zones of polymers can undergo a liquid to glass transition. If the polymer is totally amorphous, such as poly(methyl)methacrylate, this transition means a transformation to a final state exhibiting all the characteristics of glasses, including high modulus and low elongation to break.

If, however, crystalline material is also present, then the effects of the glass–rubber transition on mechanical properties are not so drastic. The polymer behaves in a similar fashion to a slightly crosslinked rubber going through its glass temperature, that is, it remains solid and tough above Tg. This analogy shows that the crystallites act like cross-links in the amorphous matrix.

Typical moduli and elongations at break for crystalline and amorphous polymers above and below their glass transition points are shown in Table 4.3. These data show clearly that the mechanical properties of semicrystalline polymers lie between the extremities of the amorphous polymers shown in their glassy and rubbery states. It is also evident that the glass–rubber transition in semicrystalline polymers leads to less drastic changes than in amorphous polymers.

Table 4.3

Typical moduli and elongation at break values for polymers in different states, above and below the glass temperature

Material	Modulus (Pa)		Elongation at break, %	
	Above T_g	Below T_g	Above T_g	Below T_g
Slightly crosslinked amorphous polymer	10^6	10^{10}	1,000%	2%
Semicrystalline polymer	10^8	10^9	>10%	<5%

This difference can be demonstrated by practical examples. At room temperature, for example, poly (ethylene glycol terephthalate) is well below its glass transition temperature, whereas polyethylene is about 130°C above its glass temperature, nevertheless, the physical properties of these two polymers are quite similar. Reversible long range elasticity in amorphous rubbers above their glass temperature exist up to very high deformations (1,000%), since inter- and intramolecular forces are weak, these materials deform readily and recovery is not hindered.

In semicrystalline polymers intramolecular forces tend to maintain a crystalline order of chain segments, but these forces are less active in the amorphous regions. Intermolecular forces lead to viscous effects hindering the recoil process. If deformation exceeds more than a few per cent the recovery is seldom total, but improves if intermolecular forces are weakened by heat or swelling agents. Semicrystalline polymers above the glass transition point have free rotation around the single bonds between the monomeric units, so that all the conditions for long range elasticity are present. Below the glass transition point rotation becomes more hindered, and both elongation and recovery processes are more difficult.

In totally amorphous polymers below the glass temperature rotation is almost completely restricted, as demonstrated by the data in Table 4.3. Elongation processes are regulated here by intermolecular forces, thus giving a crosslinked resin character to the material. The sample can break at very low deformation, and is well demonstrated by a rubber sheet, cooled by liquid nitrogen which can readily be broken into small pieces in a brittle manner.

Electron microscopic studies of highly crystalline polymers have not lent support to the idea of the existence of separate amorphous zones in the polymer, as would be expected from the fringed micelle model. Explanation of X-ray line broadening, and the presence of amorphous rings in terms of the paracrystalline model have become more readily accepted.

Accordingly, the main reasons for the appearance of amorphous diffraction rings in highly crystalline polymers are the distortions and defects in the crystalline lattice. Such defects can be caused by thermal motion, branching, chain folding, and end groups amongst other influences. This assumption has still to be proven, yet similar interpretations of amorphous zones have appeared in recent literature.

4.2.3. Spherulites

Spherulites are birefringent, usually spherical, structures observable in a series of polymers by optical microscopy during the crystallization process, beginning from nucleation through various stages of crystal growth. For example, if high molecular mass polyethylene oxide is heated to $100\,^{\circ}$C between microscope slides, and cooled back to room temperature, the appearance of spherulites can be observed in a polarization microscope at quite a low magnification (the melting point is $66\,^{\circ}$C, and crystallization usually begins at $50\,^{\circ}$C) (Fig. 4.11).

Spherulites are important structural–morphological units in polymers. It is interesting to note that while the existence of spherulites in low molecular substances has been known for a long time, such structures were first observed in polymers by Staudinger in 1937, and systematic studies have continued since 1945. X-ray studies, being technically more complicated, began about twenty-five years earlier. This time lag can perhaps be explained by the

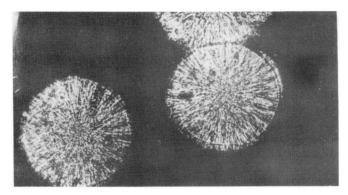

Fig. 4.11. Poly(ethylene oxide) spherulites grown from supercooled melt. Spherulite diameter is about 0.2 mm. (After Sharples 1966.)

fact that, although spherulites are present in almost all semicrystalline polymers, there are only a few in which they are large enough to be observed by optical microscopy. If, for example, thin sections of polyethylene, polyamide or polyester are studied by microscopy, only an inhomogeneous, unresolved birefingent mass can be observed (Fig. 4.12). The situation is further complicated by the fact that if thin polymer films are heated to high temperature between microscopic slides, cooling results in well-defined spherulites. Such spherulites are, however, the outcome of special heat-treatment, and in this form they are not characteristic of bulk crystallized samples (Fig. 4.13).

It is more advisable to study the properties of small sections cut from a bulk sample than to strive to describe the bulk properties using data obtained from these films. Many examples of micrographs can be found in the literature, especially on polyethylene, but the majority have been taken of thin films, thus they are not characteristic of everyday samples. It is probable, however, that these differences originate from the different densities of nucleation, and can be explained by the effects of heat treatment. Large spherulites are the result of fewer nuclei, the interference effects of neighbouring areas being weaker.

It is important to note that spherulites are not single crystals. Single crystals can be split into smaller units by cleavage, and the resulting limiting planes are related to the original structure. This is clearly impossible with spherulitic units. It follows that spherulites are aggregates of the smaller crystalline units referred to previously.

Dimensions of crystallites are in the range of 10^{-5}–10^{-8} m, and these orders of magnitude are the lowest limit of spherulitic dimensions. Upper

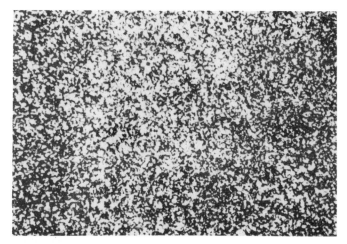

Fig. 4.12. Polarization optical micrograph of high density polyethylene. Polyethylene was melted at 200°C, then slowly crystallized at 129°C, about 100x magnification. (After Sharples 1966.)

Fig. 4.13. Polarization optical micrograph of a high density polyethylene sample. As in Fig. 4.12, but after a short heat treatment at 350°C and fast crystallization. Magnification is about 100x. (After Sharples 1966.)

limits can only be estimated for spherulites, but spherulites of 1 cm diameter have been observed, and there is no theoretical limit to their growth to even larger units. Theory suggests that in order to do so the number of nuclei must be reduced further. In practice this problem can only be studied empirically, since the factors influencing the number of crystal nuclei are not well understood.

According to electron microscopic studies the experimental lowest limit of spherulite diameter is around 10^{-6} m. This, or a slightly lower value can be regarded as the limit, since at smaller dimensions spherulitic units cannot be formed from crystallites.

Spherulites are formed in almost all crystallizable polymers. Perhaps the only exceptions are cellulose and its derivatives. In cases such as polyethylene oxide, isotactic polypropylene or isotactic polystyrene well-defined spherulites are formed with separated nuclei. In other polymers, such as polyethylene, and poly(decamethylene terephthalate), only very small birefringent units are observed. Based on such evidence, early studies of semicrystalline polymers divided them into two subgroups: spherulitic and non-spherulitic. This resulted from the fact that granular birefringence cannot be resolved by optical microscopy, but is now known to be due to the existence of several smaller spherulites. Since the practical limit of optical resolution is 10^{-6} m, it is unreasonable to assume that these structures do not exist simply because they cannot be observed optically. Methods exist when non-spherulitic semicrystalline polymers are converted into spherulitic forms by changes in the nucleating density of the system. Sometimes a gradual change of optical patterns can be observed, when granular birefringent regions are transformed into well-defined spherulites and vice versa.

In some cases the spherulites of a polymer belonging to the spherulitic group referred to above are embedded into a granular structure. This can be observed, for example, in high density polyethylene samples (Fig. 4.14).

The study of these samples has shown that, during crystallization, spherulites appear in increasing number, whereas their size does not change. In such cases kinetic analysis, which otherwise described nucleation and growth processes quite well, is not applicable.

The fact that most polymers exhibit a certain degree of birefringence gives important information in itself, since it proves the presence of oriented units which are large enough to produce depolarized light scattering. To produce such an effect, particles somewhat smaller than the wavelength of visible light $(0.3 \times 10^{-6}$ m) must be present, in good agreement with the estimated lower limit of spherulite size mentioned above.

The presence of a fibrillar structural sub-unit is typical in spherulitic structures (Fig. 4.15). Fibrils will be discussed in detail under Section 4.2.5, but some remarks should be made here. It is now clear that the formation of spherulites is the result of a fibrillar ordering, fibrils growing radially from a central nucleus, thus building up the spherulite.

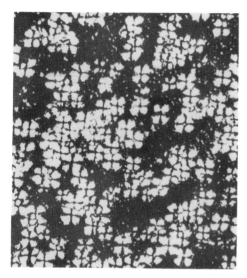

Fig. 4.14. Polarization micrograph of a polyethylene sample, where the spherulites are embedded into a granular structure, about 100x magnification. (After Sharples 1966.)

Fig. 4.15. Polypropylene spherulite with fibrillar structure. (After Keith and Padden 1963.)

In fast-grown structures the fibrils are of sub-microscopic dimensions, and therefore optically invisible (see Fig. 4.13). If, however, the growth process is slow, the fibrils are thicker, and can be studied by a normal optical microscope. If the fibrils are of such thickness, they probably consist of several microfibrils, and within them are the crystallites.

In samples of high crystallinity the fibrils can be crystallites themselves, that is, their thickness and width dimensions are equal to those of crystallites with a random distribution of defects along the fibrillar axis. In such cases the crystalline and amorphous regions do not run in a sequence, and the sample can be better characterized as a continous single crystal grown in one direction with a considerable number of defects. Although the defect phase is related to the amorphous one, the true amorphous regions can be found mainly between fibrils and at spherulite boundaries.

An important consequence of fibrillar growth is branching, this is the means by which free space is filled during continuous radial growth. According to present knowledge the angle of branching is small and not related to the crystallite axes. They are called small-angle, non-crystallographic entities in contrast to dendritic growth which can be observed, for example, in ice crystals, or in polymer single crystals. Branching fibrils frequently twist during their growth, and in fast grown samples such twistings can occur cooperatively, giving rise to a periodic change in the refractive index. The result is a complicated circular structure. In such highly ordered structures branching angles are necessarily low. A spherulite is probably formed from a single crystal in the early stages of growth, and later becomes polycrystalline through branching. A schematic diagram of this process according to Bernauer is shown in Fig. 4.16.

It is important that the process should start from a rod- or plate-like unit as the shape of the single crystalline nucleus, although geometrical shape may

Fig. 4.16. Stages of spherulite formation. (After Bernauer 1929.)

change during the growth process. When studying crystallization kinetics these changes have to be reflected by equations. In polymer samples with different densities of nucleation (with crystallizing structures of different size) the crystallization process can be different, if, instead of fibrillar or lamellar structures, spherulitic growth occurs. If the density of nucleation is very high, spherulite formation is possible in principle, but does not happen in practice. This is probably the case with cellulose and its derivatives where, if the structure exceeds the size of crystallites, rod-like structures are formed. Here the growing units are very small as indicated by the fact that in the unstretched state no birefringence is detected.

One further remark can be made here which will be important later; optical determination of molecular orientation led to the surprising result that in spherulites chain axes are perpendicular to the radius, that is, cross the radial fibrils. Refractive index is higher in the chain direction than in the perpendicular. Tangential orientation of chains leads to extinction in opposite quadrants if the spherulite is observed in a polarization microscope between crossed Nicols. This is the characteristic "Maltese cross" pattern observed in well developed spherulites (see Figs 4.23 and 4.14).

The refractive index of spherulites is usually lower in the radial than in the tangential direction, which shows that the chains themselves are oriented tangentially and not radially within the spherulite. Such spherulites are denoted as negative (Fig. 4.17).

Certain spherulites of, for example, polypropylene exhibit a higher refractive index in the radial direction, as if the chains are all directed along the radii (the so-called β-polypropylene). It can be imagined that in these spherulites the fibrils grow in a radial direction (as in normal, negative spherulites) but many branches grow perpendicularly to the fibril direction,

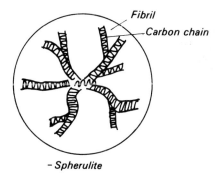

– *Spherulite*

Fig. 4.17. Negative (−) spherulite.

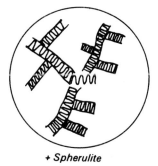

+ *Spherulite*

Fig. 4.18. Positive (+) spherulite.

and in these branchings the chains lie perpendicular to the branching direction, so that they finally exhibit a radial orientation. This concept has been proved by small angle X-ray scattering studies (Fujiwara 1968). The structure of these positive spherulites according to our present knowledge is outlined in Fig. 4.18.

In normal negative spherulites fibrils are branched at small angles, while in positive spherulites the branches run practically perpendicular to the growth direction of the fibrils. The perpendicular orientation of chains with respect to fibrils is a consequence of chain folding to be dealt with in detail in the next section.

4.2.4. Single crystals

If a polymer is crystallized from the molten state by cooling, polycrystalline samples are usually obtained, and several crystallites with individual nuclei being formed. It has been mentioned already that these spherulitic structures cannot be regarded as single crystals. Under special circumstances, however, single crystals can be grown even from polymers. Since single crystals can be studied relatively easily, they have been the subject of several investigations since their discovery in the 1950's.

Polyethylene single crystals can be grown relatively easily from xylene solutions (concentration 0.05%, crystallization temperature 70–80 °C) (Fig. 4.19.). So far more work has been devoted to polyethylene single crystals than

Fig. 4.19. Polyethylene single crystal obtained from dilute solution. Magnification is about 7,500x. Chain molecules are folded, their direction is perpendicular to the plane of the photograph. (After Sharples 1966.)

other materials, but results obtained using this system can be readily applied to several other polymers such as poly (methylene oxide), poly(ethylene oxide), poly(4-methyl-pentene-1), polyamide-6 and cellulose triacetate. A typical single crystal of polyamide-6 is shown in Fig. 4.20.

The thickness of single crystals depends on the crystallization temperature, but the order of magnitude is 10 nm (100 Å). The most striking fact is

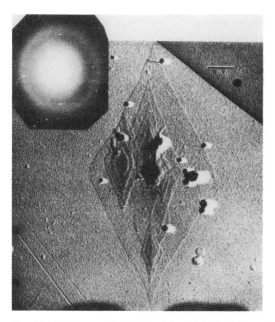

Fig. 4.20. Single crystal of polyamide-6. Obtained from dilute solution, solvent was glycerol. Lamellar thickness is 6 nm. (After Geil 1960.)

that the molecular chains are oriented along the thickness of the crystal and not along the lamellar plane. The lengths of macromolecules can reach 1000 nm (10,000 Å), and chain folding must be the mechanism. This phenomenon led to a series of new ideas in polymer morphology. It has been shown that chain folding occurs in melt crystallized samples and in fibres as well.

Chain folding has turned out to be the general form of polymer crystallization. It will be shown later that this mechanism is dominant not only in single crystals, but, with a few exceptions, in fibrils also.

The first examples of chain folding were found in the course of electron microscopic and electron diffraction studies of polymer single crystals.

Accordingly, the position of molecular chains in single crystals can be described schematically by Fig. 4.21.

It is important to clarify the mechanism of crystallization and to find out the reason for constant folding lengths under indentical crystallization conditions. One has to explain the relationship between the folding length

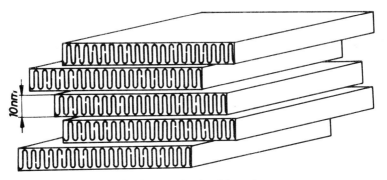

Fig. 4.21. Chain folding of polymer molecules. Schematic.

and crystallization temperature (in the case of polyethylene single crystals grown from 0.05% xylene solutions this value changes from 9 nm to 35.5 nm if the crystallization temperature is changed from 50 °C to 130 °C).

There are two current theories in the literature. One is that of Fischer and Peterlin based on thermodynamic arguments. They assumed that the folding length corresponds to the free energy minimum influenced by two factors: on the one hand the interaction of molecular chains leading to increasing free energy per unit mass, and on the other hand the surface free energy. The other theory has been developed by Frank, Tosi, Price, Lauritzen and Hoffman which analyses the chain folding phenomenon from a kinetic point of view. Accordingly, the folding length is determined by the original length of the primary nucleus. Further modifications of the theory take into account secondary crystallization in order to explain that the size of the monocrystals depends on crystallization temperature rather than the size of the primary nucleus.

It is not yet clear which of these theories holds true, and it is impossible to predict folding length in a quantitative way. It is clear, however, that the only possible configuration in monocrystals is chain folding, chains being oriented perpendicularly to the lamellae and lamellar thickness (10 nm) being much smaller than the chain length (100 nm). There are several spatial models to describe this situation. In the case of polyethylene, for example,

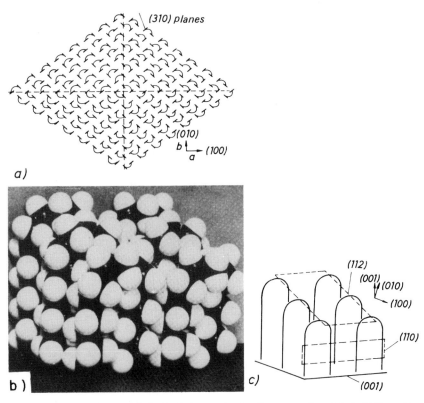

Fig. 4.22. Conformation model of polyethylene molecules. Demonstration of chain folding. *a* Top view, *b* model, *c* side view. (After Reneker and Geil 1960.)

molecules can fold so that only about 5 skeletal C-atoms take part in the folding (Fig. 4.22).

Polymer single crystals also exhibit secondary structural properties, frequently cracks or step-like structures can be observed on them. Crystals originally form hollow pyramids which collapse under the effect of surface tension when the solvent is evaporated (Fig. 4.23).

It is interesting to note that in single crystals the chains can reorganize themselves below the crystalline melting point, which demonstrates the high mobility of the structural units, for example, polyethylene single crystals grown at low temperatures, and consequently having a small folding period, can thicken on heating, that is, they can accomodate to the new equilibrium conditions. This will be especially important in the annealing processes to be discussed later (Fig. 4.24).

Fig. 4.23. Pyramidal single crystals of polyethylene. The mark of collapse can be clearly observed. (After Kalló 1979.)

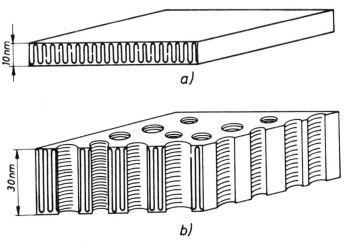

Fig. 4.24. Increase of lamellar thickness of polyethylene single crystals on annealing. *a* Before annealing, *b* after annealing.

The relationship between well-developed single crystals and melt-crystallized polymers (exhibiting a more complicated structure, albeit commercially more important) has also been studied. For example, Besset, Keller and Mitsuhashi have studied the continuous change in the crystal morphology caused by the gradual increase in concentration of the crystallizing solutions. With increasing concentrations multilayer single crystals are being formed. At higher concentrations polymer molecules crystallize through more layers so that the early stages of spherulite formation can be observed.

According to other opinions monocrystals develop from the melt so that one plane grows into ribbons or fibrils. It is important that if these fibrils join single crystals then the constituent molecules are folded and oriented perpendicular to the growth direction. This is in agreement with the observation above that in spherulites fibrillar growth occurs from the centre in the radial direction and the molecules are oriented perpendicular to the radii.

Although the relationship between single crystals and melt-crystallized structures is not totally clear, experience gained on easily available and measurable single crystals can also be applied to melt-crystallization processes.

4.2.5. Fibrils

It has already been discussed that spherulites grow from central nuclei in the form of fibrils (see Fig. 4.15). In some cases these fibrils can be observed by optical microscopy but in most polymers electron microscopy is needed to clarify the structure. The widths of these fibrils are some tens of nanometers. Fibrils are the basic unit for spherulites.

In several polymers optical microscopy does not show the substructure, and preparation of thin (about 10 nm thick) samples needed for electron microscopy is difficult technically. In such cases mechanical effects (such as ultrasound treatment) are used to clarify the structure. In oriented samples, like fibrils, cleavage is more probable along the orientation direction than perpendicular to it, since secondary bonds are easier to disrupt than primary ones. Therefore all data obtained from these disintegration methods must be considered with care because the characteristics of the original structure can easily be confused with those of the disintegration method.

It we take as an example the Fortisan fibre (a highly stretched cellulose fibre), it is found that in water, under the action of mechanical agitation, fibrillar units form which bear no relation to the original structure. On the other hand polyamide 6 and 10 disintegrate into fibrils probably identical to the original structural units, the chain molecules running perpendicular to the fibrils and probably folded (Keller 1959) (Fig. 4.25).

Studies have shown that if the fibrils are identical with those found in spherulites they can be regarded either as structural units or as bundles of them. Such fibrils are flat ribbon-like structures, or lamellae. Their thickness is some tens of nanometers, and width is in the order of micrometers. Twisting frequently occurs, which explains the optical appearance of spherulites. The

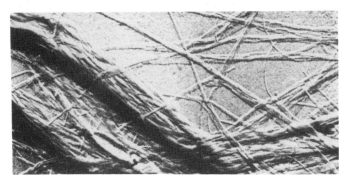

Fig. 4.25. Fibrils obtained from polyamide 6/10. Chain molecules are perpendicular to the fibre axis. Magnification is about 15,000x. (After Keller 1959.)

perpendicular orientation of chains with respect to the fibrillar axis is a consequence of chain folding discussed earlier for single crystals. It gives rise to the tangential orientation of molecules in spherulites, since fibrils grow radially from the centre. This demonstrates the utility of studies on single crystals, which are insignificant from a commercial point of view, but help in understanding the more complicated phenomena observed in melt-crystallized products.

The fact that fibrils are the structural unit of spherulites does not imply that all properties of spherulites can be understood in terms of fibrillar properties; melt-crystallized polymers can contain up to 50% of an amorphous phase. Amorphous zones can lay between the fibrils or within these units as crystalline defects. However, they are an integral part of the spherulite and their presence is important in determining the spherulitic structure and properties.

If fibrils are found among the mechanical disintegration products of a sample which has been fibrillar originally, there is a possibility that they have no direct relationship to the original structure. As an example the Fortisan fibre can be mentioned (see above) which is a highly strecthed cellulose grade, where the molecular chains run roughly parallel to the fibre direction. During mechanical disintegration, secondary bonds are disrupted easily, and

cleavage products of fibrillar shape appear but are not characteristic of the
original structure; that is, the chain orientation is perpendicular to the
original direction. In regenerated cellulose, fibrillar units can be observed
even in the absence of orientation (Fig. 4.26).

Dimensions of such fibrils lie between those of spherulites and
crystallites; their length is about 100 nm, and thickness is some 5–10 nm.
These are not crystallites as shown by the degradation of amorphous areas;

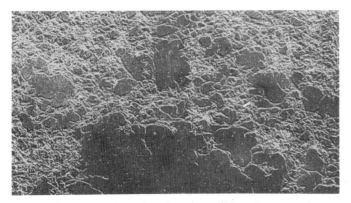

Fig. 4.26. Unoriented regenerated cellulose sample with microfibrillar structure. Sample
distintegrated by HCl solution swelling, magnification is about 15,000x. (After Dlugos and
Miche 1960.)

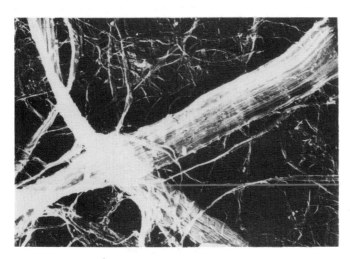

Fig. 4.27. Cellulose fibrils and partially disintegrated molecular bundles. Chain molecules are
parallel with the fibre axis. Magnification is about 15,000x.

the thickness of resulting crystallites is similar to the original value but the length reduces to about 30 nm. It seems probable that in regenerated cellulose the nucleation density is so high that the growing units can only form very short fibrils.

In natural cellulose the fibrils are much more defined, and show no tendency for spherulite formation (Fig. 4.27). The width of the fibrils is around 10 nm, but their lengths in samples of animal, bacterial or vegetable origin are usually greater than 1 μm.

It has been shown by electron diffraction measurements that in contrast to polyamide fibrils (see Fig. 4.25) the chains in cellulose run parallel to the fibrillar axis. It has to be taken into account, however, that cellulose crystallizes in nature under very different circumstances compared to, for example, melt-crystallized polyamides. Thus, for example, in vivo synthesis and crystallization can occur in parallel. Crystallization of natural cellulose shows some peculiar characteristics, for example, the fibrils shown in Fig. 4.27 cannot be produced from in vitro crystallized cellulose.

Polymers can also crystallize under circumstances favourable for fibril but not for spherulite formation as outlined in Fig. 4.28.

This phenomenon is known as transcrystallization, the name coming from the fact that crystalline nuclei are displaced on a planar surface. Such crystallization modes can be induced by stresses in the melt, or by an impurity on the wall etc.

Fibrils play an important role in almost all types of polymer crystallization. They can form branches and twists, and if the nucleation density is low and the growing units reach about one micron in size, then fibrils grow in

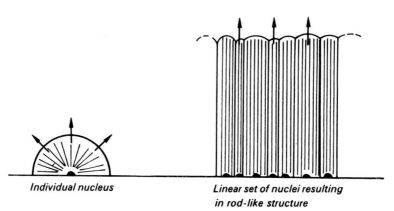

Individual nucleus   Linear set of nuclei resulting
                                         in rod-like structure

Fig. 4.28. Parallel ordering of fibrils, on transcrystallization. Growth process starts from surface nuclei. (After Sharples 1966.)

the radial direction, and spherulites are formed. In general, chains are oriented perpendicular to the fibrillar growth direction, which suggests that the growth operates through chain folding and its mechanism is similar to that observed in growing polymer single crystals from solution. In certain exceptional cases (as in cellulose) chain molecules are oriented along the fibrils.

4.2.6. Other structural units

The main structural units found in polymers crystallized under normal conditions have been reviewed, there are, however, a number of other structures which must be included in a morphological survey, such as hedrites, dendrites and extended chain lamellae.

Hedrites are polygonal structures quite similar to single crystals, as they have planar surfaces. They consist of several lamellar layers, and the total length can reach about one micron. A typical hedritic structure is shown in Fig. 4.29. It is important to note that they can form under certain conditions

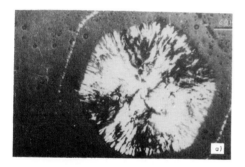

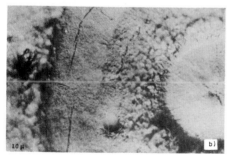

Fig. 4.29. Hedritic structure at various levels of magnification a mark $= 50~\mu$m, b mark $= 10~\mu$m. (After Geil 1963.)

both from solution and the melt. It was shown for polymethylene oxide by
Geil et al. (1960), and for isotactic polystyrene and poly(4-methyl-pentene-1)
by others. It proves directly that lamellar growth found in single crystals
grown from solutions can also appear in melts.

Pelzbauer studied the fracture surfaces of polyamide samples cooled
from the melt, and found well-developed single crystals; thus single crystal
formation from the melt is also proven.

Dendrites are intermediate products. They consist of poorly developed
single crystals being formed under imperfect crystallization conditions. Their
shape resembles that of pines. A dendritic polyethylene sample is shown in
Fig. 4.30.

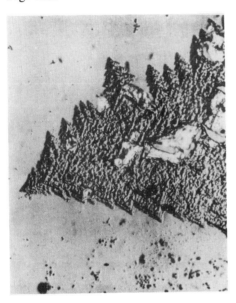

Fig. 4.30. Dendritic structure. (After Geil 1963.)

Extended chain lamellae, though not common, are also found in some
polymers.

The viscosity of polymer melts is very high, a long range migration of
macromolecules on crystallization, and a reorganization in the crystalline
state are not therefore expected. It is suprising that polyethylenes with a
molecular mass exceeding 10,000, under conditions of slow crystalization,
can form extended chain crystals.

Not only are the chains parallel, but even the chain ends are in a common
plane forming lamellae, as in low molecular paraffinic crystals. An example of

this type of crystallization is shown in Fig. 4.31 where a cryogenic fracture surface is demonstrated. Anderson (1964) has proven that, with faster cooling, lamellae are formed in which chain-folding is present. It is important that under some circumstances a significant reorganization of molecules can

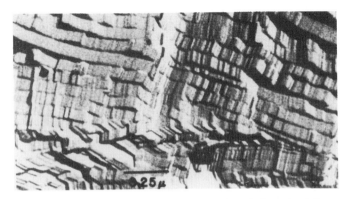

Fig. 4.31. Electron micrograph of an extended chain polyethylene crystal. The steps corresponding to the chain lengths can be observed on the fracture surface of the bulk polymer. (After Anderson 1964.)

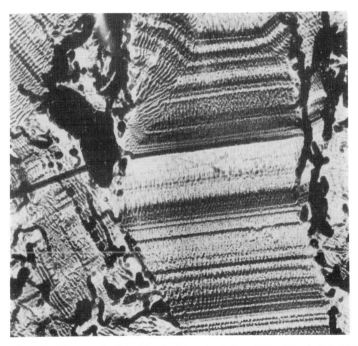

Fig. 4.32. Micrograph of the Wunderlich polyethylene. (After Wunderlich 1963.)

occur. There is evidence for the presence of a kind of molecular mass fractionation; molecules with different molecular mass forming different structural units. Wunderlich (1963) has pointed out that extended chain crystals can be grown from polyethylene if crystallization is performed under high pressure (Fig. 4.32).

According to the theory of Keith and Padden (1963, 1964), to be discussed later, during spherulitic growth molecules with lower molecular mass or stereoregularity do not contribute to fibril formation in the crystalline lattice but remain in the interfibrillar phase forming an amorphous zone.

4.2.7. A general view of polymer morphology

It is clear from the experimental data considered above, that in polymers, different structures are present simultaneously. The size of these units can cover many orders of magnitude in the same sample, for example, fibrils can be observed in several polymer samples, and these fibrils contain crystallites with dimensions two orders of magnitude lower.

However, it is not always appreciated that different methods of analysis do not necessarily give equally direct information about structural dimensions. This will be discussed in more detail later, but, for example, from the broadening of X-ray diffraction rings one can calculate the average crystallite size. In the case of cellulose this value is probably equal to the dimension of crystalline areas isolated by amorphous regions. This is supported by the electron-microscopic investigation of crystallites obtained by acidic hydrolysis. If, however, single crystals are studied, the same X-ray diffraction line broadening can be observed in structures of much greater dimensions, which can be explained by structural imperfections. This has been mentioned in connection with the paracrystalline theory of Hosemann.

When reviewing morphological structures it has to be pointed out that the same unit can appear in various contexts; for example, spherulitic fibrils are underdeveloped if the nucleation density is high, and these fibrils will be the highest structural unit in the system. If, however, the nucleation density is low and spherulites can fully develop, fibrils are only a subunit in a higher structure.

In kinetic measurements used to detect crystallization processes it is important to know the form of the growing units, and, if several types of unit are present, their nature. If, for example, spherulites develop in a crystallizing polymer simultaneously with the crystallization process, and crystal growth is fast, then nucleation and spherulitic growth are the directly measurable

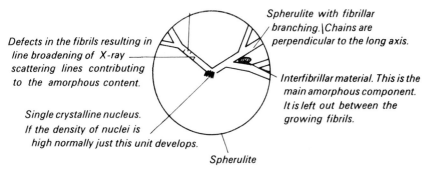

Fig. 4.33. A schematic summary of polymer structure. Appearance of various morphological units within the spherulites. Amorphous zones are formed at the interfaces of spherulites. (After Sharples 1966.)

processes, they are the rate-determining steps, not the crystallization itself. By this method, and under these circumstances, crystallization cannot be monitored.

The most problematic topic in this field is the relationship between the structure of solution grown single crystals and melt-crystallized samples. Solution grown single crystals with well-defined surfaces can be studied relatively easily, but from a practical point of view the melt-crystallization process is more important, albeit the study of these structures is more difficult.

A summary of structural units can be found in Fig. 4.33.

CHAPTER 5

Methods for studying the supermolecular structure of polymers

The growth of different polymer structures can be studied by direct microscopic observation or by the determination of volume changes associated with crystallization. The two methods yield essentially the same result. Microscopic observation, can however, be applied only if the developing spherulites are large enough for microscopic observation. As will be shown later, volume changes measured as a function of time at constant temperature give information about crystallization processes. Such an approach uses methods based on the Avrami equation.

In the course of discussion use will be made of results obtained by other methods as well, so it seems worthwhile therefore, to review briefly in advance the more important approaches.

This introductory description gives only a qualitative picture. Later, where necessary, further details will also be discussed.

5.1. DILATOMETRY

Changes in volume or density are determined using dilatometers. Such equipment is used in other branches of science, nevertheless a brief description will be given here.

The required accuracy of volume measurement is in the case of 1% crystallinity change, 1 unit in 2,000 should be measurable. This may not appear to be an easy task, nevertheless it is within the capabilites of common dilatometers. Given a 0.1 cm^3 polymer sample and using a capillary of 0.4 mm diameter, a meniscus displacement of a few tenths of a millimeter would be expected. Such a dilatometer is shown in Fig. 5.1.

The volume of the sample placed in the dilatometer is normally about 0.1 cm³, then the whole system is evacuated, and sometimes heated simultaneously. If the polymer has been precipitated from solvents a long evacuation is required to remove the solvent. After evacuation the glass tube is sealed at point *A*. Here care must be taken to avoid polymer degradation. After turning

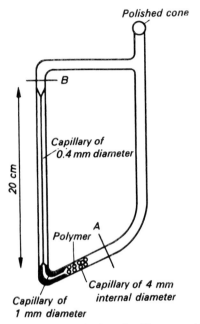

Fig. 5.1. Schematic view of a dilatometer. (After Sharples 1966.)

the system around the polished cone, it is filled with mercury by distillation. Finally the dilatometer is sealed at point *B*. If only relative measurements are made, no further calibration is required.

The mass fraction of the unchanged liquid polymer is given by:

$$\frac{m_{L}}{m_{0}} = \frac{h_{t}-h_{\infty}}{h_{0}-h_{\infty}},\tag{5.1}$$

where h_t is the height of the meniscus at time t, h_0 and h_{∞} are the heights at the beginning and at the end of crystallization respectively, m_L is the mass of the liquid polymer, m_0 the total mass of the polymer. It is to be noted that the transformed material is only partially crystalline, and information obtained by this method covers only the extent of transformation to the new state, the transition to the newly formed crystallinity but not the extent of the crystallinity itself. If crystallinity is to be determined, the absolute densities, and the density–crystallinity relationship must be known.

To determine the absolute density it is necessary to know the volume of the dilatometer, the mass of the inserted mercury and the thermal expansion coefficient of both. The sensitivity of the system decreases if the volume ratio of the fluid to that of the sample increases. This simple measurement requires accurate temperature control, temperature fluctuations exceeding 0.01 °C show up at the meniscus. A dilatometer filled with mercury only is re-commended to check thermal stability.

The time needed for thermal equilibration must also be known. This can be determined by moving the dilatometer from one bath to a second held at a different temperature—both above the melting point—and measuring the time needed for equilibration.

h_0 can be determined by measuring the equilibrium meniscus height–temperature relationship and extrapolating to the temperature where crystallization begins. Eq. (5.1) contains h_0 determined in this way.

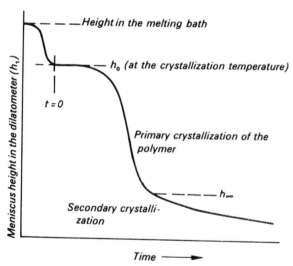

Fig. 5.2. Results of a dilatometric measurement. (After Sharples 1966.)

When studying crystallization the dilatometer is moved from a bath held at a temperature above the melting point to another one at the temperature of crystallization. The result of such an experiment is shown in Fig. 5.2.

The h_0 value shown in the Figure should be checked by the method described above. Choice of t_0 is to some extent arbitrary. In the case of samples moved from one bath to another, both above the melting point, it is easy to detect the time when the system reaches the temperature of the new bath. In the case of the system described above t_0 is of order of 2 minutes. Determination of h_∞ is complicated if the polymer exhibits secondary crystallization. During secondary crystallization h_∞ is determined from the slope. The accuracy of the analysis is strongly influenced by that achieved in h_∞ determination.

The three main sources of error in dilatometry are:

1. Presence of air bubbles. The source of this error is either insufficient evacuation of the polymer, or its degradation at the test temperature. This kind of error can usually be detected by inspection, but it should be emphasized that their presence can be highly misleading when processing the data.

2. Jump of the meniscus. This kind of error appears if the solidifying polymer seals the total cross-section of the dilatometer. It can be detected on the meniscus height–time curve during crystallization or on subsequent melting.

3. Voids and fissures in the polymer. If such defects appear on the surface, but they are small, or if they are inclusions, the liquid cannot penetrate from outside. Even microscopic investigation cannot help in detecting them. If the sample becomes opaque on solidification the presence of such voids is highly probable.

5.2. LIGHT MICROSCOPY

In the case of polymers forming large spherulites the processes of crystal nucleation and growth can be followed by optical means. Such polymers are, for example, polypropylene, polyethylene oxide, polyamide and polyethylene. In such cases optical microscopy is the must suitable method to study the crystallization process. Using a polarization microscope the two processes can be observed easily, and the effect of external conditions, such as crystallization temperature, can be measured.

The optical microscope has found application in many branches of scientific research, but here some special features will be discussed which are applicable to the study of polymer crystallization.

The polymer melt is not birefringent—except under the effect of shear deformation as, for example, in flow birefringence tests. The growing spherulites are, however, strongly birefringent, thus crystallization can be observed easily using crossed Nicols. The study of details in unpolarized light is highly recommended, since not only are the diameters of spherulites important, but the evolving substructure is also crucial, and this is frequently better discernible in nonpolarized light. The experiments can be performed with the aid of a relatively simple microscope, although an ocularmicrometer, photographic facility and a hot stage are also required.

By placing a thermocouple close to the sample, temperature control can be achieved within $\pm 0.03\,°C$. The actual temperature of the sample must of course be calibrated with materials of known melting points.

Before performing the experiment at the crystallization temperature, the sample must be melted at a higher temperature. The determination of thermal equilibration during crystallization is also important, as it can lead to serious errors in finding the zero time point of the experiment, especially when crystallization is rapid. Observations are usually performed on thin films, between two glass plates. If fast equilibration is required, thin plates must be used and the sample must be moved from the melting stage to that thermostated at the crystallization temperature as rapidly as possible.

If crystallization at the temperature of measurement is rapid, the equilibration time can be estimated by observing the time lag between placing the polymer on the stage and the appearance of birefringence.

If it turns out that the melting temperature has no effect on properties, the polymer can be melted in situ by a very simple method, using heat radiation, via a small heating filament above the sample.

Crystallization of thin films cannot be studied effectively by this method because of the possibility of degradation. For example, polypropylene crystallizing between 120 and 145°C can yield very different rates depending on degradation unless special precautions are taken. One such method for preventing the penetration of oxygen into the sample is to cover it with silicone oil of high viscosity. Even then the sample must be checked for signs of degradation. A sensitive test for this is the constancy of the growth rates under constant crystallization temperature within a given series of measurements, since the growth rate normally depends strongly on the degree of degradation.

For some polymers, such as polyethylene and polyamides, results obtained from the observation of thin films are different from those performed with bulk materials. Its exact reason is not known, but the cooling process is different in thin films and this probably influences the density of nucleation. Owing to differences of this kind cross-check experiments on sections cut from samples crystallized under bulk conditions are highly recommended. Since such sections are never as clear as crystallized thin films—if the bulk behaviour is identical with that of the films—then films are preferable for study.

5.3. LIGHT SCATTERING IN POLYMERS

Polarized monochromatic light beams passing through thin layers of a polymer containing spherulites and viewed through crossed Nicols, give a scattered, easily photographed intensity distribution resembling a four-leaf clover. This system is known as small angle light scattering (SALS). The phenomenon can be used for spherulite characterization even at spherulite diameters which cannot be resolved by polar–optical means. Fig. 5.3 shows a

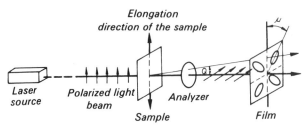

Fig. 5.3. Small angle light scattering (SALS) measuring system.

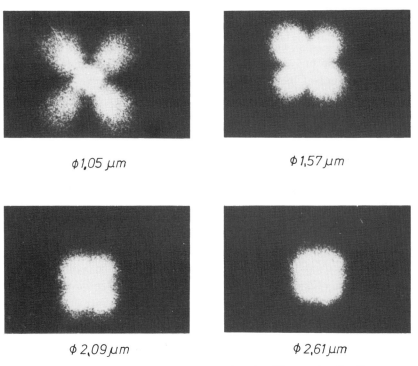

$\phi 1{,}05 \, \mu m$ $\phi 1{,}57 \, \mu m$

$\phi 2{,}09 \, \mu m$ $\phi 2{,}61 \, \mu m$

Fig. 5.4. SALS photographs of polypropylene samples with different spherulite diameters.

schematic apparatus for measuring small angle light scattering, where the light source is a laser. Fig. 5.4, for example, shows a series of SALS photographs using a He–Ne gas laser (wavelength 632.8 nm) on polypropylene samples with different average spherulite diameters. Here the light was polarized vertically and the analyzer was horizontal, these are known as H_v SALS experiments.

Studies with a vertical analyzer (with the plane of polarization parallel to the analyzer) are referred to as V_v SALS experiments.

The theoretical explanation of the SALS phenomenon has been provided by Stein and Rhodes (1960). They have shown that the scattering of homogeneous anisotropic spheres in an isotropic medium can be described by the equation:

$$I_{Hv} = AV_0^2 \left[\frac{3}{U^3} \right]^2 \times (a_r - a_t) \cos^2 \left(\frac{\Theta}{2} \right) \sin \mu \cos \mu \times$$

$$\times (4 \sin U - U \cos U - 3 \mathrm{Si}\, U)^2 \tag{5.2}$$

where V_0 is the volume of the anisotropic sphere, α_t and α_r the tangential and radial polarizabilities of the sphere respectively, Θ is the radial and μ the azimuthal scattering angle. A is a proportionality factor, $\mathrm{Si}U$ is defined by the integral:

$$\mathrm{Si}U = \int_0^\mu \frac{\sin x}{x} dx \tag{5.3}$$

and U is:

$$U = \frac{4\pi R_0}{\lambda'} \sin \frac{\Theta}{2}. \tag{5.4}$$

In this equation R_0 is the radius of the sphere, Θ the angle between the scattered and direct beams, and λ' the wavelength of the light in the medium studied.

The scattered intensity in H_v SALS experiments shows a maximum at $\Theta/2$ where the value of U is 4.09. Thus the average radius of the anisotropic spheres (the underformed spherulites) can be calculated from the measured angle of maximum intensity ($\Theta/2$) using the equation:

$$4.09 = \frac{4\pi R_0}{\lambda'} \sin \frac{\Theta}{2}. \tag{5.5}$$

In practice spherulites of diameters ranging between 0.1 and 50 micrometers can be studied using the equation.

With samples crystallized at different temperatures, the greater the spherulite diameter, the smaller the clover-shaped diffraction pattern (see Fig. 5.4).

Heat treatment strongly influences the spherulite size. Fig. 5.5 shows an example of the relationship between spherulite size and crystallization temperature for a polypropylene sample cooled from 170°C.

The method can be applied to follow the deformation of the spherulites under tensile deformation (Samuels 1974).

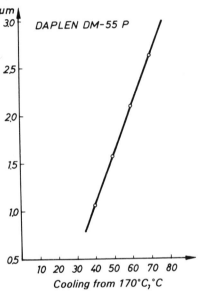

Fig. 5.5. Spherulite diameter of polypropylene (Daplen DM 55) samples as function of crystallization temperature.

5.4. ELECTRON MICROSCOPY

Electron microscopy is a very powerful tool for the study of polymer morphology at the optical level. The devices are divided into two groups: transmission (TEM) and scanning (SEM) electron microscope. These two techniques are used for different purposes and their resolving powers differ substantially.

The resolving power of normal light microscopes is 0.2 μm, that of TEM devices is 1 nm, whereas in SEM systems 10 nm would be expected. The depth of focus for scanning electron microscopes is about 300 times larger than that of optical instruments, so that even at relatively low magnification SEM shows considerably more relief than is possible with other systems.

We have to note, however, that there exist electron microscopes with resolving powers of 0.2 nm (TEM) or 0.5 nm (SEM) (capable of detecting single atoms), nevertheless in polymer research only supermolecular structures above 2–10 nm have been discovered so far.

5.4.1. Transmission electron microscopy (TEM)

Transmission electron microscopy requires 10^{-7}–10^{-8} m thick samples which do not degrade in the high vacuum prevailing in the microscope.

Thin single crystals can be studied by TEM with relative ease, they are mounted on carbon film and contrasted by evaporating platinum or other metals. In the case of samples with no contrast an electron beam can be passed through the sample at predetermined points so that electron diffraction patterns can be detected. Thus the position of molecular chains in relation to the visual position of the crystals can be determined. Electron irradiation frequently damages thin polymer samples, therefore performing measurements of this kind is quite difficult. It is necessary to take into account that the crystal structure itself may be modified under the action of irradiation.

In the case of thicker samples the difficulties are even greater, which explains the scarcity of knowledge about bulk structures.

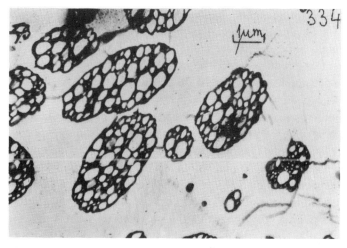

Fig. 5.6. Structure of high impact polystyrene. (After Kalló 1979.)

Surface replicas can be made but the surface is not always representative of the internal structure. Replicas are frequently made of fracture surfaces obtained at low temperatures. This is especially useful in materials exhibiting rubber elasticity at room temperature. There are also special techniques where the polymer sample is prepared by mixing it with materials of a similar viscosity which can later be dissolved from the polymer. The remaining structure corresponds to the bulk polymer structure. By the application of ultrasound small pieces of material can be produced from polymer samples which are suitable for study by transmission electron microscopy. It has to be taken into account, however, that during preparation, structures different from those of the original sample can be formed, especially in samples which are oriented or contain particles capable of orientation.

TEM is very useful in studying finely distributed additives. In the case of high impact PS, for example, the structure of rubbery additives can be explored if the electron density of the elastomer is increased by binding heavy metal ions to it. Fig. 5.6 shows a micrograph of this kind where even the internal void structure of the additive is easily discernible.

5.4.2. Scanning electron microscopy (SEM)

SEM does not need thin sections. With such a high depth of focus these instruments can produce very informative micrographs of surfaces or fracture surfaces of samples up to several cm^2 in area. Samples are prepared for SEM by evaporating metals onto the surfaces to be viewed. Samples must resist high vacuum during metal coating and in the SEM. For example whole bodies of insects can be put in the chamber for examination.

As an example SEM micrographs of PVC powders prepared by suspension, emulsion and bulk polymerization are presented. The pictures of these three types are significantly different. Suspension PVC powder (Fig. 5.7) has an external crust controlling the plasticizer uptake. Inside this crust, within the secondary globules, of 100 μm size, are the primary ones (0.5 –1μm) which can be observed in section by transmission electron microscopy.

Emulsion PVC powder has no external crust, the size of secondary globules lies between 10 and 20 μm, within which the primary ones (0.1– 0.5 μm) are observable even on SEM micrographs (Fig. 5.8). It can be seen that the size distribution is not uniform in the sample, which can lead to processing problems. Bulk polymerized PVC powders also have no crust, the primary globules (0.2–1 μm) are relatively homogeneous, local crust formation, can however, cause processing problems (Fig. 5.9).

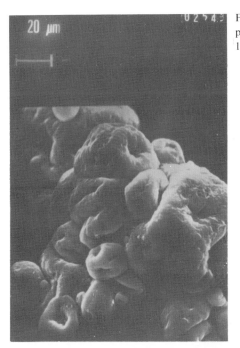

Fig. 5.7. SEM micrograph of a PVC powder suspension sample. (After Kalló 1979.)

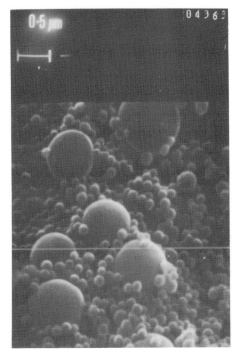

Fig. 5.8. SEM micrograph of a PVC powder emulsion sample. (After Kalló 1979.)

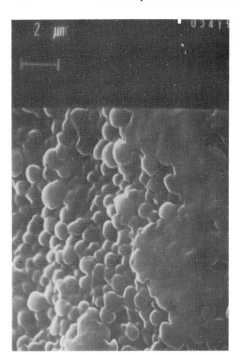

Fig. 5.9. SEM micrograph of a bulk polymerized PVC powder sample. (After Kalló
1979.)

Fig. 5.10. SEM micrograph of the ion-
etched surface of an impact modified
PVC-sheet. (After Kalló 1979.)

Surfaces or fracture surfaces of multi-component polymer systems frequently show no characteristic structure. Structures can be developed in these cases, for example, by etching. Chemical etching often gives no result because of secondary crystallization. Ion sputtering frequently gives good results where chemical etching cannot be used. As an example the ion sputtered surface of high impact PVC is shown in Fig. 5.10. PVC degrades faster under the effect of ion bombardment than the additive (chlorinated polyethylene), thus the position and distribution of the high-impact additive is represented in the micrograph by the particles emerging from the surface (Kalló 1979).

5.5. X-RAY DIFFRACTION

Diffraction of monochromatic X-rays is used in the study of both molecular and supermolecular structures. The method will be dealt with in detail in Chapter 7., here only the main applications are outlined.

5.5.1. Normal (wide angle) diffraction

The X-ray diffraction pattern of an isotropic polymer exhibits a series of rings, that for oriented samples is the fibre diagram, consisting of crescent arcs. These diffraction patterns contain a variety of information.

From the diffraction pattern it is possible to determine the position of the atoms building up the molecule in the crystalline phase, the size of the crystalline unit cell (from which the density of the perfect crystal can be calculated) and the position of chain segments within the unit cell. This kind of information refers to atomic arrangements and can be obtained by classical methods of structural analysis. These methods do not differ from those used in structural analysis of inorganic or low molecular organic crystals, they will therefore not be discussed in detail here.

Polymer texture, crystallinity ratio, size of crystalline particles and molecular orientation can also be determined from diffraction patterns. These methods are used extensively for polymers, thus they will be discussed in detail later.

On the supermolecular structural level X-ray diffraction can detect crystallites in spherulites and the amorphous nature of extraspherulitic areas. Electron diffraction study of areas chosen by electron microscopy can reveal the arrangement of carbon chains.

5.5.2. Small angle diffraction

Investigating diffraction at small angles (below 5°) reflections are often found produced by periodic structures with an identity period of 5–100 nm. This phenomenon can be observed both in natural and synthetic polymers, but the origin and nature of such large ordered units is still the subject of research.

In the case of oriented polymers (for example, fibres and films) periodically alternating crystalline and amorphous structures are expected to exist in certain directions. In fact the presence of such structures differing in electron density can be demonstrated by electron microscopy.

Polymer single crystals also exhibit similar long periods in small angle X-ray diffraction studies. Periods obtained in this case correspond to the thickness of the crystal observable in electron microscopy, that is, to the folding length of the chains.

It is worth mentioning in connection with supermolecular structures that in the case of imperfect structures, where electron microscopy is not applicable, long periods can be determined by small angle X-ray diffraction. They do not necessarily mean folding lengths, yet chain folding in polymer structures is the norm.

Small angle X-ray scattering will be discussed further after the consideration of polymer molecular textures, here only details concerning supermolecular structures were necessary.

5.6. THERMAL ANALYSIS

Thermoanalytical measurements have for many years been performed in calorimeters. Nowadays automatic equipment is available. The most important methods are differential thermoanalysis (DTA) and differential scanning calorimetry (DSC).

5.6.1. Differential thermoanalysis (DTA)

In this method the temperature difference is detected between the sample under investigation and a known reference sample when the two substances are cooled or heated with equal energy. The two samples have a single, common heating system. The method is adiabatic, the heat evolved heats the sample. Heating rates can vary between 0.1 and 1,000 °C/min, usually between 1–2 °C/min. The temperature difference (ΔT) between the sample and the reference is detected by an accurate temperature sensor as a function of temperature (or time). The equipment automatically records temperature and weight difference (if any) (*TG*), and the derivative of the latter quantity (*DTG*).

Variation of the sample temperature depends on thermal conductivity and heat capacity. The latter changes at phase transitions, thus transformation points can easily be detected by this method.

During melting and crystallization the sample absorbs or emits heat, leading to endothermic and exothermic peaks respectively. These effects are demonstrated in Fig. 5.11 which shows a typical thermogram. Polymers melt

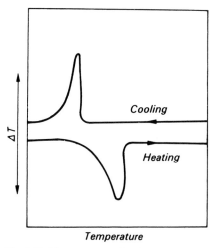

Fig. 5.11. Thermograms of polymers. Schematic.

within a fairly wide temperature range, the temperature interval for crystallization is much narrower, but a fair degree of overcooling is required to promote crystallization.

Since temperature changes during crystallization due to the continuous cooling or heating, this method is not suitable for accurate crystallization kinetics measurements.

By calculating the area under melting peaks the crystallinity of samples can be estimated if suitable calibration is used. These kind of measurements, however, can be performed with much higher accuracy using DSC.

If a quenched sample (for example a polyester) is studied between room temperature and the melting point, further transitions may appear. Secondary transition temperatures appear as a shift in the base line. This temperature is often referred to as the glass temperature (T_g). Some crystallization can appear below the melting point but above the glass transition temperature. This is known as cold crystallization which appears if the sample is in a liquid-like state between the secondary transition and the melting point.

In the course of melting and subsequent cooling the crystallites reorganize, and a new, much more stable crystalline structure develops. In oriented samples an additional heat effect due to chain disorientation can be observed below the melting point.

5.6.2. Differential scanning calorimetry (DSC)

This method is also based on the comparison of a sample with an inert standard, where, however, provision is made for keeping the two samples at the same temperature. Thus no temperature difference develops between the sample and the reference, and the experiment is isothermal. If any temperature difference does appear between the sample and the standard, the instrument compensates by changing the heat flux. The heat flux difference, the thermal energy difference per unit time is measured:

$$\Delta \mathring{H} = d\Delta H/dt \tag{5.6}$$

The two samples have two independent heating systems.

In this method also, T_g appears as a shift in the baseline, and melting and crystallization as exothermic and endothermic peaks respectively. The majority of scientific publications plot the endothermic peak as negative, and the exothermic as positive. DSC is better for quantitative work than DTA. Heat exchange involved in the transition can be determined by measuring the area under the peaks with about 2% uncertainty. The following formula is applied:

$$\Delta H = \frac{KRA}{mS} \tag{5.7}$$

where K is a constant characteristic of the equipment, m the mass of the sample, R the sensitivity used, S the chart speed, and A the area under the peak.

Heating and cooling rates can vary within wide limits (for example, 0.3–320°C/min), a typical value is 10°/min. Isothermal crystallization can also be studied here, the crystallization heat is measured as a function of time. The method is suitable for determining the heat capacity of samples. In this case

$d\Delta H/dt$ is determined within the desired temperature limits. Displacement of the recorder is proportional to the heat absorption of the sample:

$$\frac{d\Delta H}{dt} = mC_p \frac{dT}{dt},$$ (5.8)

where m is the mass of the sample, C_p the specific heat of the sample, and dT/dt the programmed speed of temperature change. Using Eq. (5.8) C_p can be determined directly.

More accurate measurements can be performed if repeated with a sample of known mass and specific heat (for example, sapphire standards) where the unknown heat capacity is calculated from the ratio of the two equations. (Heating rate is kept constant thus eliminating any potential error in the calculation.)

DSC is important in determining the crystallinity of polymers. If the heat of crystallization is known for the totally crystalline polymer, the crystallinity ratio can be determined from the heat absorbed during melting and the mass of the sample:

$$x\% = \frac{\Delta H_{meas}}{\Delta H_{cryst}} \cdot 100.$$ (5.9)

The heat of crystallization for totally crystalline polymers will be discussed when polymer crystallinity is considered. However, H_{cryst} values for three frequently used polymers are:

polyethylene 293 J/g,
polypropylene 138 J/g,
polyamide-6 188 J/g.

Thermoanalytical methods can be used to determine crystallinity only if the polymer melts on heating. Cellulose, for example, is crystalline according to X-ray data, but it degrades below its melting point.

DSC is very useful in distinguishing between chemically similar polymers produced by different technologies. The melting point of high density (low pressure) polyethylene (HDPE) is for example 408 K (135 °C), its crystallinity is around 68%, the melting point of low density (high pressure) polyethylene (LDPE) on the other hand is 386 K (113 °C), and its crystallinity is around 35%. The melting point and crystallinity of linear low density polyethylene (LLDPE) grades lie between these limits depending on the amount and chemical composition of the co-monomer(s) used. In the case of block copolymers (such as PE–PP block copolymers) the different blocks give separate melting peaks. The position of the peaks indicates the type of polyethylene block. Fig. 5.12 shows the DSC curve for a high density, low pressure polyethylene, while the diagram of a low density, high pressure PE sample is shown in Fig. 5.13. Comparing the two Figures the difference of the melting points and crystallinity is quite clear.

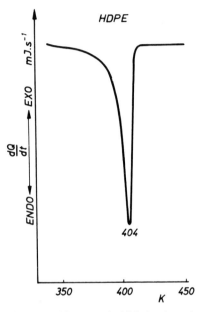

Fig. 5.12. DSC curve of a high density polyethylene sample. Crystallinity 57%.

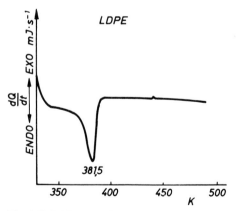

Fig. 5.13. DSC curve of a low density polyethylene sample. Crystallinity 35%.

The weight of a sample is about 5 mg in either powder, film or fibre form. Temperature of small samples can change rapidly, so that heat exchange problems emerging with thick samples can be avoided.

Fig. 5.14 illustrates the behaviour of block copolymers. Separate melting transition of polyethylene and polypropylene blocks and the corresponding melting heats are clearly visible.

Returning to the differences between the melting and crystallization behaviour mentioned in connection with the DTA technique, Fig. 5.15 shows an experiment performed on biaxially oriented polypropylene film. On first heating two endothermic peaks can be observed, the first indicates the disappearance of orientation, the second the melting of crystallites. On

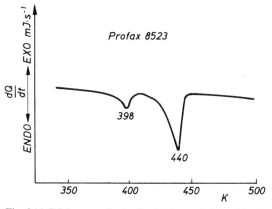

Fig. 5.14. DSC curve of a Profax 8523 type PE–PP block copolymer.

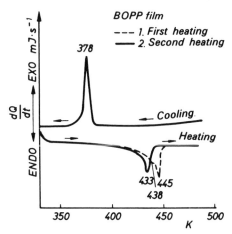

Fig. 5.15. DSC curve of a biaxially oriented PP film.

cooling crystallization appears as an exothermic peak at a lower temperature and with much narrower transition interval than in the case of melting. On repeated melting a single endothermic peak appears, orientation is now absent. The area under the melting curves is, however, constant and equal to that under the crystallization peak, whereas crystallinity of polypropylene does not change on simply melting and recooling the polymer. This is worth mentioning since even in the literature, it is frequently suggested that thermal treatment or cooling can influence crystallinity considerably, for example, it is often claimed that amorphous PP can be obtained by quenching in water. It must be stressed that any crystallinity change will be reflected in changes of melting heat and X-ray diffraction lines. Heat treatment leads to a change in the size of ordered domains or to stress relaxation at a molecular level. Quenching—expecially when using polypropylene—leads to smaller crystallite sizes and to frozen in stresses. In the case of LDPE even the size of crystallites cannot be influenced by quenching, only frozen in deformations. The thermal behaviour of LLDPE shows strong dependency on its thermal history. As branching influences the crystallization rate, very different

Fig. 5.16. SEM micrograph of a PP sheet annealed under a particular thermal program for several days.

structures appear (Bodor and Dalcolmo 1987). These remarks refer to the different levels of molecular order coexisting in, and characteristic of, polymers which can be determined by thermal analysis using versatile experimental and theoretical approximations, and hence related to other material properties.

Annealing has a considerable effect on the supermolecular structure: spherulite size and the sharpness of their interfaces increases. As an example Fig. 5.16 shows an SEM micrograph of PP spherulites annealed for a long time. Fig. 5.17 shows the DSC thermogram of the same sample.

Well-developed spherulites in PP do not correspond to single crystals, since molecular chains are organized, beginning from an internal core following the rules of chain folding referred to earlier. Cleavage planes of the spherulites do not coincide with crystallographical planes. According to the DSC trace the crystallinity of the sample increased as well.

DSC-thermograms indicate with fine sensitivity the thermal history of a sample. In the case of synthetic fibres annealing relieves internal stresses up to the annealing temperature. Investigating such a sample the stresses relax further at the temperature of the previous annealing, and an exothermic effect appears. This effect is shown in Fig. 5.17. In order to obtain large spherulites the sample has been annealed just below its melting point, and the result is a double melting peak.

Thermoanalytical methods, especially DSC, give a great deal of information about relatively small samples within a short time. They can be used to determine the melting point, melting temperature interval, crystallinity, phase transitions, crystallization and curing kinetics, specific heat and internal stresses in polymers.

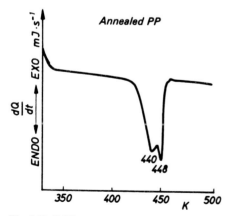

Fig. 5.17. DSC curve of the PP sample shown in Fig. 5.16.

CHAPTER 6

Kinetics of the development of supermolecular structures in polymers

Most studies of polymer crystallization kinetics so far have dealt with the average kinetic process observable by dilatometry, that is, crystallization as a combination of nucleation and crystal growth.

The density difference between crystalline polymers and their super-cooled melts at the same temperature is readily measurable, the volume difference can be as high as 10%. Thus dilatometry is suitable for monitoring the time dependence of crystallization of bulk polymers.

6.1. BASIC PRINCIPLES OF KINETIC ANALYSIS

The general theory of phase transition kinetics was developed by Avrami in 1940. Later Evans published similar equations in 1945 for the growth of circles and spheres, describing the growth of corrosion centres on metal surfaces and the development of grains in metal blocks.

These principles were applied by Morgan, Flory and Mandelkern to the crystallization of polymers at the beginning of the 1950's.

Structural development can be best understood in terms of elementary processes. The basic processes will be discussed in the following. Later, the validity of these concepts will be checked. All crystallization processes consist of nucleation and crystal growth, and before going further, it is useful to consider a number of definitions:

(i) Spontaneous nucleation. On cooling molten polymers to a fixed temperature nuclei appear randomly throughout the volume of the polymer.

(ii) Time dependence of nucleation. Nuclei can appear instantaneously on cooling or during crystallization in the non-transformed phase. In the first

case the time dependence is of zero order, in the latter it is a first order quantity, so that:

$$p = Nt \qquad (6.1)$$

where p is the number of nuclei per unit volume at time t, N the nucleus formation constant, the number of nuclei formed in unit time.

(iii) Nature of the growth process. Growth can proceed in one, two or three dimensions resulting in needles, discs or spheres. Growth stops when neighbouring chains are impinged upon.

(iv) Time dependence of growth. Linear dimensions of growing bodies increase in proportion to time, and generally the following equation is assumed:

$$r = Gt \qquad (6.2)$$

where r is the relevant linear dimension (for example, in the case of spherical growth the radius of the sphere), G is the growth constant.

(v) Density of the growing units. The transformed phase is not totally crystalline, it is, however, generally assumed that the crystallite content and thus the density of the growing units is constant. Let us denote the density of the untransformed liquid phase by ρ_L, and that of the growing solid phase by ρ_s. As a consequence of the notation the density of the totally solidified material is ρ_s.

The main goal of the Morgan–Flory–Mandelkern mathematical formalism is to establish the relationship between the average density of the crystallizing system and the time of crystallization. Characteristic parameters of the growth process can be determined from these equations. Most important of these is the Avrami exponent which will be discussed in detail later.

Here we mention only briefly that the functional form is complicated, and the exponent value depends on the dimensionality of the growth process, and on the kinetic order of nucleation. In the most general case, that is, where three-dimensional growth is combined with first order nucleation, the Avrami exponent, n is $3 + 1 = 4$.

Besides the Avrami exponent the rate constant can also be determined from the equation, which is a combined function of constants describing nucleus formation and growth. Values of these constants can in principle be proven by independent microscopic measurements.

Equations describing the early stages of growth are simple. It is known from experience that these equations can be applied up to 30% conversion. Later processes can be described by more complicated equations where coincidence of increments has to be taken into account. As will be shown later simple equations can be used to check the validity of far more complex ones.

In the following sections the three-dimensional spherulite formation will be described, based on random nucleation. Derivation of formulae for other cases—for example for needle-like growth—is similar, so that in these cases only the final formula will be described without derivation.

6.2. DESCRIPTION OF THE EARLY STAGES OF CRYSTALLIZATION

If spherulite nucleation and growth proceeds for x minutes, during a given time interval, dx:

$$\frac{N \cdot m_0 dx}{\rho_L}$$

(6.3)

nuclei are formed, where m_0 is the mass of the crystallizing material at $t = 0$, N the nucleation rate constant in unit volume as defined previously, and ρ_L the density of the liquid phase.

If crystal growth results in a spherulite of radius r, the amount of transformed material is:

$$\frac{4r^3\pi}{3} \rho_s.$$

(6.4)

Let us investigate the degree of crystallization, m_s/m_0 at time t. The amount of transformed material from the beginning of nucleation $(t - x)$ is the product of the number of nuclei and the transformed mass per nucleus:

$$dm_s = m_0 \frac{N}{\rho_L} dx \frac{4}{3} \pi G^3 (t - x)^3 \rho_s$$

(6.5)

where G is the growth constant (see Eg. (6.2)).

The amount of material transformed to spherulites up to time t is:

$$m_s = \int_{x=0}^{x=t} \frac{4\pi N G^3 m_0 \rho_s (t - x)^3 dx}{3\rho_L}$$

(6.6)

and so:

$$\frac{m_s}{m_0} = \frac{\pi N G^3 \rho_s t^4}{3\rho_L}$$

(6.7)

The mass ratio of the liquid material, m_L/m_0 after time t can be expressed in the following way:

$$\frac{m_L}{m_0} = 1 - \frac{m_s}{m_0},$$ (6.8)

thus Eq. (6.7) can be given by

$$\frac{m_L}{m_0} = 1 - \frac{\pi N G^3 \rho_s t^4}{3\rho_L}.$$ (6.9)

The important consequence of Eq. (6.7) is that in the case of random nucleation and three-dimensional spherulite growth the mass fraction of the crystalline phase increases in proportion to the fourth power of time in the initial stage. The time exponent—the Avrami exponent—is 4 in this case. As mentioned earlier it consists of two terms: the dimensionality of growth which in this case is 3; and the time dependence of nucleation, which is unity here.

In practice, of course, the conclusion is reversed; if the experiment yields $n = 4$, nucleation and growth is likely to proceed according to the mechanism discussed.

6.3. DESCRIPTION OF THE TOTAL CRYSTALLIZATION PROCESS INCLUDING COLLISIONS

There are two main methods by which Eqs (6.7) and (6.9) may be extended in order to describe the later stages of crystallization where the growing bodies meet or collide.

Evan's method assumes that the centres of growing discs or spheres are fixed. This assumption is not, however, valid for continuously shrinking systems. The treatment proposed by Flory and Mandelkern gives more exact results, and discussion will be limited to this approach. It has to be noted, however, that differences between results obtained by the two methods are small.

The mass of crystallizing material (dm_s) in a time interval dx after time x is proportional to that mass dm_s' which would crystallize in the absence of collisions. The proportionality constant is equal to the fraction remaining in the liquid phase:

$$dm_s = \left(1 - \frac{m_s}{m_0}\right) dm_s'.$$ (6.10)

This means that at the beginning of crystallization the probability of collisions is small, and the value of the proportionality factor is close to 1. As transformation proceeds the probability of collisions increases continuously, and the proportionality constant approaches zero. The value of dm_s/dm_s' is of course zero when crystallization is complete.

The assumed amount of crystallized material without collisions is:

$$m_s = m_0 \frac{\pi N G^3 \rho_s t^4}{3\rho_L}, \tag{6.7}$$

so that

$$dm_s = \frac{4\pi m_0 N G^3 \rho_s x^3 dx}{3\rho_L}. \tag{6.11}$$

Rearranging Eq. (6.10) and substituting Eq. (6.11) gives:

$$\frac{dm_s}{m_0\left(1 - \dfrac{m_s}{m_0}\right)} = \frac{4\pi N G^3 \rho_s x^3 dx}{3\rho_L}. \tag{6.12}$$

Integrating between $x = 0$ and $x = t$:

$$\ln\left(\frac{m_0 - m_s}{m_0}\right) = \frac{-\pi N G^3 \rho_s t^4}{3\rho_L}. \tag{6.13}$$

Comparing with Eq. (6.8) and expressing in an exponential form suggests that:

$$\frac{m_L}{m_0} = \exp(-zt^4), \tag{6.14}$$

where z is the rate constant with a value of $\dfrac{\pi N G^3 \rho_s}{3\rho_L}$.

This is the Avrami equation which describes the crystallization process in polymers. Expanding Eq. (6.14) into series form the first term approximates to Eq. (6.9) being correct at short times when the effect of collisions is insignificant.

6.4. ANALYSIS OF DILATOMETRY DATA

Application of derived equations becomes possible if connection is made between dilatometric data and crystallization.

The crystallinity ratio is:

$$\frac{m_0 - m_L}{m_0}. \tag{6.15}$$

Whereas the total volume at time t is given by:

$$V_t = \frac{m_L}{\rho_L} + \frac{m_0 - m_L}{\rho_s} = \frac{m_0}{\rho_s} + \frac{m_L}{\rho_L} - \frac{m_L}{\rho_s} \tag{6.16}$$

where m_0, the initial mass of the system is constant. Final volume is $V_\infty = m_0/\rho_s$, where ρ_s is the density of the growing spherulite, and therefore the final density. Similarly $m_0/V_0 = \rho_L$:

$$V_t = V_\infty + m_L \left(\frac{V_0}{m_0} - \frac{V_\infty}{m_0} \right), \tag{6.17}$$

$$V_t - V_\infty = \frac{m_L}{m_0} (V_0 - V_\infty), \tag{6.18}$$

$$\frac{m_L}{m_0} = \frac{V_t - V_\infty}{V_0 - V_\infty}. \tag{6.19}$$

If the meniscus height in the capillary is measured instead of the volume:

$$\frac{m_L}{m_0} = \frac{h_t - h_\infty}{h_0 - h_\infty} = \exp(-zt^4). \tag{6.20}$$

For other types of nucleation and growth mechanism similar Avrami equations can be derived with the general form:

$$\frac{h_t - h_\infty}{h_0 - h_\infty} = \exp(-zt^n) \tag{6.21}$$

Table 6.1

Exponents n of the Avrami equation for different nucleation and growth mechanisms

n	
$3+1=4$	Spherulitic growth + random nucleation
$3+0=3$	Spherulitic growth + instantaneous nucleation
$2+1=3$	Disc-like growth + random nucleation
$2+0=2$	Disc-like growth + instantaneous nucleation
$1+1=2$	Rod-like growth + random nucleation
$1+0=1$	Rod-like growth + instantaneous nucleation

After Sharples 1966.

where n is the Avrami exponent. Various values of n are summarized in Table 6.1.

$$\text{Plotting } \ln\left\{-\ln\left[\frac{h_t - h_\infty}{h_0 - h_\infty}\right]\right\} \qquad \text{versus } \ln t,$$

n can be calculated from the slope. The intercept on the other hand gives the rate constant:

$$\ln\left(\frac{h_t - h_\infty}{h_0 - h_\infty}\right) = -zt^n, \qquad (6.22)$$

$$\ln\left[-\ln\left(\frac{h_t - h_\infty}{h_0 - h_\infty}\right)\right] = \ln z + n \ln t. \qquad (6.23)$$

Two further parameters can be derived from the Avrami equations: induction time and the half-time of the process. Induction time is the time where m_L/m_0 differs appreciably from unity. Half-time also characteristic of crystallization can be calculated from Eq. (6.14) ($m_L/m_0 = 0.5$):

$$t_{1/2} = \left(\frac{\ln 2}{z}\right)^{1/2}. \qquad (6.24)$$

It depends on both z and n, therefore it can be used to compare z rate constants of different processes only if n is invariant.

6.5. EXPERIMENTAL DATA

Keller et al. in 1954, and Flory and Mandelkern independently in the same year proved that crystallization in crystallizable polymers proceeds in accordance with the Avrami equation.

Fig. 6.1 shows the crystallization of polyethylene terephthalate as a function of time at three different temperatures. Continuous curves are calculated from the Avrami equation assuming:

$n = 2$ at 110°C,

$n = 4$ at 236 and 240°C

z has been chosen to give the best fit to the experimental data. The curves demonstrate the acceleration of crystallization at the beginning of the phase transition, and the slowing down of the process as a result of collisions. It is quite clear that different crystallization conditions lead to changes in the mechanism, reflected in the change of the Avrami exponent as a function of temperature.

Fig. 6.2 shows the crystallization of fractionated poly(ethylene adipate) after Takayanagi (1957), plotting $\ln(-\ln m_L/m_0)$ versus $\lg t$.

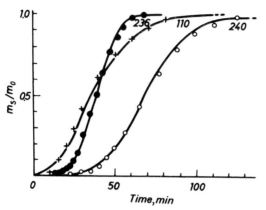

Fig. 6.1. Crystallization isotherms of poly(ethylene glycol terephthalate). (After Keller et al. 1954.)

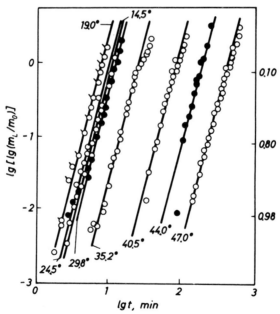

Fig. 6.2. Avrami plots of fractionated poly(ethylene adipate) samples. Crystallized at different temperatures. (After Takayanagi 1957.)

6.6. DEVIATIONS FROM THE AVRAMI EQUATION

Avrami equations have been used frequently to verify assumed n values. Gent has shown that in such cases plots with different n values (such as $n = 3$ and $n = 4$) are only slightly different, and deviations are comparable with the experimental error. Fig. 6.3 shows as an example the crystallization of different rubber samples at $-26\,°C$ together with calculated values using $n = 3$ and $n = 4$. Errors come from h_0 and h_∞ determination, and the $\ln(-\ln m_L/m_0)$ $-\ln t$ plot is a highly distorted scale where reasonable agreement can be obtained with experimental data even if it is limited to a small portion of the total process. Takayanagi has demonstrated this using unfractionated poly(ethylene adipate) samples (Fig. 6.4). According to these results the Avrami equation is valid only over in a limited range of the crystallization process. m_L/m_0 values are shown on the right hand side of the plot, so it can be seen that only about a half of the crystallization process can be described by the Avrami equation.

These deviations do not mean that data obtained from the Avrami plots are invalid. On the contrary, if the cause of this deviation is found, further information is gained about the system.

Processes leading to deviations can be simultaneous or consecutive.

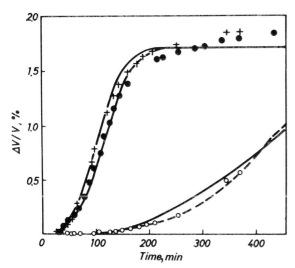

Fig. 6.3. Crystallization of various rubber samples at $-26\,°C$. Continuous line corresponds to $n = 4$, dashed line to $n = 3$. (After Gent 1954.)

An example of a simultaneous process is the appearance of two types of spherulites (the existence of such structures is known in the case of polypropylene and nylon-66) or the development of similar units from different types of nuclei (for example instantaneous and random).

Consecutive processes can be spherulite growth from needle-like or disc-like nuclei. All these processes lead to an apparent time dependence of the Avrami exponent.

There are indications that sometimes the Avrami analysis gives false information. In the case of poly(ethylene oxide) crystallization can be described by $n = 2$, which seems to be compatible with growth of disc-like structures from instantaneously formed nuclei or with the growth of needle-like structures from randomly formed nuclei (see Table 6.1). Microscopic observations on the other hand prove unequivocally that crystallization is a spherulitic growth process from instantaneously formed nuclei (see Fig. 4.11). The Avrami exponent in this case should be 3. Such discrepancies for directly observable structures cast some doubt on those structures, which cannot be detected independently.

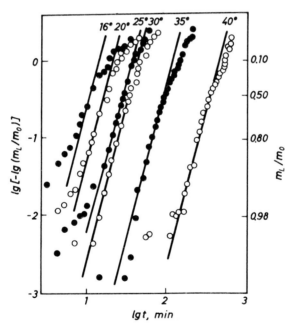

Fig. 6.4. Avrami plot of unfractionated poly(ethylene adipate) samples. Agreement is only partial, as is shown by the right hand side m_L/m_0 values, it extends to about 50% crystallization. (After Takayanagi 1957.)

Further complications appear with systems where n is non-integer. Such an example is poly(decamethylene terephthalate) shown in Fig. 6.5.

Deviations from the Avrami equation can originate from:

1. the simultaneous appearance of different growth mechanisms (for example needle- and disc-like);

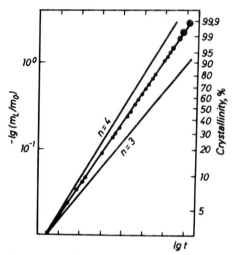

Fig. 6.5. Crystallization of poly(decamethylene terephthalate) at 112.95 °C. The Avrami equation seems to be valid, but with fractional exponent $n = 3.587 \pm 0.008$. (After Banks et al. 1964.)

2. impurities influencing crystal growth;

3. the density of the growing phase not being uniform (higher in internal regions) so that it is time dependent.

In the latter case the Avrami equation must be modified by a term taking into account the time dependence of ρ_s:

$$\frac{m_L}{m_0} = \exp\left(-zt^4 At^{-m}\right), \tag{6.25}$$

where At^{-m} describes the time dependence of ρ_s.

If $m = 0$, Eq. (6.25) is equal to the original Avrami equation. If $m = 0$ the resulting Avrami exponent cannot be regarded as an unambiguous characteristic of crystal nucleation and growth.

4. molecular mass distribution can influence crystallization kinetics. Studies by Banks on poly(ethylene oxide) were repeated by Hay et al. in 1969. According to their experiments the Avrami exponent can be altered by fractionation. Values of $n = 2$, 2.5 and 3 were obtained. The existence and significance of such fractional exponent is still unexplained. If $n = 4$ is obtained in an experiment, branching of fibrils occurs at small angles. Under these conditions growth packing is so dense that the density of the crystallized phase does not change, consequently the assumption that nucleation is random and growth is spherulitic appears to be justified.

The Avrami analysis yields interesting results concerning the time dependence of the crystallization process and is applicable even when the growing units cannot be observed by microscope. Whereas some of the conclusions may be disputed the analysis is largely correct and forms a useful means of tackling situations where other practical methods cannot cope.

The effect of external variables such as melting temperature, molecular mass and plasticizers can be studied quickly be the Avrami method. Often the determination of half-times is enough to solve technological problems.

Cases when the Avrami equation is applicable but with fractional exponents can lead to a better understanding of growth mechanisms. Phenomena of this kind are under investigation, and the analysis of these results is expected to provide new information about fibrillar morphology.

6.7. SECONDARY CRYSTALLIZATION

Crystallization does not always end as predicted by the Avrami equation, which can only be applied to primary crystallization, after this process there occurs a secondary crystallization which lasts for a considerable period of

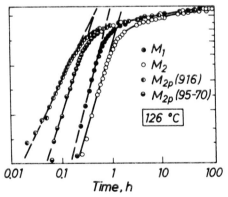

Fig. 6.6. Crystallization of various polyethylene samples as a function of time. Deviation from the Avrami equation can be seen in the final stage of transformation. (After Rabesiaka and Kovács 1961.)

time. Fig. 6.6 demonstrates a secondary crystallization process of this kind, where Avrami plots are presented versus time for 4 polyethylene samples.

In the case of secondary crystallization the crystallinity–time relationship can be given by the formula:

$$x = C + D \ln(t - t_0) \tag{6.26}$$

where x is the crystalline mass fraction at time t, t_0 the beginning of the secondary crystallization process, C and D empirical constants.

Eq. 6.26 is, of course, only an approximation, since x does have a limiting value, it cannot for example be greater, than 100%.

In some polymers, such as poly(ethylene terephthalate) or poly(ethylene adipate) no secondary crystallization occurs, in other cases (polyethylene or polyamides), however, it takes up almost one half of the total crystallization process, and thus cannot be neglected.

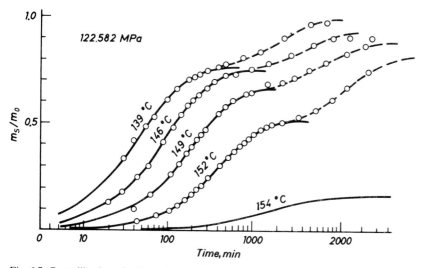

Fig. 6.7. Crystallization of polyethylene at elevated pressures. (After Matsuoka 1960.)

Fig. 6.7 shows the crystallization of polyethylene at high pressures. In this case not only the Avrami plots, but the isotherms themselves show the presence of secondary crystallization. Both the increase of crystalline fraction and a continuous perfection of the already existing crystallites have been offered as explanations of the underlying mechanisms. The second hypothesis seems to be the most probable, the perfection process can be monitored by the increase of long-range order, the chain folding distance, so that the development of larger and more perfect crystals can be demonstrated.

Secondary crystallization plays a prominent role in fibre-producing technologies (annealing).

6.8. NUCLEATION

6.8.1. Types of nucleation

Two types of nucleation have been observed: homogeneous and heterogeneous. Homogeneous nucleation means that polymer chains can aggregate spontaneously below the melting point. After reaching a critical size nuclei of this sort can begin participating in crystal growth processes. Distribution of these nuclei is random, and their generation is usually a first order function of time (Eq. (5.1)).

Heterogeneous nuclei start from impurities either distributed in the polymer or located at surfaces, frequently the wall of the container. In this case nuclei form simultaneously as soon as the sample reaches the crystallization temperature, and time dependence of nucleation is a zero order function.

6.8.2. Size of growing units

The number of nuclei determines the maximum size of crystallizing units. In the case of heterogeneous nucleation the final average volume of crystallized units, Φ can be calculated by dividing the final volume, V_∞ by the number of nuclei, $N'V_0$ where N' is the number of nuclei per unit volume, and V_0 the initial volume:

$$\Phi = \frac{V_\infty}{N'V_0} \approx \frac{1}{N'}. \qquad (6.27)$$

If the nucleation is homogeneous, new nuclei form during the growth of previously formed units, the relation being more complicated, approximately:

$$\Phi = \left(\frac{G}{N}\right)^{3/4}, \qquad (6.28)$$

where N is the nucleation constant per unit volume (see Eq. (6.1)), and G the growth constant (Eq. (6.2)).

The number of nuclei, and hence that of formed crystallites has been shown to influence the physical properties of many crystalline polymers. The size of spherulites in the crystallized product can vary due to artifical nucleating agents.

6.8.3. Homogeneous nucleation

The relationship between nucleation rate and temperature according to Turnbull and Fischer is:

$$N = N_0 \exp\left(-\frac{E_D}{kT} - \frac{\Delta G^*}{kT} \right), \qquad (6.29)$$

where E_D is the activation energy at the surface of the nucleus, G^* the Gibbs energy needed for the formation of critical nuclei; ΔG^* is proportional to $T_m^2/\Delta T^2$ where T_m is the melting point, ΔT the overcooling (nucleation rate increases with increasing overcooling). N_0 is a material constant, rate of nucleation if the exponent is zero, k the Boltzmann constant, and T the absolute temperature.

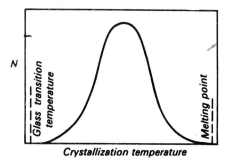

Fig. 6.8. Dependence of nucleus formation rate on the degree of undercooling.

At lower temperature E_D/kT becomes dominant, the activation energy, E_D, decreases in proportion to temperature, thus nucleation slows down. Fig. 6.8 shows the rate of nucleation as a function of temperature.

6.8.4. Heterogeneous nucleation

Impurities, nuclei of external origin, frequently play a decisive role in polymer crystallization. In the case of fibre-forming polymers, such as polypropylene and polyamide, crystallization sets out from heterogeneous nuclei. If
nucleation is independent of time, nucleation is usually assumed to be
heterogeneous. If, however, nucleation is time dependent—for example
exhibits a linear function of time—, it does not necessarily mean that
nucleation is homogeneous. There are numerous data showing that even time
dependent nucleation can be heterogeneous.

    Price et al. carried out interesting experiments to prove the presence of
heterogeneous nucleation. Small polyethylene particles were dispersed in an
inert liquid medium and melted by increasing the temperature of the medium.
If the dispersion is fine enough and nucleation is heterogeneous, some of the
molten polyethylene droplets are not likely to contain a nucleating centre.
The polyethylene involved in the experiment crystallized rapidly at 125°C in
bulk form. This corresponds to about 13°C overcooling. In dispersed form,
according to microscopic observation, some of the droplets crystallized
between 120°C and 125°C. The majority, however, did not crystallize until
90°C, corresponding to about 50°C overcooling. This value is 35°C higher
than the normal overcooling (Fig. 6.9).

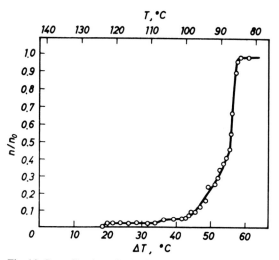

Fig. 6.9. Crystallization of polyethylene droplets as a function of the crystallization temperature.
n/n_0 denotes the ratio of crystallized droplets. Polyethylene droplets were dispersed in an inert
medium. (After Cormia et al. 1962.)

This experiment demonstrates that homogeneous nucleation plays practically no role in bulk crystallization of polymers.

Heterogeneous nuclei do not necessarily differ chemically from the polymer melt, a difference appears perhaps only in physical order and melting point. Obviously these units survive even at temperatures somewhat higher than the melting point of the polymer, and reappear instantaneously on cooling.

6.8.5. Artifical nucleating agents

Nucleation density in polymers may be controlled. This is important from a commercial point of view as well since it means that the final size of the growing units can be predetermined, and this in turn allows some control over the properties of the final product.

Artifical additives controlling nucleation have been discovered empirically, and interestingly, their structure has no relation to that of nuclei formed spontaneously in polymers.

Table 6.2

The effect of nucleating agents on the spherulite size of polyamide-6

Nucleating agent	%	Spherulite diameter, μm	
		150 °C	5 °C
		Crystallization temperature	
Base polymer	—	50—60	15—20
Polyamide 6/6	0.2	10—15	5—10
	1.0	4—5	4—5
Poly(ethylene terephthalate)	0.2	10—15	5—10
	1.0	4—5	4—5
$Pb_3(PO_4)_2$	0.05	10—15	8—10
	0.1	4—5	4—5
	0.5	4—5	4—5
NaH_2PO_4	0.05	5—10	3—7
	0.1	3—7	3—5
	0.5	5—7	3—5
$Na_7P_5O_{16}$	0.05	5—6	4—6
	0.1	4—5	3—5
	0.5	6—9	5—8
TiO_2	0.05	20—25	5—10

After Inoue 1963.

Various polymers show different degrees of compatibility with artifical nucleating agents. It is relatively easy to insert artifical nucleating agents into polyamides, but in the case of polyethylene it is more difficult. For polyolefins nucleating agents can be colloidal silica-gel or a variety of salts (for example aluminium sulphate). In the case of polyethylene nuclei can be formed by dispersing fine polypropylene powder in the polymer.

Inoue (1963) studied the effect of different nucleating agents on nylon-6. His results are summarized in Table 6.2.

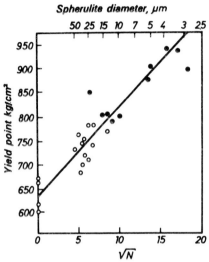

Fig. 6.10. The effect of spherulite diameter on the flow point in polyamide 6/6, N is the number of spherulite interfaces (After Starkweather and Brooks 1959.)

Artifical nucleating agents are being used more and more frequently by the plastics and synthetic fibre industry. The relationship between spherulite size controlled in this way and the yield stress is demonstrated by Fig. 6.10.

6.9. THE GROWTH PROCESS DURING CRYSTALLIZATION

6.9.1. General mechanism

The crystal growth process, as we have already seen, can result in a variety of structures. In describing the growth process we shall limit ourselves to more characteristic mechanisms for the sake of simplicity.

Crystal growth of low molecular compounds is usually described in the following way: molecules reach the crystal surface by diffusion, where they adsorb and reach the minimal free energy position by migration, that is, the part of the surface where growth is not yet finished. This process is shown in Fig. 6.11a. This relatively fast process stops for a while when the layer is saturated. Then the next step is the rate limiting process, formation of a new, secondary nucleus on the filled layer (Fig. 6.11b). The rate of formation of secondary nuclei is a similar function of overcooling as in the case of primary nuclei (Eq. (6.29)). There is only one difference; here the nucleus is only two dimensional, and so the logarithm of rate is proportional not to $1/\Delta T^2$, but to $1/\Delta T$. So that:

$$\text{growth rate} = \exp\left(-\frac{A}{T} - \frac{B}{T\Delta T^m}\right) \tag{6.30}$$

where $m = 1$ if nucleus formation is two-dimensional. Owing to the exponential dependence, slight overcooling does not result in significant growth. Later, however, even a relatively small change in ΔT increases the growth rate sharply (see Fig. 6.8).

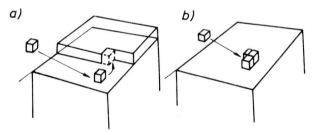

Fig. 6.11. Crystal growth process in a low molecular material. *a* The molecule diffuses to the growing surface then migrates to the low free energy points; *b* those molecules which are to form a new, secondary nucleus, before the new growing step begins. (After Mullin 1961.)

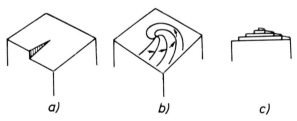

Fig. 6.12. Spiral form crystal growth by screw dislocation mechanism. *a* Crystallization starting at crystal defects; *b* molecules crystallizing in a spiral form (top view); *c* spiral growth (side view). (After Mullin 1961.)

In practice many low molecular compounds crystallize according to Eq. (6.30), but there are many exceptions as well. Frank has shown that crystal defects on the growing surface are low energy points where crystallizing molecules can easily bind. Precipitating molecules regenerate the crystal defect, and the crystal grows in the form of a spiral without any need for secondary nucleus formation (Fig. 6.12). These growth processes based on screw dislocations are less sensitive to temperature. Their growth rate can be described by the empirical relation:

$$\text{growth rate} = A \, \Delta T^2 \tag{6.31}$$

where A is a material constant.

In the case of low molecular materials there exists a third possible mechanism, where the diffusion of molecules to the growing surface is the rate determining step. This mechanism will not be treated in detail here.

6.9.2. Growth of polymer single crystals

Polyethylene single crystals grown from dilute solutions form by chain folding, that is, folding of the molecular chains at uniform, regular distances, the backbone carbon chains being essentially perpendicular to the crystal planes. According to Holland and Lindenmayer crystal growth is proportional to time, that is, the process is a first order function of time. This means that the rate-determining step is not diffusion, since if that were the case the exponent would be smaller than unity. It has also been shown that the growth rate depends on temperature according to Eq. (6.30), that is, it proceeds via secondary nucleation.

6.9.3. Relationship between single crystals and spherulites

The study of single crystals has led to the discovery of chain folding as a general crystallization form for polymers. From solutions of greater than 0.1% concentration, instead of single crystals, spherulite-like structures form on cooling. In these cases, growth is of course more complex. Similar complications can occur in dilute solutions if unfractionated samples are used, if the solution is stirred, or perhaps if temperature control is not satisfactory. The growth process itself can change during crystallization, as single crystals are highly irregular in the early stage. Other examples are known, where morphology of growing units or even crystallization kinetics change during the crystallization process.

6.9.4. Crystallization from the melt

In order to interpret structures crystallized from melts it is first necessary to know how the growth rate depends on temperature. As we have already seen the two probable mechanisms—secondary nucleation and spiral growth—vary characteristically with temperature (Eqs (6.30) and (6.31)).

It has been established earlier that rate of crystallization in rubber increases on cooling. The maximum rate is reached at $-23\,°C$, below this temperature the growth rate decreases and reaches zero close to $-50\,°C$ (Wood and Bekkedahl 1946).

Crystallization rates for synthetic polymers during slight overcooling depend on temperature according to Eq. (6.30), that is, secondary nucleation occurs.

Fig. 6.13 shows as an example the spherulite diameter as a function of time for polypropylene crystallized at $125\,°C$.

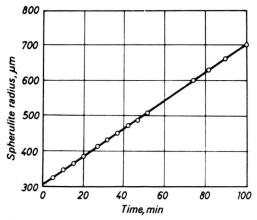

Fig. 6.13. Growth of a polypropylene spherulite at $125\,°C$. (After Keith and Padden 1964.)

Crystal growth proceeds according to a secondary nucleation mechanism. Some uncertainty springs from the fact that the melting point is not sharp, it is therefore difficult to decide whether m is 1 or 2 in the equation, that is, the dimensionality of the secondary nucleus is 2 or 3. This question is still open to debate, to settle it one way or the other a much more precise method would be needed for melting point measurements. In any event, for the study of polymers in spite of the high viscosity of the polymer melt, the rate determining process of spherulite growth is not diffusion, as this would lead to a $t^{1/2}$ type of time dependence.

The radial growth rate always exhibits a maximum as a function of temperature, an example is that of polyamide-6 shown in Fig. 6.14.

Growth thus proceeds according to Eq. (6.30), via a secondary nucleation mechanism.

It is interesting to investigate how this temperature dependence influences the final size of growing units. According to Eq. (6.28) this

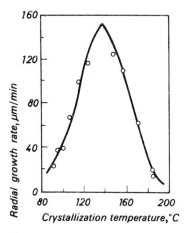

Fig. 6.14. Temperature dependence of the spherulite growth rate in polyamide-6. (After Magill 1961.)

parameter depends on the ratio of nucleation constants (G/N), during homogeneous nucleation. So that final dependence is determined by the temperature dependence of G relative to N, both being related to nucleation (primary and secondary nucleation respectively).

Secondary nucleation must be more rapid than primary, especially if the secondary nucleus is only two-dimensional. It is now generally accepted that since both G and N represent similar processes, their temperature dependence is probably also similar, their ratio being independent of temperature. In contrast to the general opinion that a higher crystallization temperature leads to large spherulites, in some cases a constant G/N ratio has been observed, when spherulitic size is independent of temperature. If this is not the case, it can probably be explained by non-random, instantaneous nucleation. If nucleation is instantaneous (heterogeneous), spherulitic size is independent of growth rate (see Eq. (6.27)). Nucleation density increases with decreasing temperature leading to a decrease in the average spherulitic size.

6.9.5. Mechanism of spherulitic growth

After establishing that secondary nucleation is the rate determining step in spherulitic growth, we have to take a closer look at how this leads to the formation of the fairly complicated structures described earlier.

Spherulitic growth was first observed in low molecular compounds during the last century.

Keith and Padden generalized the earlier results obtained on low molecular materials and published their ideas about the general mechanism of spherulitic growth.

Two important requirements of this mechanism are the high viscosity of the crystallization medium, and the presence of some impurity not being able to participate in the growth process. In the case of low molecular materials this component can be added to the system or can be present originally, it can even be the solvent itself. In polymer melts there are always branched or atactic components present less amenable for crystallization as compared to the pure polymer, so they do not build into the crystal during the growth process.

Some peculiar characteristics of spherulites could be studied by slowing down the growth process, for example, by reducing the overcooling. It should be observed (see Fig. 4.15) that fibrils grow from the central nucleus, normally by small angle branching, and sometimes by twinning.

Keith and Padden explained the growth process in the following way: impurities do not build into the growing parts, and, as a result, narrow lobes penetrating deep into the melt develop, which grow further only if they meet crystallizable material. This phenomenon is well-known in metals where these lobes allow cellular structures to develop. In polymers, because of the high viscosity these lobes become fibrils.

The diameter of these fibrils is determined by two factors, diffusion of non-crystallizing molecules from the growing surface to the melt, and by the growth rate. A parameter, δ has been introduced which is the ratio of diffusion and growth constants (D/G). This gives a rough approximation of the fibril diameter. The δ value for polymers is 1μm or less. Growing fibrils can expand, but if they become too wide (wider than δ), the growing surface becomes unstable. This leads to scission and branching, but the angle between the new units is usually small. This mechanism explains satisfactorily spherulitic growth and also describes the observed properties. It follows from these observations that δ should increase with decreasing G. In fact it has been observed that the fibril diameter does increase with decreasing growth rate.

A growing surface meets an unchanging environment during the growth process, so that the growth rate will be constant (Fig. 6.15) provided that impurities are not present in high concentrations, or that they do not become concentrated in the first stage of the growing process. If impurities are concentrated, they occur between the growing units, and if they are too close to each other, the growth process slows down. Under normal circumstances this is readily measurable, as shown in Fig. 6.15.

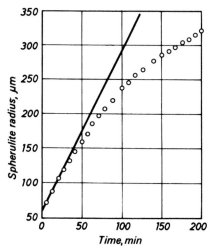

Fig. 6.15. Radial growth of a polypropylene spherulite in the presence of a large amount of impurity. Impurity in this case is atactic polypropylene. (After Keith and Padden 1964.)

Direct experiments to study the incorporation of radioactive impurities have also been performed. Results show that these indicators concentrate at the interface of spherulites at the end of the crystallization process. Thus spherulites appear to form single crystalline nuclei, or from several layers of single crystals. Nuclei become unstable and begin to grow if δ reaches the $1\,\mu m$ limit. Subsequently growth is fibrillar with branching, which leads eventually to spherical symmetry. Fibrils are in fact elongated single crystals consisting of folded chains. Molecular chains are approximately perpendicular to the growth direction, which explains why the chains in the spherulite are tangential.

Interfibrillar regions abundant in impurities either do not crystallize or do so at a much lower speed. If crystallization is slow, the Avrami kinetics become more complicated, and this phenomenon is probably responsible for secondary crystallization.

6.9.6. The effect of annealing on spherulitic structure

The structural effects of annealing in polymers, especially in synthetic fibres, are not completely understood. Annealing close to the melting point leads to partial melting and increases chain mobility. Since the melting point is $T_m = \Delta H / \Delta S$ and the melting heat of polymers is not substantially different from their monomers, it is reasonable to assume that the melting heat differences among polymer molecules themselves are even less, and that melting entropies are likely to be different. This means that melting point differences are not due to differences in molecular mass, but to differences in molecular order. The melting point of incompletely crystalline polymer areas is significantly lower than that of the crystalline regions.

Partial melting is accompanied by recrystallization. In polyethylene, for example, annealing close to the melting point first causes partial melting followed by further crystallization (Fig. 6.16).

Very close to the melting point, T_m, the specific volume of the sample increases rapidly with time, and then remains constant even after very slight cooling. At lower temperatures the plateau value of specific volume begins to

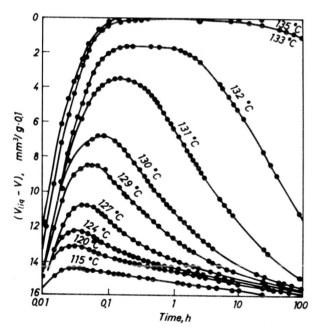

Fig. 6.16. The effect of annealing on the specific volume of polyethylene as a function of time. (After Gubler et al. 1963.)

decrease after some hours because of recrystallization, and the final value can be even lower than the original. At even lower temperatures the plateau disappears, the initial increase in specific volume being followed by a decrease giving a peak value at a relatively short time.

The annealing process consists of partial melting and recrystallization, and its result is influenced by earlier crystallization, and furthermore by the melting point of crystalline areas already present.

From a morphological point of view melting is not simply the inverse process of crystallization. Crystallization means nucleation and growth, while melting occurs simultaneously at all crystallized parts of the spherulite. At interfaces, of course, where incompatible impurities are concentrated, melting proceeds faster.

The processes accompanying annealing can be studied using mono-crystals. Annealing of single crystals also leads to secondary crystallization, indicated by the increase in density after a transient decrease (Fig. 6.17).

The chain-folding length detected by small angle X-ray scattering experiments increases during recrystallization. This has already been discussed in another context with Fig. 4.24. According to Fischer's results this parameter for annealed polyethylene increases from 10 nm to 50 nm. The increase is very rapid above 110°C (Fig. 6.18).

In melt-crystallized samples detection of folding lengths by electron microscopy is difficult, since it is necessary to use fracture surfaces or chemical

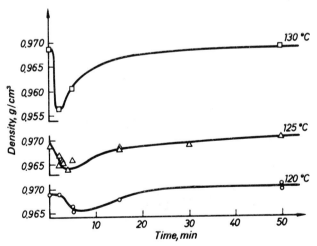

Fig. 6.17. Time dependence of the density of polyethylene single crystals at various annealing temperatures.

etching. These methods can themselves alter the structure, but small angle X-ray scattering yields results similar to single crystals, consequently presumably the effect of annealing is similar for bulk polymers.

As the density change in annealing is similar to that during secondary crystallization, the two processes are usually assumed to be similar.

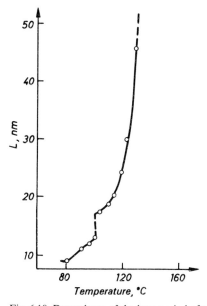

Fig. 6.18. Dependence of the long period of polyethylene single crystals on the temperature of annealing.

6.10. MELTING OF POLYMERS

High molecular crystalline materials do not melt at a single, well-defined temperature, but over a fairly wide temperature interval.

For the melting point of these materials the Thomson equation is valid:

$$T_m = T_m^0 \left(1 - \frac{2\sigma_e}{\Delta H_m D} \right) \tag{6.32}$$

where T_m is the melting point related to a lamellar thickness, D; T_m^0 is the melting point for an infinite crystal, σ_e the surface energy of the base plane involving chain folding, ΔH_m the melting enthalpy per unit volume, and D the actual lamellar thickness.

Assuming that the heat uptake during the melting process is proportional to the lamellar fraction calculated from the Thomson equation,
Wlochowitz and Eder (1984) developed a method to determine the lamellar
thickness, D, distribution. The essence of this method is that from the
temperature dependence of the DSC melting curves a lamellar thickness
distribution curve is calculated using the Thomson equation. Fig. 6.19 shows

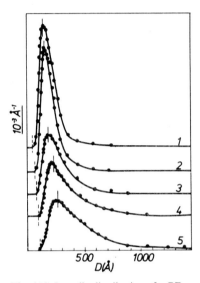

Fig. 6.19. Lamella distribution of a PE sample, crystallized at different temperatures.

the lamellar distribution curve calculated by this method for a polyethylene sample crystallized at different temperatures. Constants used for the
calculation are as follows:

$$T_m^0 = 415 \text{ K}, \qquad \sigma_e = 60.9 \times 10^{-3} \text{ J/m}^2, \qquad \Delta H_m = 2.88 \times 10^{-8} \text{ J/m}^3.$$

Vertical lines shown on the Figure at the peaks denote folding periods
calculated from small angle X-ray diffraction measurements. This lamellar
thickness value can be calculated from the long period L, using the formula:

$$D = xL \tag{6.33}$$

where x is the crystalline fraction.

CHAPTER 7

Molecular structure of polymers

7.1. OPTICAL AND X-RAY LEVEL STUDIES

So far the solid state order of polymers observable by optical or electron microscopy has been discussed at length. This level of arrangement is called supermolecular to distinguish it from the molecular structure which describes the relative position and orientation of molecular chains. Molecular ordering begins at about 1.0 nm, and is studied mainly by X-ray diffraction. These two levels of ordering are usually referred to as optical and X-ray levels. It should be noted that the X-ray level includes the determination of atomic positions within the crystals (crystallographic analysis), and that it corresponds to molecular and atomic ordering, the main method of investigation being X-ray diffraction with, to a lesser extent, electron diffraction.

Determination of atomic order in crystals is the main goal of crystallography, and this is essentially the same in inorganic, organic and macromolecular materials. The crystallographic area will not therefore be dealt with in detail, only from the point of view of molecular structure. Molecular ordering, however, will be thoroughly discussed, and the technique normally used is X-ray diffraction.

7.2. RELATIONSHIP BETWEEN LIGHT AND X-RAY DIFFRACTION

The basis of X-ray diffraction investigations is the fact that, under the action of X-rays, electrons in the atoms become excited and irradiate energy, thus forming secondary radiation centres. (The radiation is scattered on the electronic shells of the atoms.) These secondary waves interfere depending upon the relative positions of the scattering centres. From the position and intensity of these interferences, the type and position of the electronic clouds (of the atoms) can be determined.

Characteristic distances observable by optical microscopy are around 100 nm, by electron microscopy 0.15–2.5 nm, and by X-ray diffraction 0.01 nm. The majority of our knowledge about atomic positions and intra-molecular distances is gained from X-ray diffraction measurements.

X-ray diffraction is the analogue of light diffraction. If a monochromatic light beam is passed through a grating of grating constant, a, interference lines appear on a photographic plate displaced at a distance, t, from the grating. If the distance l of the interference line from the centre, and the wavelength are known, the optical grating constant, a, can be calculated.

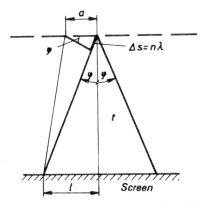

Fig. 7.1. The schematic representation of light diffraction on a grating.

Fig. 7.1. clearly shows that if interference appears at a distance, l, from the centre line, the path difference, ΔS, between the neighbouring scattering centres must be equal to $n\lambda$. From the path difference, and the scattering angle, φ, the grating constant, a can be calculated:

$$\Delta S = n\lambda = a \sin \varphi, \tag{7.1}$$

$$\tan \varphi = \frac{l}{t}. \tag{7.2}$$

At small angles:

$$\tan \varphi \approx \sin \varphi \approx \frac{l}{t}, \tag{7.3}$$

$$a = \frac{n\lambda}{\sin \varphi} = \frac{n\lambda t}{l}. \tag{7.4}$$

It should be noted that the measurable quantities, l and t, are in the order of centimeters and meters respectively, whereas the calculated grating constant and the wavelength are in the order of micrometers. So, for example, if $l = 100$ mm, $t = 3000$ mm, $\lambda = 0.0005$ mm (0.5 μm) and $n = 1$, that is, from the first interference line, then:

$$a = \frac{3000 \cdot 0.0005}{100} = 0.015\,\text{mm}.$$

According to the theory of diffraction the wavelength of the applied radiation, independent of its kind must be close to the dimensions to be observed. This means that X-radiation with a wavelenth of 0.1 nm gives information concerning dimensions of around 0.1 nm. This relationship can be given in a more exact way, such that:

$$R = \frac{\lambda}{4\pi \sin \Theta}, \tag{7.5}$$

where Θ is the diffraction (or Bragg) angle, and is equal to half of the angle between the incident and observed radiation, whereas R is the dimension to be studied (Frenkel 1968).

Using this Debye criterion one can predict precisely the wavelength necessary to study a certain kind of structural order in advance, provided that the order of magnitude of the periodicity is known. In practice it means that spherulites are studied using visible light, while structures falling below X-ray level are studied by electron diffraction.

X-ray studies are very important in fibre production. Investigating the conditions for fibre formation, Happey (1951) formulated the following three criteria:

1. The polymeric raw material has to be soluble or fusible.

2. The polymer has to be orientable.

3. The orientation has to be uniaxial.

Points 2. and 3. can be studied easily by X-ray diffraction.

Processes during fibre drawing and annealing can be followed by X-rays. The majority of our knowledge about chain structure is gained from X-ray studies.

Before going on to a discussion of X-ray techniques, mention should be made of different ideas concerning crystallinity in fibres. Kargin (1940) had the opinion that the presence of crystallinity in fibres, especially in cellulose fibres, is doubtful. This opinion was based on the theoretical expectation that interference lines should become sharper with decreasing wavelengths (for example, in electron diffraction). Such an effect has not been observed in practice and has led to continuing discussion in the Soviet Union. The opposite opinion, which is accepted by this author, has been stated by Kitaigorodsky. According to him, the only criterion for crystallinity is the presence of three dimensional order as indicated by X-ray diffraction (Kitaigorodsky 1952). It should be said, however, that this is a sufficient, but not necessary condition. The presence of crystallinity has been demonstrated for example in polyamides where the crystallites were too small to produce well-resolved interference lines, yet their presence has been unambiguously demonstrated by differential scanning calorimetry (Bodor 1969).

7.3. BASIC NOTIONS OF X-RAY DIFFRACTION

The basic equation of X-ray diffraction is the Bragg formula:

$$n\lambda = 2d \sin \Theta, \tag{7.6}$$

where n is the order of reflection, λ the applied wavelength, d the distance between the atomic planes, and Θ the Bragg angle, that is, half the angle between incident and diffracted beams.

It should be noted that d is an interplanar distance between atomic planes, not an interatomic distance. This difference is explained by Fig. 7.2. Atomic planes are usually characterized by the Miller indices. These indices describe the intercepts of the given plane with the crystallographic axes and are related to the unit cell dimension. (The unit cell is the smallest repeating

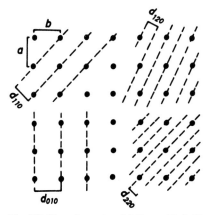

Fig. 7.2. Two dimensional lattice with lattice constants a and b, and with atomic planes of different indices.

volume unit.) Fig. 7.2. shows the [110], [010], [120] and [220] planes. The first index refers to axis a, the second one to axis b, the third one is zero, as axis c is zero in the case of two-dimensional lattices. The [110] plane is thus a crystalline direction which intersects the a and b axes at a distance of $1a$, and $1b$ from the origin of the unit cell. All other lines running parallel with this direction are equidistant from one another, d_{110}, and are all characterized by the same subscripts [110]. Similarly, d_{120} denotes the distance between planes which intersect the crystalline axes at $1a$ and $2b$ respectively. Atomic planes have a similar meaning in three dimensions. As an example the [111] plane of a cubic unit cell is shown in Fig. 7.3.

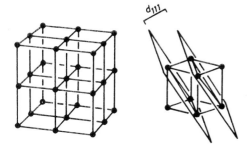

Fig. 7.3. A three dimensional lattice and the d_{III} atomic plane.

The positions of diffraction spots are determined by interplanar spacings obeying the Bragg equation; their intensity being determined by the number and type of atoms lying on the plane. The aim of structural analysis is to identify the lattice which gives a particular diffraction pattern, from the positions and intensities of the diffraction spots. The lattice cannot always be reconstructed from the diffraction pattern unequivocally, and even in cases where the molecule is not considered complex, it requires cumbersome calculations.

Not all interplanar distances can be related to a diffraction spot. Depending on the spacial grouping of the crystal, systematic extinctions may occur, for example, in the case of certain spacial groups only even reflections appear along the b axis (Kakudo and Kasai 1972).

Figs 7.2 and 7.3 clearly show that in this representation one plane is plotted several times, and that planes with higher indices are crowded around the origin of the unit cell. It seems reasonable therefore to introduce the so-called reciprocal lattice, which gives a clearer representation, where all planes with the same indices refer to a single point only, and, as we shall see later, are directly related to the X-ray diffraction pattern.

The reciprocal lattice is such a set of points, where all atomic planes are denoted by a separate point. Such a point lies on a line perpendicular to the atomic plane, at a distance of λ/d from the origin. Thus, for example, the reciprocal lattice of a two dimensional lattice is shown in Fig. 7.4. The lattice

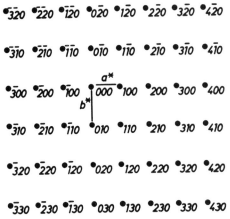

Fig. 7.4. The $hk0$ section of the orthorhombic reciprocal lattice.

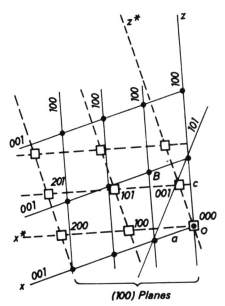

Fig. 7.5. The $hk0$ planes of the monoclinic reciprocal lattice together with the unit cell.

constants of the reciprocal lattice are usually denoted by stars: a^*, b^*, c^*. (d^* = λ/d, and in orthogonal systems $a^* = \lambda/a$, $b^* = \lambda/b$, $c^* = \lambda/c$).

The relationship between the reciprocal lattice and the crystallographic planes in a monoclinic crystal is shown in Fig. 7.5. The points identify the real lattice points, while the squares represent the reciprocal lattice. The points with higher indices are farther from the origin, and vectors directed from the origin to the reciprocal lattice points are perpendicular to the crystal planes. In non-orthogonal systems the directions of crystallographic and reciprocal lattice axes are different.

The reciprocal lattice will be investigated more thoroughly in subsequent sections. What is to be emphasised here is that it is not simply a convenient device, but is pivotal in the diffraction phenomenon. For example, the diffraction pattern obtained from a single crystal using a Buerger camera,

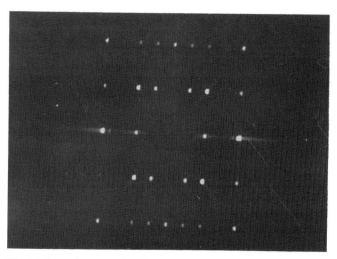

Fig. 7.6. One reciprocal lattice plane of the ε-aminocaproic acid on a photograph obtained with a Buerger camera.

is an undistorted picture of the reciprocal lattice (Fig. 7.6). This camera and the corresponding procedure is used for the study of monocrystals, but the principle is also important in understanding polymer X-ray diffraction.

7.4. CONDITIONS FOR INTERFERENCE AND GEOMETRY
OF THE INTERFERENCE PATTERN

For the sake of simplicity it is useful to study a linear array of atoms, and find the conditions for interference with this geometry. If the lattice constant (the identity period) of this one dimensional lattice is a, and the incident radiation is perpendicular to the linear array, then interference will occur as described by the following equation (see also Chapter 7.2 on light diffraction):

$$\Delta S = n\lambda = a \sin \varphi, \tag{7.1}$$

or with the auxiliary angle:

$$n\lambda = a \cos \Phi \tag{7.7}$$

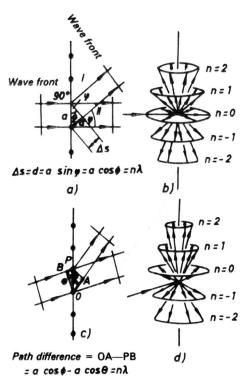

Fig. 7.7. Scattering of radiation from a point array. *a* Conditions for interference if the incident radiation is perpendicular to the point array; *b* the layer lines of the interference pattern; *c* conditions for interference if the incident radiation is not perpendicular to the point array; *d* layer lines of the interference pattern in case *c*.

where Φ is the angle between the point array and the diffraction direction (Fig. 7.7*a*).

One can see that the first, or zero order diffraction appears in the direction of the incident beam, symmetrically around the point array; the first ($n = 1$ or -1) and second ($n = 2$ or -2) order diffractions ranking conically upwards and downwards from the direction of the incident radiation (Fig. 7.7*b*).

If the incident radiation is not perpendicular to the point array, then the condition for diffraction is modified according to Fig. 7.7*c*:

$$n\lambda = a \cos \Phi - a \cos \Theta. \tag{7.8}$$

In this case diffraction is no longer symmetrical, but the diffraction cones appear similarly to the previous case (Fig. 7.7*d*).

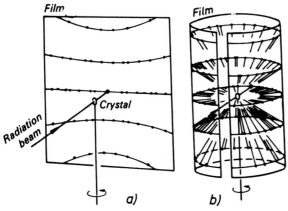

Fig. 7.8. Appearance of diffraction cones *a* on flat film, *b* in a cylindrical camera.

When flat recording films are used, the diffraction cones are distorted to hyperbolic curves (Fig. 7.8*a*). Should cylindrical films be used (Fig. 7.8*b*) the intersection of the cones with the cylindrical surface yields circles, on flat film these give straight lines. These lines are referred to as layer-lines, and are labelled according to the notation of Fig. 7.7.

In practice X-rays are scattered by three-dimensional crystalline lattices, not one dimensional point-arrays. Three-dimensional lattices can be represented by three-dimensional atomic arrays, "synthesized" from three one-dimensional arrays. (Mathematically it can be regarded as the convolution of three one-dimensional point arrays.) All the linear arrays produce their diffraction cones, and reflections appear where the three cones intersect one another (Fig. 7.9).

A diffraction photograph for single crystals is outlined in Fig. 7.10. For a better understanding of single crystal diffraction photographs we have to get

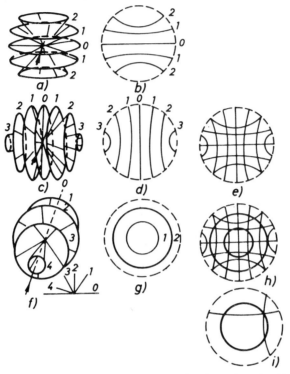

Fig. 7.9. Superposition of scattering cones in a three-dimensional lattice; a, c, f scattering cones of an individual point array; b, d, g intersection of cones with the flat film; e superposition of two and h three point-array diffraction cones on a flat film; i condition for the appearance of an interference reflection, when the three cones intersect in a single point.

acquainted with the relationship between interference conditions and the reciprocal lattice and with the reflection sphere.

As we have seen, the condition for interference-spot formation is the simultaneous intersection of three conical surfaces at one point. These conditions can be determined mathematically using this geometric picture and analytical geometry, but the solution is more simple if the reciprocal lattice is used. It was Ewald (1921) who first demontrated the simple relationship between the reciprocal lattice and the Bragg equation.

Take a circle with unit radius and centre C (Fig. 7.11). If radiation enters the circle at point P, and, passing through point C, reaches the crystal placed at point 0, and if the angle between the atomic planes and the radiation is equal to Θ, the diffraction will appear at 2Θ, so long as the conditions for diffraction are fulfilled. It can be shown that these conditions are met if one point of the reciprocal lattice with origin 0 touches the circle.

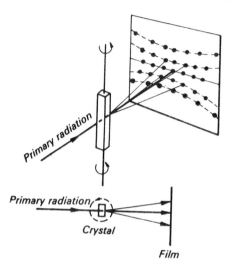

Fig. 7.10. Schematic of a single crystal X-ray analysis.

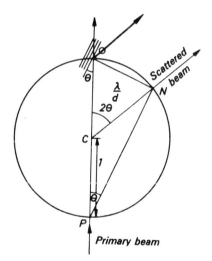

Fig. 7.11. Relationship between the reciprocal lattice and the Bragg-law.

Let a reciprocal lattice point on the circle be N. This point, according to the definition of the reciprocal lattice lies on a straight line perpendicular to the atomic plane, at a distance $\frac{\lambda}{d}$ from the origin. The angle ONP, according to the Thales theorem, is a right angle. The ratio $\overline{ON}/\overline{OP}$ is equal to $\frac{\lambda}{2d}$ and $\overline{NP}$ is parallel to the crystallographic planes, since $\overline{ON}$ is perpendicular to both. Angle $\overline{NPO}$ is thus equal to Θ, that is, the angle between the incident radiation and the crystallographic planes:

$$\sin \Theta = \frac{\lambda}{2d} \qquad\qquad\qquad (7.6)$$

according to the Bragg equation.

It has been shown that such crystallographic planes which lie on the reflection circle fulfill the conditions of diffraction. The reflection circle has been introduced for the sake of simplicity, but the proof is equally valid for a sphere with unit radius. Such spheres, because of the properties referred to above, are known as reflection spheres. The direction of the reflected beam runs parallel to the line drawn from the centre of the sphere to the reciprocal lattice point. Physically the lattice scatters radiation from point 0, but calculations are easier if performed from the centre of the reflection sphere. It can be shown that $\overline{CN}$ is parallel to the scattered radiation; since the angle $\overline{OPN}$ is equal to Θ and $\overline{CP}$ is equal to $\overline{CN}$, thus the angle $\overline{CPN}$ and $\overline{PNC}$ are equal, and angles $\overline{NCP}$ and $\overline{OCN}$ are $\pi - 2\Theta$ and 2Θ respectively.

Diffraction can thus be regarded as the coincidence of the reflection sphere surface with the reciprocal lattice points. The positions of the incident radiation and crystal define the position of the reflection sphere.

7.5. TYPES OF X-RAY DIFFRACTION PHOTOGRAPH. FIBRE DIAGRAM

As we have seen, the atoms of a single crystal give interference only under specific conditions. These are that the Bragg conditions should be fulfilled, or in other words, the reciprocal lattice point should touch the surface of the Ewald-sphere. Therefore, if a fixed single crystal is illuminated with a monochromatic X-ray beam, very few, if any, interference spots will be observed on a film placed around the crystal. In order to get several reflections for data processing the following techniques may be applied:

1. The position of the single crystal should be fixed, and the radiation not monochromatic, but continuous (Laue photographs).

2. The radiation is monochromatic, but the position of the crystal is changed systematically (rotation or precession photographs).

3. The radiation is monochromatic, but the sample is in powder form instead of single crystal, and as a result, reflecting planes with all orientations occur (powder-photographs).

The last two techniques can be applied to evaluate interplanar distances using the Bragg equation.

Rotating the crystal also rotates the reciprocal lattice, since the direction of the reciprocal lattice points is always perpendicular to the crystalline planes. As the reciprocal lattice rotates, its points touch the surface of the reflection sphere. In this case reflection occurs, and its direction is given by a straight line drawn from the centre of the reflection sphere to the particular point. This is shown in Fig. 7.12.

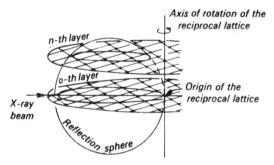

Fig. 7.12. Relationship between the reflection sphere and the reciprocal lattice. Reflection is observed where the reciprocal lattice point touches the surface of the reflection sphere.

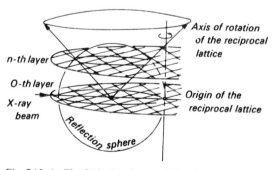

Fig. 7.13. As Fig. 7.12, showing the diffraction cone.

All layers of the reciprocal lattice (the zero layer and the n-th layer are shown in Fig. 7.12) intersect the reflection sphere in a circle. Connecting the circumferences of these circles with the centre of the reflection sphere forms the reflection cones shown in Fig. 7.8. Obviously, a cone related to a particular layer line originates from the same reciprocal lattice layer. Labelling of the layer-line and reciprocal lattice layer are indicated in Fig. 7.13.

Fig. 7.14 shows diffraction photographs taken of single crystals, and indicates intersections of the rotating three-dimensional reciprocal lattice with a plane. The plane in this case represents a part of the reflection sphere in the vicinity of the origin. In practice, as was suggested earlier, the layer lines deform to hyperbolae on flat sheet film. Linear layer lines can be obtained using cylindrical cameras, and Fig. 7.14 shows how the reflection spots can be assigned to reciprocal lattice points.

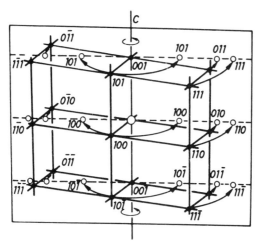

Fig. 7.14. The reciprocal lattice crosses the plane representing a small part of the surface of the reflection sphere during its rotation.

Fibres usually contain axially oriented, highly anisometric crystallites along the fibre axis. There is no orientation perpendicular to the fibre axis, that is, all positions obtainable by the rotation of the single crystal are present. The X-ray diffraction pattern for a fibre is analogous to the precession (or rotation) photograph of a single crystal (Fig. 7.15).

When taking a fibre photograph, the fibre is placed in the camera, and illuminated through a small hole (diameter 0.5–0.8 mm) by monochromatic X-ray radiation, and the diffraction pattern is exposed on a sheet film (Fig. 7.16).

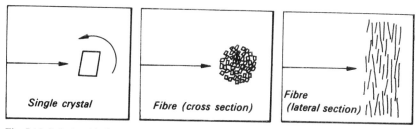

Fig. 7.15. Relationship between the fibre diagram and X-ray diffractogram of a rotated single crystal.

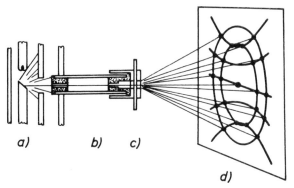

Fig. 7.16. Schematic of a camera used for fibre studies. *a* Anode of the X-ray tube, *b* collimating system, *c* sample, *d* film.

In practice X-ray tubes with copper anodes are normally used, and the characteristic wavelength is 0.154 nm. In fact this characteristic radiation consists of two components with slightly different wavelengths, leading to a broadening, and sometimes to a doubling of interference spots. For more precise work monochromators or special filters (Ni-filter, of optimum thickness 0.0085 mm) are used to remove one of the components. Diffraction photographs can also be obtained with X-ray tubes having molybdenum anodes, here the characteristic wavelength is 0.071 nm. In this case zirconium filters are used (of optimum thickness 0.037 mm).

Since the maximum value of $\sin \Theta$ is 1, it follows from the Bragg equation (Eq. (7.6)) that diffraction can be obtained from planes satisfying the inequality:

$$\frac{n\lambda}{2d} \leqq 1. \tag{7.9}$$

If the characteristic wavelength is 0.154 nm and $d = 0.1$ nm, only the first diffraction line appears, and an interplanar distance lower than 0.077 nm cannot be detected with this radiation.

Within the reciprocal lattice concept it means that those reciprocal lattice points which lie outside the reflection sphere can never be in the reflection position. If the wavelength gets lower, the reciprocal lattice shrinks, and further points, hitherto outside the reflection sphere can reflect the radiation.

In practice λ is known, and n is taken to be unity, so for example, the second line related to the [111] plane is called [222]; the reflection angle, 2Θ is measured, and the distance between the atomic planes can be calculated.

7.6. POWDER DIFFRACTION. DIFFRACTOMETERS

The essential difference between diffractograms obtained from monocrystals and powders lies in that in the latter case reflections are obtained from several microcrystals with different orientations. Here the d_{hkl}^* reciprocal lattice point related to the d_{hkl} plane no longer appears in a single point. Owing to the various orientations, directions perpendicular to the d_{hkl} plane can appear in any spatial direction. The reciprocal lattice point, d_{hkl}^* smears out, producing a spherical surface. The radius of this sphere is equal to d_{hkl}, and the centre is the origin of the reciprocal lattice. The Ewald reflection sphere intersects this

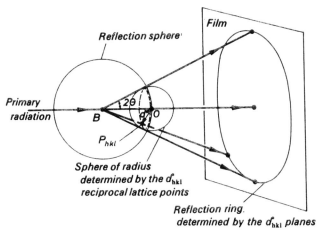

Fig. 7.17. Appearance of a powder diffractogram at the intersection of the reflection sphere and d_{hkl}^*.

sphere of radius d_{hkl} in a circle (Fig. 7.17) and this gives the reflection position. Diffraction is obtained along the conical surface determined by the intersection of the spheres, shown in Fig. 7.17. The situation is similar for other crystallographic planes, so, for example, from planes d_1-d_5 a diffraction pattern consisting of five concentric cones can be obtained, as shown in Fig. 7.18. Cones denoted as d_4 and d_5 refer to scattering angles greater than 90°, the direction of the cones is therefore reversed.

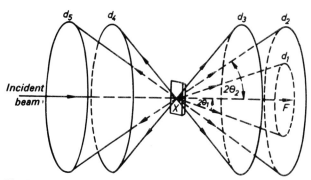

Fig. 7.18. Schematic of the powder diffractogram.

If the powder is not sufficiently fine, the diffraction ring becomes "granular", indicating uneven distribution of crystalline orientation.

In powder diffraction measurements cylindrical cameras are used in order to detect the full range of scattering angles. Since diffraction rings are symmetrical, a section is sufficient for their description. Debye–Scherrer cameras take such photographs, and the diffraction patterns are known as Debye–Scherrer photographs (Fig. 7.19). The Debye–Scherrer cameras used in practice are designed so that their circumference makes the calculation of reflection angles somewhat easier. The angle between two, symmetrically displaced interference sites is equal to 4Θ. If the circumference of the camera is equal to 180 mm, the distance measured in mm between the symmetrical diffraction lines, being related to the 4Θ scattering angle is equal to 2Θ (in degrees).

Cameras with a circumference of 360 mm are also available, and then 1 degree corresponds to 1 mm.

As an alternative to photographic film X-rays can be detected by Geiger–Müller or other counters. A schematic diagram for a diffractometer of this type is shown in Fig. 7.20. Such diffractometers are very useful in the measurement of powder diffractograms. The counter, however, moves along a single radius, and cannot be used to study orientation, at least not in this simple form.

Unoriented crystalline polymer samples exhibit powderlike diffractograms, since the crystalline parts contain crystallites with various orientations. When studying such samples powder cameras or diffractometers are used. In most polymers processable reflections lie in the 10–40° angular range, and higher order reflections cannot easily be evaluated owing to the line broadening effect of small crystallites, or crystalline defects. In the angular range mentioned, planar films can be used. These are easier to handle than cylindrical cameras.

The advantages of cylindrical cameras are that reflections can be detected over a greater angular range than with a planar film, and that layer lines appear as straight lines. On the other hand, however, planar films can be

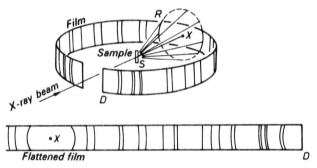

Fig. 7.19. The Debye–Scherrer camera and diffractogram.

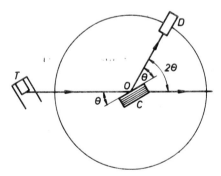

Fig. 7.20. Schematic of an X-ray diffractometer. T–X-ray tube, C—sample to be studied, D—detector (proportional counter).

handled more easily, and the evaluation of photographs from oriented samples is simplified. In practice first a planar camera is applied and then by inspection one can decide whether further photography is necessary using more complicated cameras.

7.7. PROCESSING OF DIFFRACTION DATA, MAIN ASPECTS OF EVALUATION

Important information obtainable from diffraction photographs are:
- identification of polymers;
- determination of the identity period;
- lateral ordering of polymeric chains;
- orientation of crystallites;
- crystallite size;
- crystalline content;
- unit cell data (number of monometric units in the cell, cell dimensions, absolute crystalline density, etc.);
- determination of the point group (from the systematic absence of reflections);
- atomic positions within the unit cell (fine structure analysis based on measured intensities).

Here six points will be discussed in detail, while problems grouped under the last two points will be omitted. The reasons for limiting the discussion are that the latter problems constitute themes for several books by themselves; and further, these methods are identical for all types of crystalline materials whether ionic or covalent, low- or high molecular mass; and finally, this type of structural analysis need be performed only once for a given material. Investigations listed under the first six headings monitor changes of the material under preparation, or processing of the polymers, and are therefore helpful in understanding and explaining technological processes at the molecular level.

7.8. IDENTIFICATION BY X-RAY DIFFRACTOGRAMS

Interplanar distances detected by X-ray diffraction are characteristic of individual materials in a similar way to fingerprints. Most crystalline substances, including crystalline polymers, can be identified based on their powder-diffractogram. Standard identification systems are available, for example card libraries which contain the three most intense reflections. A lot of important data can be found in a book from Kitaigorodsky (1952). This book contains the d interplanar distances and relative intensities for several chemical substances.

Distances between atomic reflection planes (d-values) can be calculated from a diffraction pattern knowing the reflection angles and the wavelength, and using the Bragg formula (7.6).

In the case of polymers, especially oriented polymeric fibres, identification of polymer and orientation are possible from the fibre-diagram simply by inspection. Even crystallographic modifications (in the case of polymorphic materials) are easy to determine.

Evaluation by inspection requires experience and practice, and some of the basic characteristics will be dealt with later. In particular, indentity period and lateral ordering are useful in the identification of oriented polymers.

7.9. IDENTITY PERIOD

The identity period is the repeat distance characteristic of one of crystallographic axes. In the case of fibres, where uniaxial orientation is present, this quantity is especially characteristic. As shown later the identity period gives some information about the chemical composition as well. The principle for measurement has already been discussed, the identity period is calculated from cylindrical or plane diffraction photographs using Eq. (7.1). If the incident radiation is perpendicular to the identity period of the sample studied, then, according to Fig. 7.7a:

$$I = \frac{n\lambda}{\sin \varphi}. \tag{7.1}$$

The identity period can be calculated from any point on any layer-line using Eq. (7.1), and equal periodic distances are obtained, provided, of course, that appropriate n values have been chosen, φ is the angle between the scattered radiation and the equator of the reflection sphere.

As an example, a diffraction photograph taken of a rotated single crystal of ε-aminocaproic acid using a cylindrical camera of 180 mm diameter is shown in Fig. 7.21. According to these measurements the first layer-line is at a distance of 7.5 mm from the equator, the second is 17.5 mm, and the third is 35 mm. From these data, and knowing the radius of the camera, the corresponding tan φ values are 0.2617, 0.6106 and 1.2212 respectively. Further studies show that the order of these layer lines are $n = 1$, 2, and 3.

Since in this case the wavelength $\lambda = 0.54$ nm, identity periods, I, are 0.609, 0.593 and 0.598 nm respectively. These are in good agreement with the literature value published by Bernal (0.59 nm).

Determination of identity periods from planar or fibre diffraction photographs is similar. It is important to remember, however, that according to Fig. 7.8 scattering cones intersect with planar film in hyperbolic curves. The

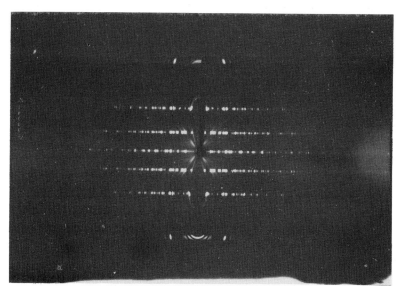

Fig. 7.21. X-ray photograph of a rotated ε-aminocaproic acid single crystal. Cylindrical camera.

calculation is outlined in Fig. 7.22. It includes the case when at the apex of the hyperbolic curve a meridian reflection, A_m, appears. In this case:

$$\tan \varphi = \frac{\overline{A_m C}}{\overline{OC}} \tag{7.10}$$

From which $\sin \varphi$ and I can be calculated using Eq. (7.1). Point A_m is on the second layer line in this case.

If there is no meridian reflection, then the identity period can be determined from any other reflection spot appearing on layer lines other than $n = 0$, for example in the case of point A_1 the calculation goes as follows:

$$\sin \varphi = \frac{\overline{AB}}{\overline{AO}} = \frac{\overline{AB}}{(\overline{AC^2} + \overline{OC^2})^{1/2}}. \tag{7.11}$$

and so:

$$I = \frac{2\lambda(\overline{AC}^2 + \overline{OC}^2)^{1/2}}{\overline{AB}}.$$

(7.12)

If the crystal is rotated, or in the case of static (non-rotating), fibre-diagrams, reflections generally appear four times (points A_1, A_2, A_3 and A_4 in Fig. 7.22). On the layer line belonging to $n = 0$ (the equator-line) the reflections

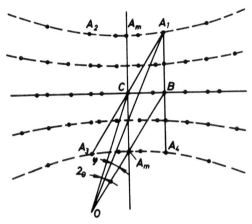

Fig. 7.22. Determination of the identity period.

of crystalline planes parallel to the rotational axis of the crystal appear, so identical reflections appear twice only. Meridian reflections originating from planes parallel to the fibre axis also appear only twice. If the crystalline axis running parallel to the fibre axis is the c-axis, then the equatorial reflections have indices of type [hk0], while meridian reflections belong to the [00l] type. If the fibre direction is parallel to the b-axis, the reflection appearing on the equator can be characterized by subscripts [h0l], and those on the meridian line by [0k0]. This is indicated in Fig. 7.14.

The identity period should not be confused with the interplanar spacings belonging to the reflections spots. These two quantities are equal for the [001] and [010] meridian reflexes only. For point A_1 for example:

$$\tan 2\Theta = \frac{\overline{A_1 C}}{\overline{OC}}$$

(7.13)

where 2Θ is the scattering angle between the direct and scattering radiation and, according to Eq. (7.6):

$$d = \frac{\lambda}{2\sin\Theta} = \frac{\lambda}{2\sin\left(\frac{1}{2}\arctan\dfrac{\overline{A_1 C}}{\overline{OC}}\right)}. \tag{7.14}$$

In practice distances $\overline{AC}$ and $\overline{AB}$ are measured as half the distance between points A_1 and A_3 and points A_1 and A_4 respectively.

The fibre diagram for a stretched polyethylene sample is shown in Fig. 7.23. The distance between film and sample was 50 mm, and the identity

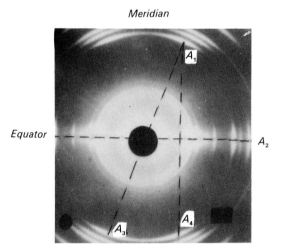

Fig. 7.23. X-ray photograph of a stretched PE fibre.

period calculated from the reflections, appearing on the layer lines is 0.254 nm. It is important to note that for the first layer line $n = 1$, which can be established only after the identification of the reflections. This means that in the drawn polyethylene fibre the molecular chains are in an oriented state. Taking into account the tetrahedral structure of the C–C chain, and assuming 0.15 nm for the interatomic spacing the experimental identity period is equal to the repetition distance of the zig-zag structure (Fig. 7.24).

For PVC the identity period is 0.51 nm, consequently in PVC every second chlorine atom is in an identical position (Fig. 7.25a). In poly(vinyl alcohol) the identity period is 0.252 nm (Fig. 7.25b), and in poly(vinylidene chloride) it is 0.47 nm (Fig. 7.25c).

The fibre diffraction diagram of polyamide-6 is shown in Fig. 7.26. Arrows show the meridian reflections appearing on the 2nd, 4th and 6th layer lines, φ-values related to these reflections are 10.8°, 20.8° and 38.7° respectively, these values are calculated from the reflection positions and the 50 mm sample to film distance. From these angles the calculated identity periods are 1.64, 1.73 and 1.72 nm respectively, in good agreement with the literature value (1.724 nm). The identity period of polyamide-66 is 1.72 nm. These values met expectations: in polycaprolactam (polyamide-6) two

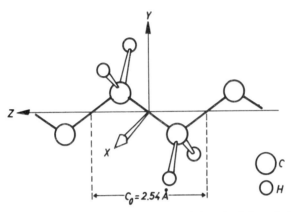

Fig. 7.24. Identity (or repetition) period of the polyethylene chain.

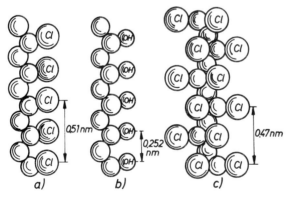

Fig. 7.25. Identity period of poly(vinyl chloride), poly(vinyl alcohol) and poly(vinylidene chloride) from left to right. (After Mark 1943.)

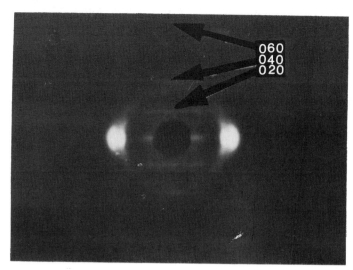

Fig. 7.26. Fibre diffraction photograph of polyamide-6.

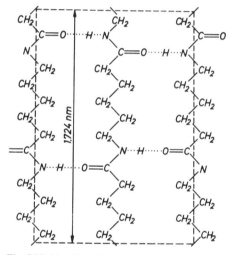

Fig. 7.27. Identity period of polyamide-6.

chemical units constitute the identity period in the fully extended state, whereas in hexamethylene adipamide (polyamide-66) only one is involved (Figs 7.27 and 7.28).

When helical structures are formed the identity period is shorter than expected. For example in poly(isobutylene) the identity period is equal to

1.863 nm, and it contains 8 isobutylene units. If the chain were fully extended, the identity period should be 3.0 nm. The reason for this short identity period is that the molecular chain is in helical form (Fig. 7.29).

Similarly the identity period of polypropylene is only 0.65 nm instead of 0.75 nm, the value expected for 3 monomeric units (Fig. 7.30). The identity period of poly(vinyl acetate) is 0.51 nm, but after hydrolysis that of the poly(vinyl alcohol) product is only 0.252 nm. This shows that only every second acetate group counts as being in the identical position. This is obviously true for its derivative (the poly(vinyl alcohol) molecule) too, but because of its lower volume, the —OH group can form in the crystal lattice in an identical manner regardless of its position, this is the reason for the halved identity period. Putting it another way, the position of the acetate group is

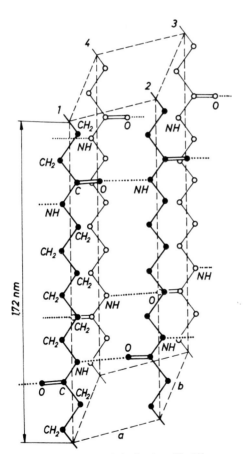

Fig. 7.28. Identity period of polyamide-6/6.

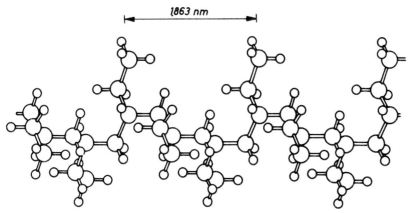

1863 nm

Fig. 7.29. Identity period of polyisobutylene.

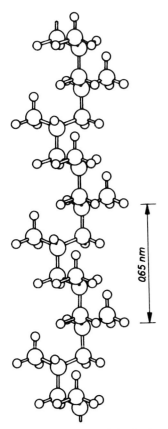

065 nm

Fig. 7.30. Identity period of polypropylene.

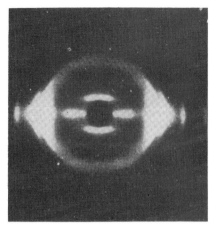
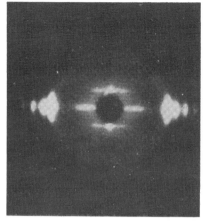

Fig. 7.31. X-ray photograph showing a fibre   Fig. 7.32. X-ray photograph showing a fibre
of poly(ω-hydroxy undecanoic acid). (After   of poly(ω-hydroxy decanoic acid). (After
Baker 1945.   Baker 1945.)

important from the point of view of crystallization, that of the OH group is not, because of its lower volume.

In polyesters one of the O atoms of the carboxyl group builds into the chain, the basic chain can be regarded as a paraffinic chain with some oxygen substituted groups, the measured identity period is equal to the repetition length of the $C=0$ groups being in identical position. For example, from the diffraction photograph of ω-hydroxy undecanoic acid the identity period is 1.5 nm (calculated value 1.49 nm), while that of ω-hydroxy decanoic acid is 2.73 nm (calculated 2.71 nm). In the latter case every second molecule is in identical position (Figs. 7.31 and 7.32).

The X-ray diffraction pattern is compatible with the order outlined in Fig. 7.33. The appearance of a meridian line instead of a single spot in Fig. 7.31 shows that the chains are not totally oriented along the fibre axis. In Fig. 7.32, however, there is one sharp spot which indicates perfect orientation. Generalizing these findings, one can say that in the case of ω-hydroxy acids containing even numbers of carbon atoms the unit cell is doubled, its geometry is orthogonal, and the fibres run parallel with the chain axis. The identity period of ω-hydroxy acids containing odd numbers of carbon atoms involves only one molecule, the unit cell is not orthogonal (usually monoclinic), and there is some angle between the chain and fibre orientation.

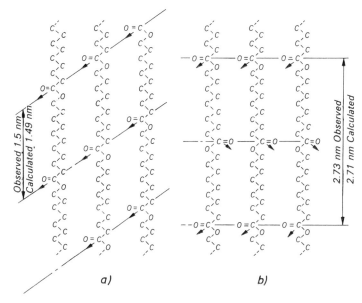

Fig. 7.33. Identity periods in fibres of poly(ω-hydroxy undecanoic acid) and poly(ω-hydroxy decanoic acid). (After Fuller and Baker 1943.)

The layered structure of polar groups in polyamides and polyesters and in its copolymers is quite general. If the number of polar groups is increased then the planar zig-zag structure is replaced by a helical one. The appearance of helical structures is accompanied by a shortened identity period. The material properties are related to the type of layered structure, so, for example, the melting point increases with increasing helicity (Fuller and Baker 1943).

7.10. LATERAL ORDER OF MOLECULAR CHAINS

On fibre diffractograms, reflections perpendicular to the fibre axis are not hyperbolic curves, they are displaced along straight lines, this is the zero layer-line, or equator line. From the position of these reflections the interplanar spacings, d related to the reflections are easily calculated using the Bragg equation. Equatorial reflections are characteristic of oriented polymers. The next Table contains the interplanar spacings showing lateral order in some of the more important polymers. These values can be determined from fibre-diffractograms (Table 7.1). It can be seen from the Table that the lateral

Table 7.1

Lateral order of some important polymers

Polymer	Index and intensity											
	A_1	Int.	A_2	Int.	A_3	Int.	A_4	Int.	A_5	Int.	A_6	Int.
Polyethylene	0.418	VS	0.371	VS	0.297	W	0.263	VW	0.247	S		
Polypropylene-α	0.626	VS	0.519	VS	0.477	VS	0.419	S	4.04	VS		
Polypropylene-β			0.549	VS	—	—	0.420	S	0.385	M	0.361	M
Polypropylene-γ	0.637	VS	0.529	VS	0.441	VS	0.419	S	0.405	S		
Polyamide-6	0.443	VS	0.378	M	0.236	M						
Poly(vinyl alcohol)	0.454	VS	0.389	VS	0.320	S	0.279	S				
Poly(vinyl chloride)	0.532	M	0.474	M	0.371	M	0.324	M	0.244	M		
Poly(vinylidene chloride)	0.565	M	0.341	M	0.316	M	0.283	S	0.244	M	0.218	W
Polyisobutylene	0.598	VS	0.347	W	0.298	M	0.273	W	0.244	M		

VS = very strong, S = strong, M = medium, W = weak, VW = very weak.

Numerical values are given in nm units.

After Baker 1945.

volume required by poly(vinyl alcohol) is higher than that for polyethylene. Interplanar spacings measured in polyethylene are characteristic of paraffinic chains. If side groups are present then interplanar spacings change, depending on the number of side groups, their bulk, and any deviations from the planar zig-zag structure.

Appearance of polymorphic modifications can be detected and monitored by the study of lateral ordering. For example, the identity periods of cellulose-I and cellulose-II modifications are both 1.03 nm. d values measured on the equator, are however, different. In polyamides, α and β modifications can be detected on the meridian and the equator. Three different modifications of polypropylene have been described by Turner-Jones (1964) (Fig. 7.34).

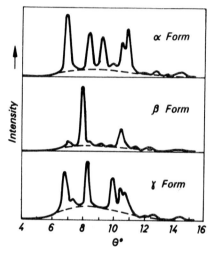

Fig. 7.34. X-ray diffractograms of various modifications of polypropylene. (After Turner-Jones et al. 1964.)

The identity period and the lateral order can be used in the identification of polymers, as mentioned earlier.

Chemical reactions on fibres destroy their crystalline structure, and, as a result, in principle, chemical reactions can be monitored by X-ray analysis. It has to be taken into account, however, that chemical reactions occur first and predominantly in amorphous parts, and any changes are not reflected in the X-ray diffraction patterns.

X-ray diffraction is not an adequate analytical tool if copolymerization occurs. The reason for this is that comonomers are normally close to each other in chemical composition and in volume (isomorphous substitution), and therefore fit into the same crystal lattice. If the volumes of the two monomeric units are different, the higher volume monomer hinders the crystallization of the smaller one, and this can be detected by X-ray diffraction.

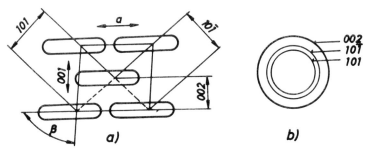

Fig. 7.35. The unit cell of cellulose and powder diffractogram. (After Howsmon 1950.)

Swelling of polymers can be studied by X-ray diffraction. One can decide, for example, whether swelling is limited to the amorphous phase, or if it also involves the crystalline phase. If the crystalline part is involved, then phase transition can occur, or the unit cell distorts. In the case of cellulose under the influence of strong acids (such as 72% H_2SO_4 or 83% H_3PO_4 solutions) the lateral order changes continuously, while under the effect of swelling agents shifts attributable to these agents can be detected. As an example let us consider the swelling phenomenon induced by diamines. Fig. 7.35a shows the unit cell of cellulose in the a-c plane, while Fig. 7.35b shows the X-ray diffractogram of the unoriented cellulose fibre. Change of the [101] distance

Table 7.2

The effect of diamines on the [101] plane distance

Natural cellulose	0.61 nm
+Hydrazine	1.03 nm
+Ethylene diamine	1.23 nm
+Tetramethylene diamine	1.46 nm

After Howsmon 1950.

under the effect of diamines is listed in Table 7.2. It can be seen that this interplanar distance may more than double in the swelling media. Such swelling processes can be monitored by X-ray diffraction quite successfully.

7.11. ORIENTATION OF MACROMOLECULES

7.11.1. Mono- and biaxial orientation: pole diagrams

Orientation can occur in both amorphous and crystalline substances. An example of amorphous orientation is stretched atactic polystyrene. Optical birefringence can be detected using a simple polarisation microscope, and chain orientation is easily recognized from X-ray diffraction patterns. In this case molecular segments exhibit a well-defined orientation without forming crystalline parts.

In semicrystalline samples orientation can be detected in both amorphous and crystalline parts. These two kinds of orientation can be measured independently, and sometimes differ substantially.

Orientation can be characterized by, for example, the directionality of molecular segments. If there is no favoured direction, then orientation vectors exhibit a uniform distribution on a spherical surface (Fig. 7.36a). Fibres in the unstretched state have a similar structure. If however, the fibre is stretched, molecular segments become oriented along the fibre axis, and two poles are observed in this direction (Fig. 7.36b). This is known as uniaxial orientation.

If orientation is biaxial as, for example, in biaxially oriented films or in stretched and rolled fibres, the poles are also deformed in the lateral direction (Fig. 7.36c).

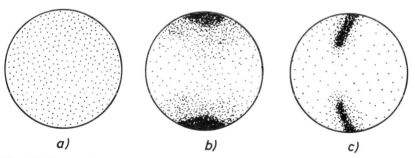

a)   *b)*   *c)*

Fig. 7.36. Characterization of orientation in polymers. *a* No orientation, *b* uniaxial orientation, *c* biaxial (planar) orientation. (After Ke 1964.)

In practice, instead of spherical surfaces pole diagrams are used; these are used widely in the crystallographic study of polycrystalline materials, particularly in metallurgy. Pole diagrams are the projections of spherical surfaces where points of equal intensity are connected by continuous smooth lines. In order to describe the orientation of all three crystalline axes two pole diagrams are needed (Fig. 7.37). If the orientation is uniaxial a single pole diagram is sufficient. In this case, instead of a pole diagram, the number of oriented molecular chains can be plotted versus the angle between the chain and the fibre axis (Fig. 7.38).

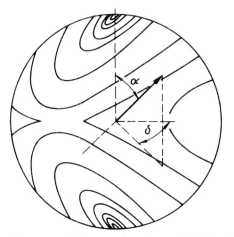

Fig. 7.37. Pole diagram of a polymer. (After Ke 1964.)

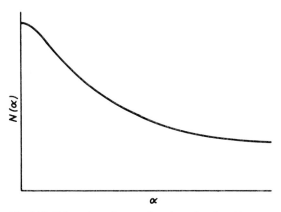

Fig. 7.38. Orientation plot as a function of angle with respect to the fibre direction. (After Ke 1964.)

For semicrystalline materials two diagrams would serve better, since the orientation of the crystalline parts can differ substantially from that of the amorphous parts.

7.11.2. Experimental methods used to determine molecular orientation in polymers

The five most important methods for orientation measurements are:
   1. X-ray diffraction;
   2. optical birefringence;
   3. velocity of ultrasound;
   4. polarized infrared radiation;
   5. magnetic anisotropy.

X-ray studies can describe the angular distribution of orientation, so this is the most popular type of measurement employed. It is very important for the study of fibre drawing processes where drawing is performed on fibre bundles, and it is not sufficient to know simply the average orientation. It is necessary to know whether orientation within the bundle is uniform or not, since the interior fibres may be stretched to a greater extent than those on the outside.

Optical birefringence gives the average of crystalline and amorphous orientation, pure crystalline orientation can only be measured by X-ray diffraction. A combination of these two methods—after cumbersome calculations—yields the amorphous and crystalline orientation values separately. It should be emphasised that X-ray diffractograms can also be used to determine amorphous orientation so long as the crystalline interference spots do not disturb data processing.

Frequently orientation means orientation of ordered phases. It has to be stressed, however, that both crystalline and amorphous materials can exist in both oriented and unoriented states, examples are the oriented and unoriented polypropylene fibre (crystalline) and the stretched and unstretched atactic polystyrene fibre (amorphous). The fact that a material exhibits birefringence indicates the presence of orientation, but does not necessarily imply crystallinity.

The velocity of ultrasound in polymers is very sensitive to chain orientation. Fast and accurate methods have been developed to determine the average (amorphous + crystalline) orientation (see under section 7.11.8).

Generally speaking, it is necessary to make a clear distinction between three notions, namely: orientation, crystallinity, and crystalline particle size

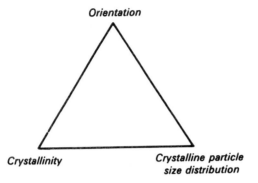

Fig. 7.39. Orientation, crystallinity and crystalline particle size as three independent properties of solid crystalline polymers.

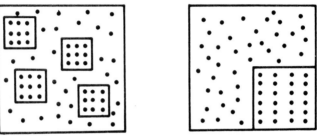

Fig. 7.40. Models of two polymers with equal crystallinity but different crystalline particle size.

distribution. These parameters can change independently and are frequently confused even in the literature. In order to characterize a semicrystalline polymer adequately, all these parameters need to be determined (Fig. 7.39). The difference between orientation and crystallinity should be clear from the previous text. Independence of crystallinity and crystalline size distribution is outlined in Fig. 7.40.

7.11.3. Cold drawing and necking

When drawing synthetic fibres the decrease of the cross section is not uniform along the length of a fibre. At certain places the fibre becomes thin and elongated, this is known as necking, and the neck moves along the fibre on further stretching (Fig. 7.41).

Experience shows that this behaviour does not occur at higher temperatures, or if the stretching process is very slow. Necking is observed only if the fibre is crystalline, or if it crystallizes on drawing.

The necking phenomenon can be explained as follows: parallel molecular bundles formed on drawing show very high resistance to further drawing, and at the interface of stretched and unstretched molecular chains (at the neck) the drawing force acts at a single point. Deformation is accompanied by a strong exothermal effect, the local temperature increase can reach 20–30°C, which makes local deformation easier. If the process is very slow, it becomes isothermal, rather than adiabatic, the heat evolved is dissipated by heat conduction, and no necking is observed. Similarly necking does not occur at higher temperatures, owing to increased molecular mobility the fibre can be stretched easily at any point (Fig. 7.42). Fibres can usually be stretched to about 4–5 times their original length, this is known as the natural drawing ratio. At higher extensions the fibre usually breaks. There are, however, some exceptions where low-volume side groups make chain slippage possible (polyethylene, polypropylene).

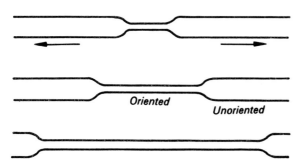

Fig. 7.41. The necking phenomenon in fibres.

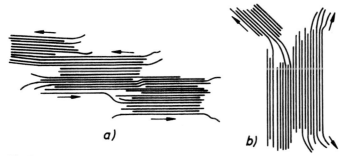

Fig. 7.42. Molecular orientation during necking.

Necking is a factor to be taken into account when studying orientation in polymers. The diameters of insufficiently drawn fibres can reach four times the diameter of uniformly stretched parts. Such "nodes" can easily be missed when samples are studied by microscopy. Frequently local or residual stresses are confused with orientation. Orientation means a directional ordering of molecular chains, while stresses are forces which tend to reorganize the polymer, sometimes leading to fracture. In strongly oriented systems (such as fibres) stresses can be relieved by annealing. High degrees of orientation can be achieved only if the parallel orientation of the chains is not accompanied by high stresses. Such "ultraoriented" fibres can be produced for example in the gel state.

7.11.4. Determination of orientation from X-ray diffractograms

On the diffractogram of an unoriented sample each interplanar distance yields a circular interference line. The effect of orientation is to transform these circles into an arc or sickle-shape. The diffraction pattern obtained from a perfectly

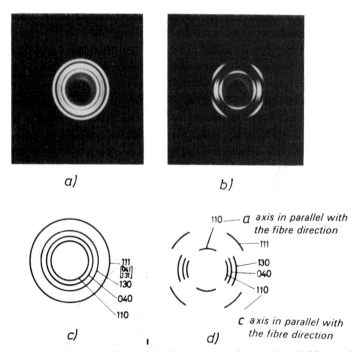

Fig. 7.43. Diffraction photograph of an unoriented and oriented PP sample. *a* Unoriented sample, *b* oriented sample, *c* schematic of Fig. *a* together with the Miller indices of reflection lines, *d* schematic of Fig. *b* together with the Miller indices of interference lines, orientation is vertical.

oriented sample is equivalent (at least in principle) to that present in a single crystal. So, depending on the space grouping, every interplanar distance leading to diffraction produces a point pair on the equator and meridian-line, and four points on higher layer-lines. In practice, orientation always leads to the appearance of arc-like patterns. As an example Fig. 7.43 shows the diffraction patterns for both unoriented and oriented polypropylene fibres together with their simplified line-diagrams. Changes induced by orientation are quite clear: the unoriented sample shows that all orientations of crystalline planes are equally probable: under the effect of orienting forces, all chains tend to orient in a single direction, and, similarly to single crystals, diffraction is obtained in one direction only.

If, instead of a point-like reflection, tangential arc-like patterns are obtained, then either the molecular chains are not perfectly oriented, or if they are oriented, their direction does not coincide with that of the fibre axis. The fibre axis is parallel to the molecular chains when drawn fibres are studied. In most unit cells this is the c axis, which is perpendicular to the a^*–b^* plane of the reciprocal lattice. In this case [00l] type reflections appear on the meridian and [hk0] type on the equator line. If the direction of molecular chains is in the b axis then [0k0] type reflections appear on the meridian and [h0l] type on the equator. This is the case, for example, with cellulose.

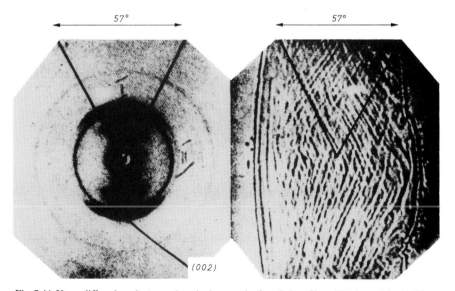

Fig. 7.44. X-ray diffraction photograph and micrograph of a cellulose fibre. (Kitaigorodsky 1952.)

In natural cellulose fibres microscopy suggests that the structure of the cellulose wall exhibits a spiral structure where the typical angle is 57° (Fig. 7.44). The same angle can be detected on X-ray diffractograms. It should be noted here that the X-ray diffractogram is rotated by 90°, interference arcs falling below the 57° angle appear perpendicular to the fibre axis, the arc shown is the [002] reflection of cellulose.

The effect of chain orientation on diffraction is demonstrated by the reflection sphere and the reciprocal lattice in Fig. 7.45, which is a modification of Fig. 7.17. Reciprocal lattice vectors, d^*, are not oriented in one direction as in single crystals, being not distributed uniformly on the surface of a sphere with radius d^* as in powder diffraction, but displaced around part of the spherical surface.

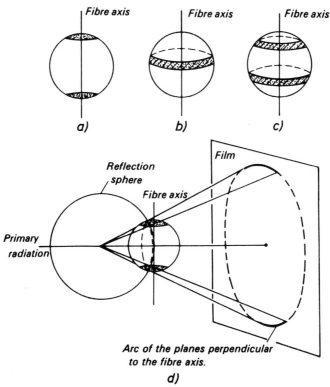

Fig. 7.45. Appearance of orientation on flat diffraction photographs in terms of the reciprocal lattice–reflection sphere model. Formation of crystalline planes characterized by reciprocal lattice vector d^*: a perpendicular to the fibre axis, b parallel to the fibre axis, c general position and d their intersection with the fibre surface of the reflection sphere which results in observable reflections.

Let us take a closer look at the differences between Figs 7.45a, b and c. Planes perpendicular to the fibre axis appear on the meridian, their distribution shows the perfection of the orientation, so, if reflections of this type appear, they can be used to characterize orientation.

Planes parallel to the fibre axis, in the case of uniaxial orientation, appear on the equator of a sphere defined by the $d*$ reciprocal lattice vector, and the points are distributed along a "belt", or band. Other, general reflections appear perpendicular to the [hkl] crystallographic plane, according to its orientation. $d*$ reciprocal lattice points drawn perpendicular to the planes at the origin are concentrated on parts of the spherical surface with radius $d*$, and a reflection appears if the end-points, p of the reciprocal lattice points coincide with the surface of the reflection sphere. This is shown in Fig. 7.45d, where the sphere with radius $d*$ (shown in Fig. 7.17) is replaced by the reciprocal lattice vectors of the atomic planes oriented perpendicular to the fibre axis, in accordance with Fig. 7.45a. The degree of scattering shows the extent of orientation. As can be seen from Fig. 7.45 the width of the arc associated with the [hkl] reflection is determined by that area where the reflection sphere and the d^*_{hkl} reciprocal lattice points touch each other.

It has to be stressed again that not all of the crystallographic planes give a reflection spot. This is limited by the reflection conditions, which are, in turn, determined by the spatial structure of the crystal.

If, for example, the planes are perfectly oriented, and the repetition period of the planes is 0.5 nm, then, according to the Bragg equation, and assuming CuK_α radiation:

$$\sin \Theta = \frac{\lambda}{2d} = \frac{0.154}{1.0} = 0.154, \qquad\qquad (7.14)$$

and hence

$$\Theta = 8° 50'.$$

So that on the meridian only those planes which have an angle with the radiation direction of at least 8° 50', (that is, are imperfectly oriented), give a reflection. In order to study such planes tilted diffraction photographs are necessary.

If this is explained in terms of the reciprocal lattice and reflection sphere, it means that a reflection sphere of unit radius rotating around the origin of the reciprocal lattice touches the vertical axis only at the origin of the reciprocal lattice, so that the reciprocal lattice points lying on the axis of rotation never get into a reflection position. If such reflections are to be studied, then some other axis has to be chosen, or the tilted scattering technique has to be used.

It can be seen from Fig. 7.46 that the toroidal body obtained by the rotation of the reflection sphere contains those reciprocal lattice points which can reflect the radiation, so points lying on the axis of rotation do not give reflections, with the exception of the origin, and provided that the orientation

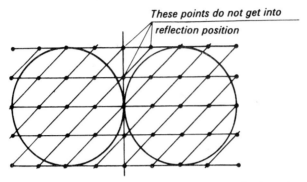

Fig. 7.46. The scattering toroid determined by the reflection sphere and reciprocal lattice.

is perfect. Orientation measurements, therefore, begin with the determination of the diffraction arc length of the equatorial reflections measured in degrees.

Fibre diagrams are made with flat film cameras and the optical density of the exposed film is determined. Optical density is defined as:

$$D = \log \frac{I_0}{I},\tag{7.15}$$

where D is the optical density, I_0 and I are the intensities of the incident, and transmitted radiation respectively. It is well-known that the optical density of photographic film is (within certain limits) proportional to the amount of incident radiation so that the angular distribution of optical density gives a measure of chain positions.

Optical densities are measured by microdensitometers or microphotometers, point by point, and the optical density is plotted versus the angle between the scattered light and the equator.

7.11.5. Types of orientation found in X-ray diffractograms

It has been mentioned that on X-ray diffractograms of unoriented samples, independent of the axis chosen, a full diffraction circle is obtained (Fig. 7.47a). In drawn fibres uniaxial orientation is present, so that orientation will be detected from the A and C directions, but not from the B direction (along the chain axis). This means that a fibre can be rotated in any direction along the fibre axis, and the molecular order remains unchanged, in this direction there is no orientation (see Fig. 7.47b).

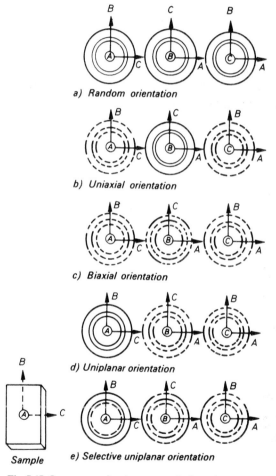

a) Random orientation

b) Uniaxial orientation

c) Biaxial orientation

d) Uniplanar orientation

Sample

e) Selective uniplanar orientation

Fig. 7.47. Occurrence of various types of orientation on the X-ray diffraction photographs. (After Howsmon 1950.)

If the fibre is calandered, or a film is drawn in two perpendicular directions, biaxial orientation occurs, and the sample shows orientation even from direction *B* (Fig. 7.47*c*).

With film it frequently occurs that when X-rays are incident perpendicular to the surface, an unoriented structure is observed, whilst investigation of the edges indicates high orientation.

This is uniplanar orientation (Fig. 7.47*d*). If in such samples, orientation is missing in certain crystallographic directions, but appears in others, the orientation is known as selective uniplanar orientation (Fig. 7.47*e*).

7.11.6. Experimental results of measuring orientation

Sisson and Clark (1933) measured the intensity of the [002] reflection in cellulose as a function of the angle between the scattered radiation and the equator. Optical densities naturally depend on the exposure time, and in order to remove this dependence the optical density curves have been

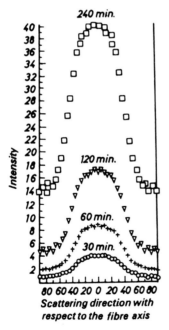

Fig. 7.48. The intensity of 002 line in cellulose as a function of the angle between the fibre axis and scattering direction at various exposure times. (After Sisson and Clark 1933.)

normalized to 100 units as the maximum (the percentage distribution is shown in Figs 7.48 and 7.49).

The degree of orientation has been characterized by the angle, where the intensity decreases to 40% of the maximum value. Barkley and Woodyard (1938) found a mathematical relationship between the degree of orientation, A, measured in this way and the ultimate strength, S, such that:

$$S = 193.5 - 3.18\,A \tag{7.16}$$

where S is measured in lbs/sq.inch. Experimental points are given in Fig. 7.50.

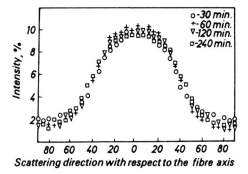

Fig. 7.49. The normalized representation of the results shown in Fig. 48. (After Sisson and Clark 1933.)

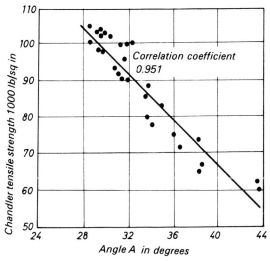

Fig. 7.50. Correlation between tensile strength and orientation of cellulose fibres.

Meredith (1944) decreased the standard deviation of the data by plotting 40% not of the optical density, but of the X-ray intensity. This was achieved by making calibration lines with different exposure times on each photographic film, so that if the optical density was not strictly proportional to the intensity of the scattered radiation, it could be corrected. Nowadays the linearity of most photographic films is good over wider ranges of exposure times, so this method is no longer necessary.

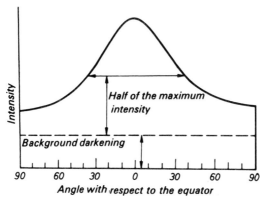

Fig. 7.51. Orientation halfwidth.

A widely used orientation parameter is the half-width of the angular distribution curve, where the scattered intensity decreases to one half of the maximum value (Fig. 7.51). When calculating the half intensity, the background optical density level is taken into account.

The orientation parameter, A, is high if the orientation is low and vice versa. It is therefore preferable to use the reciprocal value of the orientation factor:

$$O = \frac{180}{A},$$ (7.17)

or alternatively, as introduced by Krimm and Tobolsky:

$$O = \frac{180 - A}{180}.$$ (7.18)

These parameters increase with orientation, and therefore describe the orientation in a more graphic way. The following measure of orientation was originally introduced in optical measurements, but is now used widely elsewhere:

$$f = \frac{3\overline{(\cos^2 \alpha)} - 1}{2}.$$ (7.19)

f is referred to as the orientation factor, where α is the angle of orientation, that is, the angle between the axis normal to the plane and the fibre axis, the bar denotes averaging. If orientation is ideal (or perfect) then $\alpha = 0$, and $\cos^2 \alpha = 1$, $f = 1$. If orientation is perpendicular to the fibre axis, then $\alpha = 90°$, and $f = -0.5$. If there is no specific orientation, that is, the orientation of the planes is randomly distributed in all directions, then $\cos^2 \alpha = 1/3$, and $f = 0$.

In order to give a full description of orientation, three orientation functions are necessary. In the case of uniaxial orientation the third function

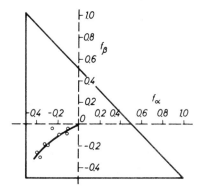

Fig. 7.52. Orientation function diagram of the cold drawing of polyethylene. (After Ke 1964.)

may be calculated from the other two values and data of the unit cell. For an orthogonal unit cell for example:

$$f_\alpha + f_\beta + f_\gamma = 0.$$ (7.20)

From the uniaxial orientation functions, an orientation function diagram can be constructed. As an example, the orientation function diagram for the cold drawing process of polyethylene is shown in Fig. 7.52. At the origin of this diagram there is no orientation. Along any of the axes the unity value means perfect orientation. As one can see from the diagram, in polyethylene orientation appears along the c-axis.

Roldan (1962) introduced the following orientation parameter for polyamides:

$$f = \left[1 - \left(\frac{I_{av}}{I_{equ}} \right) \right] K \tag{7.21}$$

where I_{av} is the average intensity of the measured line of the rotated sample, while I_{equ} is the maximum intensity of the same line on the equator, K is a correction factor which takes into account line broadening, its value being close to unity.

If $I_{av} = I_{equ}$ (that is, there is no orientation), then $f = 0$. If orientation increases and $I_{equ} \gg I_{av}$ then f approximates to unity.

When orientation is strong, it can be characterized by a distribution curve, which gives the probability density of oriented chains as a function of the angle between the chain and the equator (the polar angle) (see Fig. 7.38). If, however, the orientation is slight, then the ratio of intensities measured on the equator and the meridian gives a parameter which reflects slight changes in orientation.

It should be noted that the arcs of the fibre diagram do not indicate whether the orientation is accompanied by a spiral arrangement of the molecular chains. Because of the cylindrical form of fibres only a projection of the spiral structure appears. We have seen, for example, that in cellulose there are oriented spiral-like structures, and the angle between these structures is 57°. Arcs on the X-ray pattern indicate this 57° as an interval, as if within these limits all directions were possible. In fact these spirals exhibit a constant and uniform angle, but the X-ray diffractogram shows only the projections. The

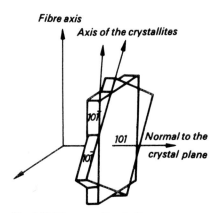

Fig. 7.53. The tape effect in fibres. (After Kast 1953.)

existence of spiral or helical structures can be inferred not from the arc shapes on the X-ray diffraction pattern, but from the shorter identity period. Diffraction of helical structures has been comprehensively discussed by Vajnstein (1966).

Hermans was the first to point out that, from the point of view of orientation, it is important which interference line is chosen for the measurement. Investigation of a single line gives unequivocal results only if the oriented units are cylindrical rods. Let us see, for example, the scattering of the [101] plane in Fig. 7.53. The crystallite can change its orientation with respect to the fibre axis, but this has no effect perpendicular to the plane. If the orientation behaviour is similar to this case, usually "tape-effect" is meant, as the orienting units are tape-like structures. Generally, more reflections are studied, and only if their dependence on orientation is similar can the orientation be characterized by a single line.

Hermans and Kratky were the first to study the change in X-ray diffraction patterns as a function of orientation in cellulose. Fig. 7.54 shows an

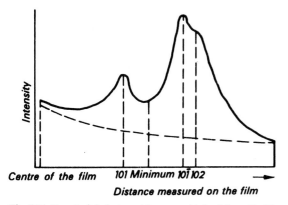

Fig. 7.54. Equatorial photometric curve obtained from the X-ray diffractogram of viscose fibre. (After Howsman 1956.)

equatorial photometric curve constructed from an X-ray diffractogram of a viscose fibre. The intensities of the [101] and [10$\bar{1}$] diffraction arcs have been measured as a function of orientation angle. These curves are different for these two reflections (Fig. 7.55), so the "tape-effect" was present. If measurements are performed on fibres freshly prepared from xanthogenate solution, the orientation curves for the two reflections are similar, when orientation is rodlike. On processing the data, it has to be taken into account that the

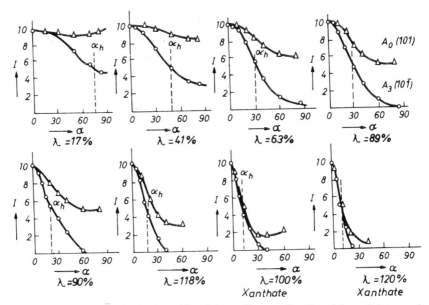

Fig. 7.55. 101 and 10$\bar{1}$ reflection intensities of viscose fibres as a function of the angle between the fibre axis and scattered radiation at different extensions. (After Kast 1953.)

amorphous parts can have their own orientation, not necessarily identical with that of the crystalline part (as discussed in earlier sections). Kast (1953) has shown, on photometric curves, that the scattered intensity of the amorphous areas also depend on the angle between the scattering direction and the equator, and that the amorphous curve shows a maximum at the equator (Fig. 7.56).

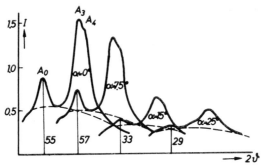

Fig. 7.56. Radial photometric curves of viscose fibres. Photometric measurements were performed under different α angles for orientation measurement. The amorphous curve also shows a maximum at the equator. (After Kast 1953.)

Kargin and Mikhailov (1940) looked for a relationship between X-ray diffraction patterns and mechanical properties of viscose fibres drawn to different extents. They found that tensile strength increases with orientation, but the mechanical product (the product of tensile strength and elongation at break) decreases. On annealing orientation decreased, but elongation and ultimate strength increased.

Ingersoll (1946) also studied the relationship between the degree of orientation and mechanical properties of viscose fibres.

Bodor and Hangos (1954) performed similar investigations under the guidance of József Vándor. They studied the relationship between the orientation $(0 = 180/A)$ of the [101] reflection and fibre properties on fresh viscose fibres produced in Hungarian Viscosa. Definite relationships were found between the degree of orientation and elongation (Fig. 7.57), tensile

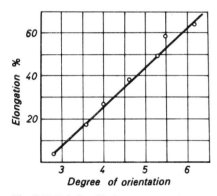

Fig. 7.57. Relationship between extension and orientation in viscose fibres. (After Bodor 1958.)

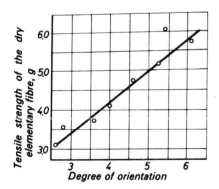

Fig. 7.58. Relationship between tensile strength and degree of orientation in viscose fibres. (After Bodor 1958.)

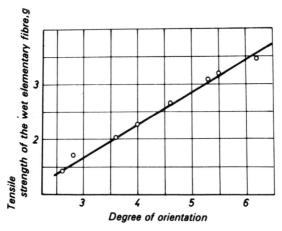

Fig. 7.59. Relationship between tensile strength in the wet state and the degree of orientation in viscose fibres. (After Bodor 1958.)

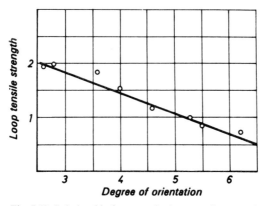

Fig. 7.60. Relationship between the loop tensile strength and degree of orientation in viscose fibres. (After Bodor 1958.)

strength (Fig. 7.58), wet tensile strength (Fig. 7.59), and loop tensile strength (Fig. 7.60).

Bodor also studied the orientation of crystalline and originally amorphous polyamide-6 fibres. The change in the X-ray diffraction pattern on drawing of polyamide fibres is shown in Fig. 7.61. Amorphous polyamide fibres can be produced if the molten filament is driven into a dry ice and acetone bath. Two sharp reflections appear on the equator of the X-ray diffraction pattern of polyamide-6 fibre, one is the [200], the other is the [020,

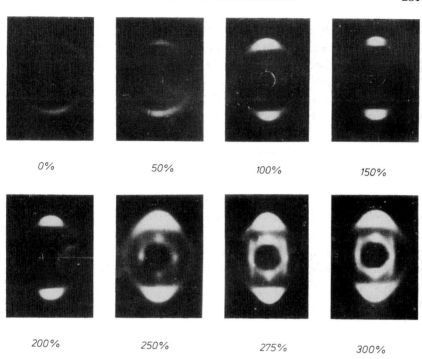

0%   50%   100%   150%

200%   250%   275%   300%

Fig. 7.61. Relationship between the orientation and extension of polyamide-6 fibres as shown by diffraction photographs. (After Bodor 1958.)

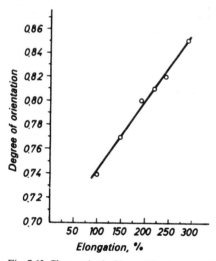

Fig. 7.62. Changes in the X-ray diffractogram of crystalline polyamide-6 during extension. (After Bodor 1958.)

220] double reflection. (The chain axis is the *c* axis). The relationship between the orientation degree of the arc and the draw ratio is linear in the case of the [020, 220] reflection, if the fibre is crystalline (Fig. 7.62). If the fibre is amorphous, this relation is somewhat different.

Orientation influences the rate of dye uptake (the rate of diffusion). If the fibres are insufficiently drawn, diffusion of dye molecules is much faster, and this leads to problems in the technology. Munden and Palmer (1950) studied dye uptake in polyamide-66, and it has been established that with a 250% draw, the dye diffusion rate decreases to 0.03% of the original value. It is interesting, however, that the equilibrium dye uptake is the same in both cases, that is, the orientation influences the diffusion rate but leaves the equilibrium value unchanged. Equilibrium is reached within about 100 hours.

Orientation can also be the result of the fibre formation process itself. Geleji et al. (1962) studied the orientation developed during fibre spinning of polypropylene fibres (Fig. 7.63). As can be seen from the photographs, a decrease in the fibre spinning temperature leads to orientation, and a similar effect is observed if the haul-off speed is increased.

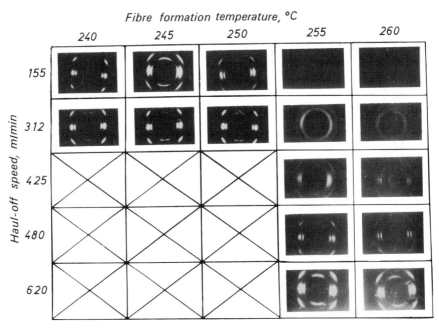

Fig. 7.63. X-ray diffraction photograph of polypropylene fibre under various conditions. Parameters are haul-off speed and fibre-formation temperature. (After Geleji et al. 1962.)

Reference has already been made to orientation in non-crystalline materials (such as polystyrene). The amorphous part of semicrystalline polymers may also be oriented, this would be detected by an increased intensity of the amorphous ring on the equator. This effect is not limited to the viscose fibre discussed above.

Other examples are found in polyethylene and polypropylene, where the intensity increment due to the amorphous ring can be separated from the crystalline interference lines, or in polyamide-6 and polyamide-66, where the two effects are superimposed on each other. It should be noted, however, that an amorphous ring does not necessarily appear on the X-ray diffractogram of every semicrystalline polymer. In the case of poly(vinylidene fluoride), for example, no halo can be detected. The probable reason for this is that an amorphous diffraction ring appears only if the scattering centres within the amorphous area are distributed randomly, and the distance between these centres does not vary too widely. It has been shown by model experiments that if the inter-particle distance between the scattering centres shows a high standard deviation (some 100%), the amorphous parts lead to background scattering only, and the amorphous ring does not appear. The main effects of orientation on fibre properties are summarized in Table 7.3.

In major cotton-producing countries the orientation of cotton fibres is measured by X-ray diffraction according to national and international standards, and these data are utilized in product qualification.

Finally a practical note: orientation can be determined more readily from flat diffraction photographs than from diffractograms taken in cylindrical cameras (for example from Debye–Scherrer diffractograms). For flat films data processing is performed along a perfect circle, whereas in the

Table 7.3

The effect of orientation on various properties of fibres

The effect of orientation		
Increases	Decreases	Does not change
Tensile strength	Elongation	Water vapour absorption
Young's modulus	Axial swelling	Elastic extension limit
Brittleness	Rate of dye uptake	Equilibrium dye uptake
Lateral swelling	Plasticity	
Refractive index	Puckering resistance	
Relaxation after elongation	Chemical reactivity	

After Howsmon 1950.

case of cylindrical films, reflections are found along the intersection of a conical and a cylindrical surface. Orientation can be measured by diffractometers too, but when used for this purpose, they require a special arrangement.

7.11.7. Orientation determination from birefringence measurements

Optically anisotropic bodies are characterized by the indicatrix. The indicatrix is a rotational ellipsoid with axes proportional to the refractive index measured in the particular direction. Birefringence is determined in the three direction of space as:

$$\Delta_1 = n_2 - n_3,$$

$$\Delta_2 = n_3 - n_1,$$

$$\Delta_3 = n_1 - n_2. \tag{7.22}$$

Of these three equations only two are independent, the third can be calculated.

In the case of uniaxial orientation $n_2 = n_3$, and only one independent equation exists. The relationship between refractive index and polarizability, α, is given by the Lorentz–Lorentz equation:

$$\frac{n^2 - 1}{n^2 + 2} = \frac{4}{3}\pi\alpha. \tag{7.23}$$

For an infinitesimal increment of refractive index it follows from Eq. (7.23) that:

$$\frac{6n}{(n^2 + 2)^2} dn = \frac{4}{3}\pi d\alpha. \tag{7.24}$$

For finite refractive index difference, Δ_1:

$$\Delta_1 = n_2 - n_3 = \frac{2\pi}{9} \frac{(n^2 + 2)^2}{n} (\alpha_2 - \alpha_3) \tag{7.25}$$

and for fibres:

$$\Delta = n_1 - n_2 = \frac{2}{9} \pi \frac{(n^2 + 2)^2}{n} (\alpha_1 - \alpha_2) \tag{7.26}$$

where n is the average refractive index.

Birefringence is measured using a microscope with crossed Nicols. Transmission of polarized light at 45° angle with respect to the fibre axis is given by:

$$T = \sin^2(\delta/2),\tag{7.27}$$

where

$$\delta = \frac{2\pi\Delta d}{\lambda}\tag{7.28}$$

d is the sample thickness, and λ the wavelength of the applied light. In practice, the path difference is corrected by a rotating wedge (known as the Berek compensator). The two extreme positions of the wedge, where no path difference is observed, can be determined since in these cases the centre of the fibre appears to be black. From these values the optical path difference, Γ, can be calculated using the tables supplied with the optics. If fibre diameter and path difference are known, the birefringence can be given by:

$$\Delta = n_1 - n_2 = \frac{\Gamma}{d}.\tag{7.29}$$

Optical birefringence gives the average orientation of crystalline and amorphous areas.

This overall orientation is the weighted average of crystalline and amorphous orientations:

$$f_{av} = x f_{cr} + (1-x) f_{am}\tag{7.30}$$

where x is the crystallinity, f_{cr} the crystalline, and f_{am} the amorphous orientation, as given by Eq. (7.19). Optical birefringence in uniaxially oriented polycrystalline polymers can be given by (Hermans 1946):

$$\Delta = x\Delta^0_{cr} f_{cr} + (1-x)\Delta^0_{am} f_{am},\tag{7.31}$$

where Δ^0_{cr} is the internal birefringence of the totally oriented crystal and Δ^0_{am} that of the amorphous parts. (In the case of isotactic polypropylene $\Delta^0_{cr} = 29.1 \times 10^{-3}$, $\Delta^0_{am} = 60.0 \times 10^{-3}$). Rearranging Eq. (7.31) gives:

$$\frac{\Delta}{x f_{cr}} = \Delta^0_{cr} + \Delta^0_{am}\left(\frac{1-x}{x}\right)\left(\frac{f_{am}}{f_{cr}}\right)\tag{7.32}$$

and a linear relationship is obtained between $\Delta/x f_{cr}$, such that:

$$\left[\frac{(1-x) f_{am}}{x f_{cr}}\right].$$

This relation can be used to determine Δ_{er}^0 and Δ_{am}^0 if values of Δ, x, f_{er} and f_{am} are known. The same equation can be used to check the validity of f_{am} obtained in the sonic modulus measurements referred to below.

7.11.8. Orientation measurements from ultrasound

This type of orientation measurement is based on the fact that the sonic modulus, E, is dependent upon the orientation of molecular chains. The relationship between molecular orientation and the velocity of sound is outlined in Fig. 7.64.

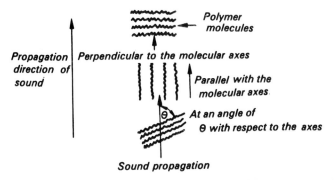

Fig. 7.64. Possible orientations of molecular chains with respect to sound direction.

Ward (1962) has shown that:

$$\frac{1}{E} = \frac{\overline{1 - \cos^2 \Theta}}{E_{lat}^0} \tag{7.33}$$

where E is the measured modulus, Θ the angle between the velocity of sound and the axis of symmetry, and E_{lat}^0 the modulus of the totally oriented fibre. Here "lat" is an abbreviation for lateral, or transverse modulus.

Samuels (1965) showed that the sonic modulus can be used to study orientation in heterogeneous two phase systems. This analysis assumes that in a homogeneous ideal mixture, density and compressibility are additive quantities. If both amorphous and crystalline phases are present, the following relationship is valid for compressibility:

$$K = xK_{cr} + (1 - x)K_{am} \tag{7.34}$$

where K is the compressibility of the mixture, K_{cr} is that of the crystalline, K_{am} is that of the amorphous parts, x is the crystalline fraction.

The relationship between compressibility, K, bulk elastic modulus, B, Young's modulus, E and Poisson's ratio, v, is:

$$K = \frac{1}{B} = \frac{3(1-2v)}{E}. \tag{7.35}$$

The ratio between the sonic modulus, E, velocity of sound, C, and density, ρ, in rigid, rod-like objects was established by Newton:

$$E = \rho C^2. \tag{7.36}$$

Combining Eqs (7.35) and (7.36) gives:

$$K = \frac{1}{B} = 3\frac{(1-2v)}{\rho C^2}. \tag{7.37}$$

According to Waltermann (1963), Poisson's ratio for polypropylene is $v = 0.33$, so that for long, rodlike polypropylene samples Eq. (7.37) reduces to:

$$K = \frac{1}{B} = \frac{1}{E} = \frac{1}{\rho C^2}. \tag{7.38}$$

Comparing Eqs (7.33)–(7.38) one can see that for the sonic modulus of an oriented polypropylene sample:

$$\frac{1}{E_{or}} = \left(\frac{x}{E^0_{lat,\ cr}}\right)(1 - \overline{\cos^2 \Theta}) + \left(\frac{1-x}{E^0_{lat,\ am}}\right)(1 - \overline{\cos^2 \Theta_{am}}), \tag{7.39}$$

In the case of unoriented samples $\cos^2 \Theta = 1/3$, and so:

$$\frac{3}{2E_u} = \frac{x}{E^0_{lat,\ cr}} + \frac{1-x}{E^0_{lat,\ am}} \tag{7.40}$$

where u refers to the unoriented state.

If sonic modulus of unoriented semicrystalline polymers is studied as a function of crystallinity, the internal lateral modulus of crystalline and amorphous areas can be obtained. Table 7.4 shows the sonic modulus, density and crystalline fraction values calculated by Samuels for some polypropylene samples. From these experimental data the following internal lateral moduli can be calculated for crystalline and amorphous parts of isotactic polypropylene:

$$E^0_{lat,\ cr} = 3.96\,\text{GPa}, \tag{7.41}$$
$$E^0_{lat,\ am} = 1.06\,\text{GPa}. \tag{7.42}$$

Table 7.4

Sonic moduli and crystallinities of polypropylene fibres

Sample	Sonic modulus (E) in GPa	Density g/cm³	Crystallinity, % after Samuels
1.	2.27	0.8875	0.410
2.	2.48	0.8936	0.485
3.	2.63	0.8980	0.540
4.	3.01	0.9061	0.643

Once these data are known, the amorphous orientation can be calculated from:

$$3/2(\Delta E^{-1}) = \frac{x f_{cr}}{E^0_{lat, cr}} + \frac{(1-x) f_{am}}{E^0_{lat, am}} \tag{7.43}$$

where f_{cr} is the crystalline, f_{am} is the amorphous orientation.
f_{cr} can be determined from X-ray data and:

$$(\Delta E^{-1}) = (E_u^{-1} - E_{or}^{-1}). \tag{7.44}$$

In practice sound velocity is determined by a "time of flight" method: that time interval is measured which is needed for the sound-pulse to cover a certain distance. If the distance is changed systematically and a path-time diagram constructed, the velocity can be calculated from the slope of the curve:

$$C = \frac{\text{path difference}}{\text{time}}. \tag{7.45}$$

If, for example, a distance of 110 mm takes 85 μs, then:

$$C = \frac{110\,\text{mm}}{85\,\mu\text{s}} = 1.69\,\text{km/s}.$$

In fibres the sonic modulus is frequently given in g/den units:

$$E = C^2 \cdot 11.3\,[\text{g/den}]. \tag{7.46}$$

The sonic modulus in Pascals for any kind of material is given by:

$$E = \rho C^2\,[\text{Pa}], \tag{7.36}$$

if the density is given in kg/m³, sound velocity in m/s and the velocity is in km/s, the result is obtained in GPa units.

Some equipments are available commercially which measure the sonic velocity by driving the fibre through two sensor heads, the distance between the heads being periodically decreased and increased. As a consequence, the time of flight also changes periodically, and the slope referred to in Eq. (7.45) can be determined.

The schematic view of such an apparatus is shown in Fig. 7.65. The time of flight versus sensor head distance is shown in Fig. 7.66.

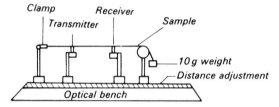

Fig. 7.65. Schematic of equipment used for the determination of ultrasound velocity. (After Samuels 1974.)

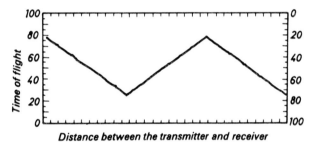

Fig. 7.66. Variation of time-of-flight for ultrasound pulse with the distance between the detectors in the ultrasound velocity determining equipment. (After Samuels 1974.)

7.11.9. Experimental results and the importance of amorphous orientation

Experimental data show that in the overall context, tensile strength, elongation and shrinkage all depend primarily on amorphous orientation. Geleji et al. (1977) studied the relationship between morphology and mechanical properties of various polypropylene fibres of different molecular mass distribution. Table 7.5 summarizes the results obtained for fibres with different draw ratios.

Table 7.5
Morphological data for PP fibres of different molecular mass as a function of extension ratio

Sample and extension ratio		Crystal-linity %	SAXS (L), nm	Crystalline particle size, (D), nm	Amorphous area (T_{am}), nm	Crystalline orientation f_{cr}		Amorphous orientation, f_{am}	Specific amorphous orientation, $f_{spec,\,am} \cdot 10^{-2}$
						$2\Theta = 14°$	$2\Theta = 17°$		
A	1:3.0	0.652	15.8	7.10	8.70	0.850	0.875	0.476	0.55
	1:3.5	0.649	14.0	7.40	6.60	0.885	0.900	0.641	0.97
	1:4.0	0.642	14.8	7.40	7.40	0.895	0.900	0.723	0.98
	1:5.0	0.636	15.6	7.40	8.20	0.900	0.920	0.861	1.05
B	1:3.0	0.657	17.0	8.90	8.10	0.850	0.870	0.464	0.57
	1:3.5	0.683	15.4	7.40	7.80	0.880	0.894	0.755	0.94
	1:4.0	0.656	14.4	6.85	7.55	0.90	0.900	0.778	1.03
	1:5.0	0.636	12.0	6.85	5.15	0.90	0.915	0.869	1.68
C	1:3.0	0.666	16.0	8.10	7.90	0.875	0.89	0.603	0.76
	1:3.5	0.660	15.0	7.40	7.60	0.88	0.895	0.743	0.98
	1:4.0	0.650	15.0	7.40	7.60	0.895	0.91	0.798	1.05
	1:5.0	0.610	14.0	7.00	7.00	0.900	0.91	0.849	1.10
D	1:3.0	0.660	15.4	8.10	7.30	0.895	0.895	0.558	0.76
	1:3.5	0.645	15.0	8.10	6.90	0.895	0.905	0.558	0.81
	1:4.0	0.628	14.2	8.10	6.10	0.895	0.915	0.828	1.35

Amorphous orientation is more sensitive to drawing than crystalline orientation. Fig. 7.67 shows the tensile strength as a function of amorphous orientation, while the initial modulus–amorphous orientation plot can be seen in Fig. 7.68.

A more thorough investigation of the data shows that amorphous orientation and the size of amorphous regions together influence the

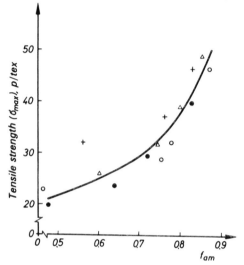

Fig. 7.67. Relationship between tensile strength and amorphous orientation for polypropylene fibres.

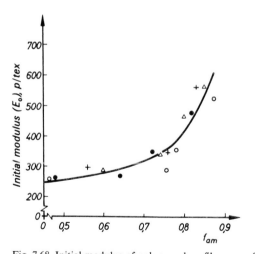

Fig. 7.68. Initial modulus of polypropylene fibres as a function of amorphous orientation.

properties. The linear dimensions of amorphous regions can be estimated from the long period, L, obtained from small angle X-ray scattering (SAXS) data and from the crystalline particle size, D. As a first approximation T_{am}, the size of the amorphous region, is the difference between the long period and

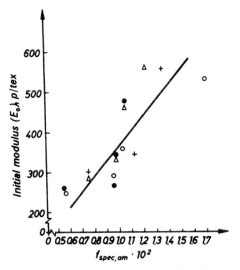

Fig. 7.69. Initial modulus of polypropylene fibres as a function of specific amorphous orientation.

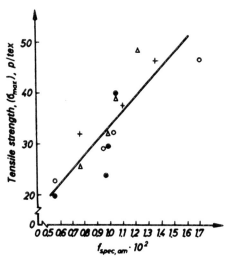

Fig. 7.70. Tensile strength of polypropylene fibres as a function of specific amorphous orientation.

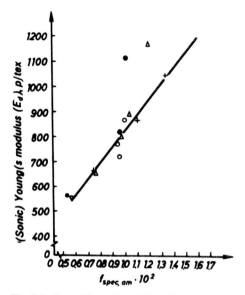

Fig. 7.71. Sonic (Young's) modulus of polypropylene fibres as a function of specific amorphous orientation.

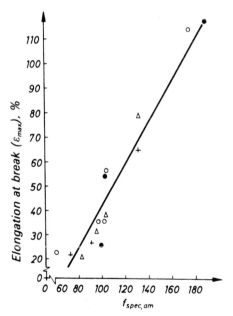

Fig. 7.72. Elongation at break of polypropylene fibres as a function of specific amorphous orientation.

crystalline particle size:

$$T_{am} = L - D. \tag{7.47}$$

If the amorphous orientation is divided by the size of the amorphous region, the specific amorphous orientation is obtained:

$$f_{spec,am} = \frac{f_{am}}{T_{am}}. \tag{7.48}$$

Initial modulus (Fig. 7.69), tensile strength (7.70), sonic (Young's) modulus (Fig. 7.71) and elongation at break (Fig. 7.72) all depend essentially linearly on the specific amorphous orientation. Amorphous orientation is limited to the amorphous regions—it is then plausible that the mechanical properties of fibres referred to above, are mainly determined by the specific amorphous orientation.

7.12. DETERMINATION OF CRYSTALLINE PARTICLE SIZE

Two properties of X-ray diffraction photographs can be used in the determination of crystalline particle size: one is line-broadening, the other is granular structure.

It was observed by Laue (1913) and Scherrer (1920) that interference lines are broadened if the size of the crystallites falls below 50 nm. If the crystallite size exceeds 10^{-6} m, then interference conditions are valid for relatively few crystallites, and the diffraction pattern becomes "granular". It can be seen even by inspection that the particle size is not fine enough for the normal Debye–Scherrer diffractogram. This "granularity" can be evaluated exactly, but the order of magnitude does not exist in polymers, so its treatment will be omitted here.

The line-broadening effect is, however, very important in polymers. The phenomenon itself is similar to that of interference lines observed for optical gratings. In the case of optical gratings, when the number of slits is sufficiently high, the interference pattern is sharp. If, however, the number of diffraction slits is decreased, the interference lines broaden somewhat. If the crystallite size falls below the limit of 50 nm, a similar line broadening occurs in X-ray diffraction (Henry et al. 1951).

Despite the fact that line broadening in polycrystalline materials cannot be ascribed to the reduced size of crystallites exclusively, internal stresses and impurities have a similar effect (Hosemann 1950, Kitaigorodsky 1952), and crystallite size can be estimated by the study of X-ray diffraction line broadening. As the effect of other factors cannot be taken into account exactly, some authors prefer to refer to "apparent crystallite size" or "particle size equivalent" (Ruscher 1958).

Apparent crystalline particle size can be calculated using the relationship:

$$D_{hkl} = \frac{K\lambda}{\beta \cos \Theta} \tag{7.49}$$

where D_{hkl} is the crystallite size normal to the hkl plane (in nm), K is a proportionality factor (its value being close to unity), λ the applied wavelength, β a measure of line broadening (in radians), and Θ the Bragg angle.

Apparent crystallite size refers to the dimension obtained by the above formula when $K=1$ is chosen (Jones 1938). Line broadening, β, is the excess line broadening as opposed to the instrument line broadening. The linewidth is measured as the halfwidth (in radians) related to half of the maximum absorption (Fig. 7.73).

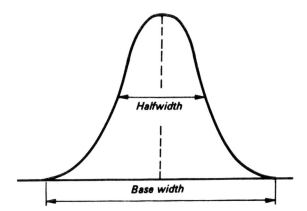

Fig. 7.73. Representation of linewidth or halfwidth.

Frequently the integral linewidth value is used, this is the ratio of the total area of the diffraction line and the maximum height. It is equal to the width of a rectangle, the height of which is equal to the height of the diffraction line, its area being equal to the total area of the diffraction line (Fig. 7.74).

Instrument broadening, β_{inst}, refers to the linewidth obtained under specific circumstances for a given camera type, slit system, and angular range, when measuring infinite crystallite size. For this purpose copper sheet or graphite are used.

Line broadening, β, is determined using the relationship:

$$\beta = \sqrt{\beta_m^2 - \beta_{inst}^2} \tag{7.50}$$

where β_m is the measured, and β_{inst} the instrument linewidth.

It should be noted that line broadening is the lateral smearing of the interference line (Fig. 7.75), while in the case of orientation it is the angular distribution along a circular arc which has to be taken into account (Fig. 7.76).

These two notions are frequently confused. The reason for this may be that in both cases a half-width is calculated as a characteristic value. For

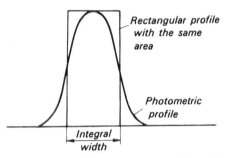

Fig. 7.74. Representation of the integral linewidth.

Fig. 7.75. Crystalline particle size determined from broadening of the crescent profile.

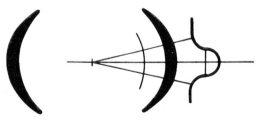

Fig. 7.76. Determination of the degree of orientation from the angular width of the diffraction arc.

orientation measurements, the angle with respect to the equator is calculated, while in the case of line broadening, the width is measured in the radial direction.

In cylindrical cameras the linewidth, expressed in radians, can be calculated simply from the linewidth, B measured on the photometric curve:

$$\beta_m = \frac{B}{fR} \tag{7.51}$$

where B is the measured linewidth (mm), f the magnification factor of the photometer, and R the radius of the camera (mm).

In the case of a flat film camera:

$$\beta_m = \frac{2\pi}{360}\left(\arctan\frac{r_1}{cf} - \arctan\frac{r_2}{cf}\right) \tag{7.52}$$

where r_1 and r_2 are the distances of the two sides of the line from the centre of the film, such that:

$$B = r_1 - r_2 \tag{7.53}$$

c is the distance between the object and the film (mm), and f the magnification factor of the photometer. In diffractometers the linewidth can be read directly in degrees, here the linewidth in radians is given by:

$$\beta_m = \frac{2\pi}{360} B^0 = 1.745 \cdot 10^{-3} B^0 \tag{7.54}$$

where B^0 is the linewidth in degrees.

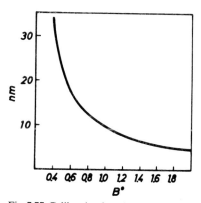

Fig. 7.77. Calibration function for crystalline particle size determination from the line broadening at $2\Theta = 14°$. Müller Mikro 111 diffractometer, slit width $1°$.

To solve frequently occurring problems quickly, graphical relationships can be established for a given line, instrument and slit system, so that the crystalline particle size can be read directly. Such an empirical function is shown in Fig. 7.77.

From a practical point of view it is important that the evaluation of linewidths is made easy in cylindrical cameras (such as in Debye–Scherrer cameras), since the formula is relatively simple in this case. Data processing is more complicated with flat cameras, and they are not therefore used for this purpose. Line broadening measurements are very simple with diffractometers.

The information content of X-ray interference lines are better exploited if the factors influencing the measured data are known individually. If experimental intensity values are denoted as $h(x)$ (where x is the spatial, here the angular coordinate), then, in order to determine the instrument distortion function, $g(x)$ spectrum analysis must be applied. Once the instrument distortion function is known, experimental curves can be improved by spectrum analysis, and the instrument effects can be removed mathematically.

This is a general problem in data processing, and several methods have been developed to solve it. Here we mention the method of Stokes (1948) which was developed before the advent of computers. This method makes use of Lipson–Beevers strips used in X-ray structural analysis.

Its essence is the determination of Fourier coefficients for the $h(x)$, experimental, and $g(x)$, instrument line broadening curves, so that the Fourier coefficients of the $f(x)$ functions, $(F(n))$ can be determined by simple algebraic division. If the Fourier coefficients of the undistorted $f(x)$ function are known, $f(x)$ can be calculated by inverse Fourier transformation.

Applying this, or some similar method, the undistorted lineshape can be found and the average crystalline particle size calculated from the Laue–Scherrer equation.

7.12.1. Determination of crystalline particle size distribution

Thus far the determination of the average value of crystalline particle size has been discussed. If more crystallographic reflections are studied, information can be gained about the average value in different [hkl] directions.

Bertaut (1949) developed a method to determine the distribution of crystalline particle size for catalysts. This method calculates the distribution function from the undistorted interference profiles obtained by Fourier analysis of the experimental curve. A similar approach has been used by

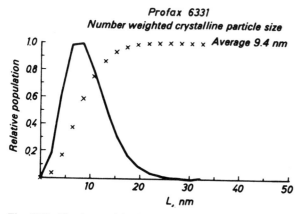

Fig. 7.78. Number weighted distribution of crystalline particle size in a Profax 6331 polypropylene sample.

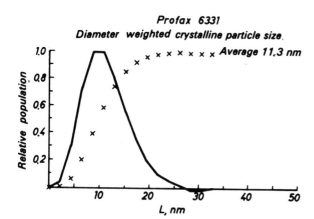

Fig. 7.79. Diameter weighted distribution of crystalline particle size in a Profax 6331 polypropylene sample.

Bodor and Füzes (1978) for polymers. The distribution of crystalline particle size is shown in Figs 7.78 and 7.79 for Profax 6331 polypropylene. In the first case weight factors are calculated from the number of particles, in the second from their diameter. Crystalline particle size distribution curves can be bimodal, an example for a polyamide sample is shown in Fig. 7.80.

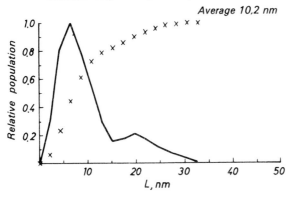

Fig. 7.80. Diameter weighted distribution of crystalline particle size in a MET 2 polyamide sample (bimodal distribution).

7.13. DETERMINATION OF CRYSTALLINITY IN POLYMERS

7.13.1. Definition of crystallinity (Crystalline mass fraction)

Earlier discussion has suggested that the partially crystalline nature of some polymers may be recognized from their X-ray diffraction patterns. By definition a crystalline material is one exhibiting three-dimensional order. The presence of crystallinity can be detected by X-ray diffraction and thermodynamic measurements.

The definition of the degree of crystallinity, that is, the ratio of crystalline and non-crystalline components, is as follows: if N different phases are present (for example one amorphous and several crystalline phases, such as hexagonal, cubic, etc.) then the mass fraction for the k-th type is given by:

$$\alpha_k = \frac{m_k}{m} \tag{7.55}$$

where m_k is the mass of the k-th phase ($j = 1$ to N) and m the total mass:

$$\sum_{j=1}^{N} m_j.$$

In the case of a two-phase model where only one crystalline (1) and the amorphous (2) phases are present, the crystalline mass fraction is:

$$\alpha_1 = x = \frac{m_1}{m},$$ (7.56)

where $m = m_1 + m_2$.

X-ray determination of crystallinity is based on measuring the intensity of X-rays scattered by crystalline, m_1, and amorphous m_2, parts of the sample, and is calculated as:

$$x = \frac{m_1}{m_1 + m_2} = \frac{m_{cr}}{m_{cr} + m_{am}}.$$ (7.57)

Since X-rays are scattered mainly by the electronic shell of the atoms, it is reasonable to assume that the scattered intensity is proportional to the number of electrons, that is, to the mass of the scattering phase. It follows that if the crystallinity is to be measured by X-ray diffraction, it is sufficient to compare the intensity of reflections scattered by the various phases, this means that the integral intensities must be calculated. Details of this calculation will be discussed later.

After suitable calibration, any physical method related to crystallinity can be used for crystalline fraction measurements, these include density, infrared absorption intensities, heat capacity, chemical reactivity or mechanical hardness.

It is important to know the number of different crystalline phases, in polyamides, for example, and frequently in polypropylene and polyethylene, several crystalline phases coexist. In these cases density measurements cannot be used since the densities of the different crystalline phases vary somewhat. Under these circumstances, besides the measured density value it is necessary to know the relative amounts of the different crystalline phases.

Crystallization processes can be monitored by thermoanalytical methods, so that the exothermal crystallization process can be detected by DSC. Sometimes crystallization results in very fine crystallites as in polyamides and their X-ray diffraction pattern is difficult to distinguish from the amorphous halo, since the line broadening effect is strong. Crystallization is, however, unambiguously detected by DSC. These samples are crystalline, despite the fact that, by current X-ray diffraction methods, interference lines overlap with amorphous scattering, and only an angular shift indicates the presence of crystallinity. The crystalline nature of these samples can be shown by a detailed analysis of the diffraction pattern, but this is indirect evidence. The DSC curves of such samples do not reveal any further crystallization, whereas the crystallization process is easily identified on DSC curves of amorphous polyamide samples on heating (Bodor 1969) under identical conditions.

Thus crystallinity can be detected by thermodynamic means, even though, owing to the small crystallite size, the interference lines are smeared out as a result of line broadening. In this sense thermodynamic criteria for the presence of crystallinity are more general than those based on X-ray scattering, that is, on the presence of three-dimensional order.

7.13.2. Resolution of scattering intensities of different origin

In order to evaluate crystallinity from X-ray diffraction curves three kinds of scattering must be distinguished. These are: background, amorphous and crystalline scattering. In practice, this means the calculation of areas under diffractometric or photometric curves.

7.13.2.1. Determination of background scattering

This scattered intensity originates partly from the thermal motion of the crystal lattice, and partly from Compton scattering which depends on the material type, but is independent of its physical state. Determination of this contribution causes problems mainly in photometric methods, since inadequate beam collimation causes the photometric curve to become ill-

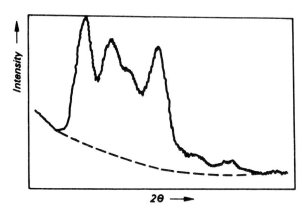

Fig. 7.81. X-ray diffractogram of polypropylene obtained in a Debye–Scherrer camera. Beam slit 0.5 mm.

defined at small angles. For diffractometric curves where photon-counters are used background noise can be calculated with greater certainty.

Included in the background scattering is that known as "white scattering" which is the result of mixed wavelength radiation from the X-ray tube, this appears close to the main radiation direction. White scattering is a function of the service time of the X-ray tube, and is caused by a thin tungsten layer being deposited on the anode from the incandescent filament. This form of scattering can be eliminated by effective monochromators.

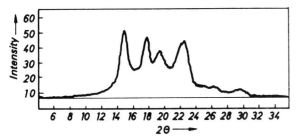

Fig. 7.82. X-ray diffractogram of polypropylene obtained by a diffractometer using a Geiger–Müller counter as a detector. Slit is 1°.

Use of diffractometers improves data processing possibilities. To demonstrate the difference, photometric and diffractometric curves of the same sample are shown in Figs 7.81 and 7.82. The wide angle diffraction range in polymers falls between 3 and 35°, and background scattering is readily measured by diffractometric methods at the extremes of this angular range. Determination of background scattering in films is problematic in the neighbourhood of the direct radiation (in the $2\Theta = 3$–$10°$ range), and much published work includes calculations only from angles exceeding $2\Theta = 10°$. This approach gives reproducible results, but neglects part of the amorphous scattering, and leads to mathematical problems, as the underestimated amorphous areas have to be corrected empirically.

7.13.2.2. Determination of the shape of amorphous scattering curves

Determination of amorphous scattering intensity is relatively simple in those angular ranges where no crystalline reflections appear, since all scattering above the background can be ascribed to amorphous areas. If, however, crystalline peaks are superimposed on the amorphous halo, amorphous scattering is usually taken as the midpoint minimum between the crystalline peaks. It will be shown later that this technique is incorrect.

Dependent upon the crystalline particle size, their interference lines overlap, so amorphous curves calculated from the contour of the inter-crystalline minima are, at least to some extent, influenced by crystalline scattering, and the crystallinity determined from these data will be incorrect. This problem will be returned to after presenting the basic ideas.

Resolution of total scattering into amorphous and crystalline parts assumes, of course, that only two phases are present in the polymer, and neglects any transitions. This approach is disputable, but is usually accepted as a first approximation.

Amorphous scattering of polymers is best measured in the molten state, or in some cases on quenched amorphous samples. The scattering pattern of amorphous polymers usually has two rings, but sometimes several more are apparent. Natta (1957) analysed amorphous scattering curves for several polymers with the conclusion that in most polymers there is a diffraction ring at $2\Theta = 19.5°$, which corresponds to the van der Waals distance. In addition there is another diffraction band with a position dependent upon the size of the side groups (in the case of polypropylene at $2\Theta = 16°$). The latter corresponds to an approximately hexagonal ordering of the molecular chains. Amorphous scattering for several polymers is plotted in Fig. 7.83.

There is evidence to suggest that amorphous scattering can be analysed better by electron than by X-ray diffraction. For example the X-ray diffraction curve for molten polyethylene exhibits only two interference lines, while in electron-diffraction measurements a third amorphous ring is observed at higher angles.

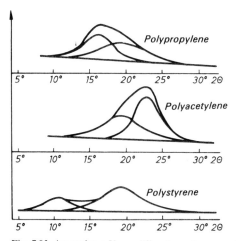

Fig. 7.83. Amorphous X-ray diffraction of some polymers. (After Natta 1959.)

Amorphous scattering is usually related to the stereo-regularity of the chain. In isotactic polypropylene, for example, a strong amorphous halo is observed at $2\Theta = 20°$, while in atactic one amorphous scattering dominates at $2\Theta = 16°$ (Bodor and Nagy 1968).

If the bandshape of the amorphous scattering is known, then the whole amorphous scattering curve can be constructed from a single experimental point, even if the crystalline diffraction lines overlap. Bandshape analysis for amorphous material is a crucial step in the determination of crystallinity, errors made at this point can lead to totally false results. All scattering in excess of the background and amorphous scattering can be ascribed to crystalline parts.

7.13.3. Calculation of crystallinity from amorphous and crystalline scattering

Data processing methods can be divided into two groups: in the first case a single diffractogram is sufficient, whereas in the second group at least two X-ray diffractograms are needed. Sometimes the first group is referred to as involving absolute methods, whilst the second uses relative ones. This distinction is not correct, since absolute and relative crystallinity values can be obtained by methods in both groupings.

7.13.3.1. Data processing from individual diffraction patterns

When crystallinity is calculated from interference patterns (or more exactly, from the areas under the diffraction curve) it is implicitly assumed that the total, integrated scattered intensity of the macroscopically isotropic sample is independent of the physical state. If the total crystalline scattering is denoted as T_{cr}, and the total amorphous scattering as T_{am}, then the crystalline fraction can be calculated from:

$$x = \frac{T_{cr}}{T_{cr} + T_{am}}. \tag{7.58}$$

Scattered intensity is proportional to the area under the scattering curve, the intensity of the individual lines, however, must be corrected by the Lorentz and other structural factors.

These geometrical correction factors take account of scattered intensity dependence on the scattering angle. If the Ewald concept is used, it is necessary to take into account the finite dimensions of the reciprocal lattice points. Their intersection with the Ewald sphere depends on the position, and velocity of the reciprocal lattice point: geometrical relations influence the intensity. In the case of rotation photographs, for example, the intensity depends on the angular velocity, that is, on the time spent by the finite volume reciprocal lattice "point" at a given position, or, in the case of powder photographs, it depends on the intersection area of two spheres.

For powder diffractograms these purely geometrical distortions can be corrected for by the Lorentz factor:

$$\frac{1}{\sin^2 \Theta \cos \Theta} . \tag{7.59}$$

As suggested earlier, diffraction can also be interpreted as a reflection from a set of parallel planes (the angle between incident and scattered radiation being halved by the normal to the [hkl] plane), and so a polarization characteristic of reflection appears, leading to reduced intensity. This phenomenon can be corrected for by the polarization factor:

$$\frac{1 + \cos^2 2\Theta}{2} . \tag{7.60}$$

When calculating the total scattered intensity, the measured intensity (or area) has to be divided by the above mentioned factors, so that:

$$T_{cr} = \Sigma f_i T_i, \tag{7.61}$$

$$T_{am} = \Sigma f_{i,am} T_{i,am} \tag{7.62}$$

where T_i is the area under the i-th line in the diffractometric curve (in arbitrary units), f_i the corresponding correction factor, $T_{i,am}$ the area under the i-th band of the amorphous diffraction curve (in the same units as T_i) and $f_{i,am}$, the corresponding correction factor.

Crystallinity determination should be based on diffractometric data taken over the full range of scattering angles from 0 to 360°.

Since this total scattered intensity is rarely available in experiments, data collection and processing is usually limited to a narrower angular range. As mentioned in relation to Fig. 7.8, the main scattering range for polymers extends to $2\Theta = 35°$. At angles higher than this value the intensity of scattering lines diminishes rapidly due to the thermal motion of lattice points, and they gradually merge into the background scattering. The ratio, Q, of intensities scattered by ideal and real crystals can be characterized by the Debye factor:

$$Q = \exp - 2B \left(\frac{\sin \Theta}{\lambda} \right)^2 \tag{7.63}$$

where B is a thermal factor, which is constant for a given crystal. B is related to the vibration amplitude, u, of the atoms, such that:

$$B = 8\pi^2 \overline{u^2}. \tag{7.64}$$

B determines the rate of intensity decrease of the reflections as a function of scattering angle. For polymers (in general for carbon compounds) B is close to 0.035 nm.

If it is assumed that the scattered amorphous and crystalline intensities outside the angular range studied are negligible (or their ratio is the same) then after correcting the measured intensities according to Eqs (7.61) and (7.62) the crystalline fraction can be calculated from Eq. (7.58). A good example of this approach to crystallinity determination was reported by Krimm and Tobolsky (1951) for polyethylene.

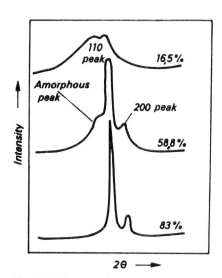

Fig. 7.84. Photometric curves of X-ray diffractograms taken from polyethylene samples of different crystallinities.

The X-ray diffractograms of three of the samples studied are shown in Fig. 7.84. The diffractograms refer to samples of different crystallinity. The two crystalline diffraction peaks are due to the [110] and [200] planes, while the amorphous ring appears close to the [110] line at a slightly smaller angle. Resolution of the diffraction pattern is based on the symmetric nature of interference lines (Fig. 7.85). Amorphous scattering was measured independently using molten samples. The centre of the amorphous scattering band corresponds to a 0.45 nm interplanar distance. The areas under the three component curves were determined planimetrically, and these intensities corrected according to the method established above. (The combined

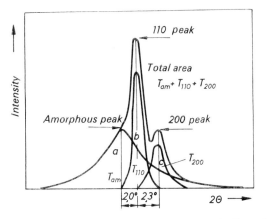

Fig. 7.85. Resolution of component curves on the X-ray diffractogram of polyethylene.

correction factors for polyethylene being $f_{am} = 0.69$, $f_{110} = 1.00$, $f_{200} = 1.43$.) The crystallinity from these three areas could then be calculated:

$$x\% = \frac{f_{110} T_{110} + f_{200} T_{200}}{f_{110} T_{110} + f_{200} T_{200} + f_{am} T_{am}} \cdot 100, \qquad (7.65)$$

$$x\% = \frac{T_{110} + 1.43 T_{200}}{0.69 T_{am} + T_{110} + 1.43 T_{200}} \cdot 100$$

where T_{am}, T_{110} and T_{200} denote the areas under the amorphous and two crystalline diffraction lines respectively (see Fig. 7.85).

Should the amorphous or crystalline scattered intensity falling outside the measured angular interval not be negligible, or if their intensity ratio is substantially different from that in the measured interval, then an empirical constant k, must be included in the crystallinity formula, such that:

$$x = \frac{T_{cr}}{T_{cr} + k T_{am}} \qquad (7.66)$$

where the value of k has to be established by measurements extending to the total angular range. The crystallinity of polypropylene has been determined by Natta et al. (1957) using this method.

Table 7.6

Characteristic lines of X-ray diffractograms of polypropylene

Interference line	Crystallographic index hkl	Diffraction angle $2\Theta°$ (Cu K_α)	Relative intensity I_{hkl}/I_{040}	Correction factor
C_1	110	14	2	3.06
C_2	040	17	1	5.18
C_3	130	18.8	0.8	6.89
C_4	111	21.2	2	10.30
	(131 041)	21.9		
Amorphous	—	16.3		6.9

After Natta 1959.

Table 7.6 contains the most important data of the diffraction lines taken into account in their calculations. Fig. 7.86 shows the diffractograms of their polypropylene samples of different crystallinity and the corresponding amorphous bands with height h_{am}.

The asymmetric amorphous band, which is the result of two overlapping peaks can be approximated by triangles. The base ends of these triangles are found at 10.8 and 25.3° respectively, while the height of the triangle, h_{am} is at 16.3°.

If measurements had been performed over the entire angular range, crystallinity could be determined using Eq. (7.58). However, acceptable results are obtained if the corrected intensity ratio is measured up to $2\Theta = 30°$. In practice measurements were performed only up to $2\Theta = 25°$, and crystallinity calculated from Eq. (7.66), where k is 0.9.

Another method which uses data measured over a restricted angular range takes the ratio of uncorrected crystalline and amorphous intensities calculated from data measured over a chosen angular interval which includes the majority of measurable reflections. Crystallinity data obtained this way has to be compared with independent results, for example, from density measurements. Such a method was used by Farrow (1961) to determine the crystallinity of polypropylene. Amorphous and crystalline ranges were distinguished by the minimum appearing between the [110] and [040] reflections. Crystallinity was calculated by Eq. (7.58) (without correction factors) and independently, using density measurements. Data calculated by these independent methods showed a systematic difference for stretched polypropylene fibres. With increasing extension ratios (and birefringence)

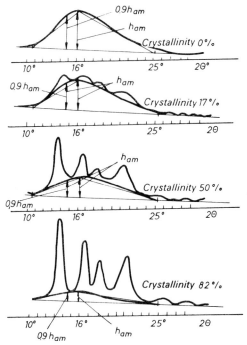

Fig. 7.86. Crystallinity determination from the X-ray diffractogram of polypropylene. (After Natta 1959.)

density values increased while the X-ray crystallinity remained constant (Fig. 7.87).

Farrow could not explain the inconsistency between the two crystallinity trends. According to present knowledge the probable cause was that the crystallite size became too small, as the average crystallite size decreased.

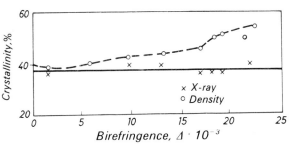

Fig. 7.87. Dependence of crystallinity on optical birefringence in polypropylene. Crystallinity data are taken from density and X-ray diffraction measurements. (After Farrow 1960.)

Owing to line broadening and smearing the amorphous area may appear to increase, or at least remain constant, so the apparent crystallinity calculated from X-ray data decreases, or remains reasonably constant.

The measurement of intensity ratios of this kind is relatively simple, they are therefore popular, but such methods yield only relative crystallinities. There is no guarantee that the intensity ratio in the chosen angular interval is equal to the corrected intensity values over the whole range. Crystallinities obtained in this way must therefore be checked, and compared with data obtained by other experimental methods.

It is interesting to note, and it proves the applicability of this approach, that there is no systematic difference between the crystallinities obtained for polypropylene by the Natta and Farrow methods. This, of course, cannot be generalized to other polymers, where scattering may be totally different.

7.13.3.2. Ruland's method for crystallinity determination

Ruland (1961) developed a method to determine crystallinity which will be discussed in some detail because of its importance. In connection with the reciprocal space the following quantity has been introduced:

$$s = \frac{d^*}{\lambda} = \frac{2 \sin \Theta}{\lambda}. \tag{7.67}$$

This is the characteristic parameter of the reciprocal lattice, which is independent of the applied wavelength. In the case of powder diffractograms reciprocal lattice points can be found on the surface of a sphere of radius s. The diffractometer sweeps the reciprocal space in one dimension, but for exact measurements, integration of the intensity should be performed in all three dimensions. If the intensity distribution is spherically symmetric, spherical polar coordinates can be used and the angular variables integrated in advance:

$$\int_0^\infty I(s) \, dv_s = 4\pi \int_0^\infty s^2 I(s) \, ds \tag{7.68}$$

and for crystalline intensity:

$$\int_0^\infty I_c(s) \, dv_s = 4\pi \int_0^\infty s^2 I_c(s) \, ds. \tag{7.69}$$

For a first approximation the crystallinity of the sample can be calculated as:

$$x = \frac{\int_0^\infty s^2 I_c(s) \, ds}{\int_0^\infty s^2 I(s) \, ds} \tag{7.70}$$

where x is somewhat lower than the exact value, since some of the crystalline scattered intensity contributes to the diffuse background owing to crystal defects and the thermal motion of lattice points.

To allow for these effects Ruland introduced a lattice defect parameter, D, into the intensity formula. This takes account of all intensity losses of crystalline reflections due to the non-ideal distribution of atoms.

The square average atom-scattering factor in the polymer studied can be calculated from:

$$\overline{f^2} = \frac{\Sigma N_i f_i^2}{\Sigma N_i}$$

(7.71)

where N_i is the number of atoms of the i-th type (determined by the stoichiometric formula), and f_i the atom-scattering factor.

Eqs (7.68) and (7.69) can then be rewritten in the form:

$$4\pi \int_0^\infty s^2 I(s)\, ds = 4\pi \int_0^\infty s^2 \overline{f^2}\, ds,$$

(7.72)

and

$$4\pi \int_0^\infty s^2 I_c(s)\, ds = 4\pi x \int_0^\infty s^2 \overline{f^2}\, D\, ds.$$

(7.73)

Division of these two expressions gives the crystallinity ratio:

$$x = \frac{\int_0^\infty s^2 I_c(s)\, ds \int_0^\infty s^2 \overline{f^2}\, ds}{\int_0^\infty s^2 I(s)\, ds \int_0^\infty s^2 \overline{f^2}\, D\, ds}.$$

(7.74)

This expression is applicable if there is no orientation, the order is three-dimensional, the integrals are valid for the limited range studied, crystalline peaks can be reliably and reproducibly resolved, and finally if the D-factor is known.

For first order lattice distortions (including thermal vibration) D can be described by a Gaussian distribution. For second order distortions the following expression is valid:

$$D = \frac{2\exp(-as^2)}{1 + \exp(-as^2)}.$$

(7.75)

Ruland found that the coefficient:

$$K = \frac{\int_{s_0}^{s_p} s^2 f^2\, ds}{\int_{s_0}^{s_p} s^2 f^2\, D\, ds}$$

(7.76)

if plotted versus s_p, the upper limit of integration, gave identical slopes when relating Gaussian curves with D curves calculated by Eq. (7.75). These results are shown in Fig. 7.89. All kinds of lattice distortion can therefore be described approximately by a single Gaussian lattice defect factor:

$$D = \exp(-ks^2),\tag{7.77}$$

where k has three components:

$$k = k_T + k_I + k_{II}\tag{7.78}$$

where k_T is the thermal contribution, and k_I and k_{II} are due to first and second order lattice distortions respectively.

Application of Ruland's method can be illustrated by the study of a series of polypropylene samples with differing thermal histories:

Sample 1. Isotactic polypropylene, quenched from the melt in room temperature water.

Sample 2. The same as Sample 1., but annealed at 105°C for 1 hour after quenching.

Sample 3. The same as Sample 1., but annealed at 160°C for 1/2 hour after quenching.

Sample 4. Heptane extract of raw polypropylene, highly atactic.

The experimental diffractometric curve was corrected for polarization effects, and the intensity data plotted versus s instead of 2Θ. Fig. 7.88 shows

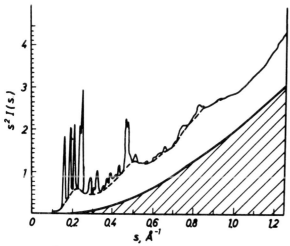

Fig. 7.88. $s^2 I(s)$–s plot of X-ray diffraction data obtained from a polypropylene sample. (After Ruland 1961.)

the $s^2 I(s)$ function plotted versus s for sample 3. The shaded area corresponds to incoherent scattering, and was calculated theoretically. The area under the dashed line and crystalline diffraction lines is due to coherent scattering. By integrating these areas, crystallinity can be calculated from Eq. (7.74), provided the constant K, in Eq. (7.76) is known.

Ruland constructed a nomogram for the determination of K. f^2 values were calculated from the chemical composition and $s_0 = 0.1$ taken as a lower limit in Eq. (7.76) gave s the K values as a function of s_P, the upper integration limit, using k as a parameter (Fig. 7.89).

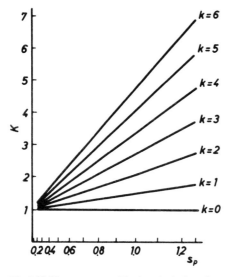

Fig. 7.89. Nomogram used in the calculation of K value using various k lattice defect factor values as parameters, independent variable is s_P, the upper integration limit. (After Ruland 1961.)

Ruland chose different k values and determined that which gave constant crystallinity results for a particular sample. Table 7.7 shows the calculated crystallinities for the four samples where $k = 0$ and 4, and using $s_P = 0.3, 0.6, 0.9$ and 1.2 as upper integration limits. He arrived at the somewhat surprising result that constant crystallinities could be obtained with $k = 4$ in all cases. It means that crystal distortion in these samples was very nearly uniform. This is a reasonable conclusion when it is taken into account that, in polypropylene, crystal distortions are small compared to the effects of thermal motion.

This assumption has been corroborated by Natta and Corradini (1960), who found the average isotropic thermal coefficient of the polypropylene

Table 7.7

Crystallinities in polypropylene as a function of k and the integration interval

| Interval | 1. Sample | | 2. Sample | | 3. Sample | | 4. Sample | |
$s_0 - s_p$	$k=0$	$k=4$	$k=0$	$k=4$	$k=0$	$k=4$	$k=0$	$k=4$
0.1—0.3	0.270	0.329	0.353	0.431	0.456	0.666	0.120	0.146
0.1—0.6	0.195	0.294	0.222	0.411	0.333	0.616	0.078	0.144
0.1—0.9	0.105	0.305	0.145	0.421	0.220	0.638	0.044	0.128
0.1—125	0.067	0.315	0.095	0.447	0.145	0.682	0.029	0.136
Average x		0.31		0.43		0.65		0.14

After Ruland 1961.

lattice to be $B = 0.075$–0.185 nm^2 (on the Å scale used by Ruland this is 0.75–1.85 Å^2). The coefficient, k, used by Ruland is related to B such that $B = 2k$, as shown by a comparison of Eqs (7.63) and (7.77):

$$Q = \exp\left[-2B\left(\frac{\sin \Theta}{\lambda}\right)^2 \right], \tag{7.63}$$

$$D = \exp\left[-4k\left(\frac{\sin \Theta}{\lambda}\right)^2 \right]. \tag{7.77}$$

As methods have been developed in fine structure analysis to determine B, k can be determined experimentally if the thermal motion is dominant.

The Ruland method is not used in routine studies, but his method has been the basis for several further investigations and simplified versions of the method are widespread.

7.13.3.3. Data processing by comparing diffraction curves

The application of this means of evaluation depends on whether the particular polymer can be obtained in the totally amorphous state or whether it is semicrystalline under all circumstances. Examples for the first group are natural rubber, isotactic polystyrene and poly(ethylene terephthalate), for the second group, cellulose, poly(tetrafluoro ethylene) or crosslinked polyethylene.

In the first case crystallinity is determined from the scattering curve of a semicrystalline sample using that of the amorphous condition, as shown in Fig. 7.90. The amorphous fraction can be measured directly if identical primary radiation and sample thickness can be achieved for both measure-

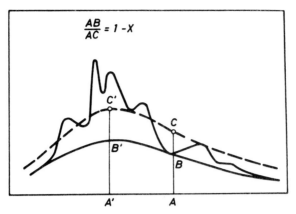

Fig. 7.90. Determination of the amorphous fraction if the fully amorphous polymer is available.

ments. In this case if the scattering curve of the totally amorphous sample is denoted by the dashed curve (running through points C and C'), and the amorphous scattered intensity of the semicrystalline sample by the continuous curve (a congruent curve running through points B and B' with similar shape). The amorphous fraction can be calculated as:

$$1 - x = \frac{\overline{AB}}{\overline{AC}}. \tag{7.79}$$

Field (1941) studied the crystallinity of various rubber samples by this technique. He measured the intensity of the amorphous halo for amorphous rubber samples under identical primary radiation conditions, as a function of the sample thickness (Fig. 7.91).

Once the calibration curve was known the d_{am} "amorphous equivalent" thickness for any semicrystalline sample with thickness d_{meas} could be determined and the crystalline fraction expressed as:

$$x = \frac{d_{meas} - d_{amorphous}}{d_{meas}}. \tag{7.80}$$

Of course, in this case standard conditions (equal X-ray power, exposure time, developing conditions etc.) have to be maintained.

Since crystalline areas develop at the expense of the amorphous component, Field developed his method somewhat further. An empirical calibration curve was constructed which related the ratio of the intensity maxima of the crystalline interference, I_{cr}, and that of the amorphous ring, I_{am}, to crystallinity (Fig. 7.92).

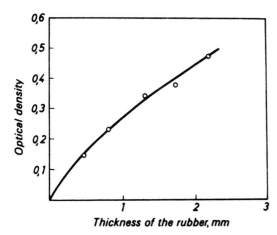

Fig. 7.91. Intensity of the amorphous ring of an amorphous rubber sample as a function of thickness. (After Field 1941.)

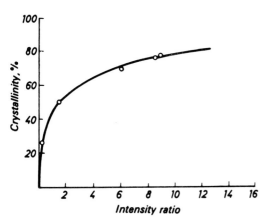

Fig. 7.92. Functional relationship between crystallinity and the I_{cr}/I_{am} ratio.

With this calibration curve the crystallinity of any rubber sample could be determined, even without standardized experimental conditions.

This method in this form is, however, appropriate only for the comparison of samples with similar crystalline particle size, since the integral crystalline interference line intensities are used as a measure of crystalline scattering. The maximum intensity is, however, a function of line broadening, and in the case of equal crystalline scattering they change in opposite directions, the area of the interference line being constant, but the height and linewidth can change.

Observed intensities can be made independent of the exposure conditions using equipment developed by Goppel (1947). The essence of his method is the application of a standard reference cell within the normal diffraction camera. The method is outlined in Fig. 7.93. The direct beam is

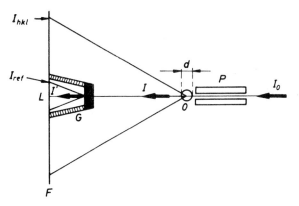

Fig. 7.93. Schematic view of the Goppel camera. I_0—intensity of the incident radiation; P—slit system; O—sample; d—thickness of the sample; I—intensity of the direct radiation passing through the sample; G—Goppel reference chamber; F—film; L—lead plate to absorb the direct radiation; I'—intensity of the direct radiation passing through the reference sample; I_{hkl}—intensity of the radiation scattered by the sample; I_{ref}—intensity of the radiation scattered by the reference sample.

viewed by both cameras and the interference patterns of the standard and investigated material appear on the same film. As a standard, Goppel used a mixture of $Al(OH)_3$ and $CaSO_4$. Using the Lambert law, where I_0 is the intensity of the incident beam, t the exposure time, d the sample thickness, and μ the linear absorption coefficient of the sample, the intensity of the amorphous halo in the totally amorphous sample can be expressed as:

$$I_a(d) = kI_0 t d e^{-\mu d}, \tag{7.81}$$

provided that Θ, the diffraction angle, is sufficiently small. The intensity of the amorphous ring in the partially crystalline sample, of total thickness d is given by:

$$I_{a,c}(d) = kI_0 t d_a e^{-\mu d} \tag{7.82}$$

where d_a is the equivalent thickness of the amorphous rubber. The intensity of one diffraction line of the reference material at small Bragg angles is:

$$I_{\mathrm{ref}} = k'I_0\, te^{-\mu_{\mathrm{ref}} d_{\mathrm{ref}}} e^{-\mu d}, \tag{7.83}$$

since the radiation passes through both the sample and the standard material before being diffracted.

Division of Eq. (7.82) by Eq. (7.83) gives the ratio which is independent of both the absorption of the rubber and the intensity of the incident radiation. It is, however, directly proportional to the effective thickness of the amorphous rubber, that is, to the amorphous rubber content of the sample:

$$\frac{I_{a,\,c}(d)}{I_{\mathrm{ref}}} = \frac{kd_a}{k'e^{-\mu_{\mathrm{ref}} d_{\mathrm{ref}}}} = k'' d_a \tag{7.84}$$

where k'' can be determined from the diffraction photograph of the totally amorphous sample, as here $d_a = d$. From a knowledge of the actual thickness and $\mathbf{k''}$, the amorphous fraction can be determined in any particular sample:

$$A\% = \frac{d_a}{d}\cdot 100. \tag{7.85}$$

Owing to technical reasons this method is limited to the flat camera technique. It cannot be applied in diffractometers with proportional counter detectors, since the reference camera cannot be positioned as required. Diffractometers are, however, much more stable devices, and the intensity of the primary radiation can be considered constant within experimental error. A further problem with this type of diffractometer is that the radiation does not pass through the sample, it is reflected from a relatively ill-defined depth, so that equal sample thicknesses in various samples cannot be ensured or checked. Since in Goppel's method only the amorphous scattering intensities are compared, and intensity measurements are sufficient, it is not necessary to integrate the individual crystalline interference lines, or to take into account the line broadening effect of crystalline particle size distribution. Because of low intensities, however, care must be taken that corrections due to incoherent scattering and the scattering of air should be made. Results obtained by the Goppel method are practically identical to crystallinities obtained from density measurements (Wildschut 1946).

If totally amorphous samples cannot be prepared, crystalline reflections must be included in the calculation of crystallinity. Such a method was developed by Hermans and Weidinger, originally for cellulose (1948), and later was extended to other polymers (1949, 1951, 1961). Diffraction photographs of various cellulose samples were prepared under standardized

conditions, and the assumption made that these samples could be characterized by different crystallinities. Fig. 7.94 shows the photometric curves of a
ramie and of a viscose fibre.

They have prepared small, macroscopically isotropic cylinders (height 2
mm, diameter 1.5 mm) from the fibres.

The left hand part of the curves (denoted as I_{ref}) is treated in the same
way as in the Goppel method, normalization being performed according to that

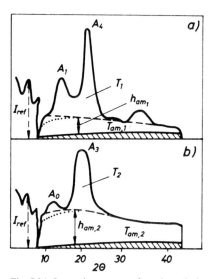

Fig. 7.94. Scattering curves of ramie and viscose fibres.

approach. The Figure shows scattering curves with amorphous areas which
exhibit maxima at $2\Theta = 18°$. The shaded areas originate from incoherent
radiation which were determined from the study of monocrystals of diamond
and sugar. From these curves characteristics were chosen which are supposed
to be proportional to the amorphous content, but the proportionality
constant is uncertain. Such characteristics are, for example, the area under the
amorphous scattering curve, T_{am}, or the maximum height of this curve, h_{am}.
Another independent datum was also chosen, which is also proportional to
crystallinity, but the proportionality constant is again unknown. Such a
datum is, for example, the total integrated intensity of the crystalline
scattering lines between $10°$ and $40°$, T_{cr}. Using two scattering curves the
unknown proportionality constants could be eliminated and their value
calculated. If the crystalline fractions of the two samples are x_1 and x_2, and
$h_{am,1}$, $h_{am,2}$ are proportional to the amorphous contents, $T_{cr,1}$ and $T_{cr,2}$ are

proportional to the crystalline content, and the following pair of equations can be written:

$$\frac{x_1}{x_2} = \frac{T_{cr,1}}{T_{cr,2}}; \quad \frac{1-x_1}{1-x_2} = \frac{h_{am,1}}{h_{am,2}}. \tag{7.86}$$

From these equations x_1 and x_2 can be determined.

Using this method the crystallinities of two samples with different crystallinities can be calculated from Eq. (7.86) and the scattering curves.

The method was developed further by Hermans and Weidinger in the sense that a single scattering diagram is sufficient if the process is standardized. That is, if the crystallinity diagram is constructed for the polymer. The essence of this is that several samples of the same chemical composition, but with different crystallinities should be studied (such as cellulose or polypropylene) and the results are plotted as T_{am} versus T_{cr} or as h_{am} versus T_{cr} diagrams. By doing so the characteristic values of the totally amorphous and crystalline samples can be established by extrapolation ($T_{am,\,100}$, $h_{am,\,100}$ and $T_{cr,\,100}$). Using these quantities:

$$T_{cr} = x\, T_{cr,100}, \tag{7.87}$$

$$T_{am} = (1-x)\, T_{am,100}, \tag{7.88}$$

$$h_{am} = (1-x)\, h_{am,100}. \tag{7.89}$$

Crystallinity can be obtained if Eq. (7.87), is divided by Eq. (7.88) or Eq. (7.89). If the measured T_{am} and T_{cr} values, and the extrapolated values are known, the crystallinity can be calculated from a single diffractogram without standardization.

$$x = \frac{T_{cr}}{T_{cr} + \dfrac{T_{cr,100}}{T_{am,100}}\, T_{am}} = \frac{1}{1 + k\,\dfrac{T_{am}}{T_{cr}}} \tag{7.90}$$

or for $h_{am,100}$

$$x = \frac{T_{cr}}{T_{cr} + \dfrac{T_{cr,100}}{h_{am,100}}\, h_{am}} = \frac{1}{1 + k'\,\dfrac{h_{am}}{T_{cr}}}. \tag{7.91}$$

Obviously Eq. (7.90) is formally equal to Eq. (7.66), when $k = T_{cr,100}/T_{am,100}$.

The crystallinity diagram (which plots the crystallinity versus the amorphous or crystalline areas) for polypropylene is shown in Fig. 7.95.

The relationship is linear with a correlation coefficient of 0.9987, and can be given by:

$$x = \frac{1}{1 + 1.297 \dfrac{T_{am}}{T_{cr}}} \, . \tag{7.92}$$

If, instead of the integral intensity of the amorphous curve, T_{am}, the height, h_{am}, is measured, then $k' = T_{cr, \, 100}/h_{am, \, 100} = 9.53$.

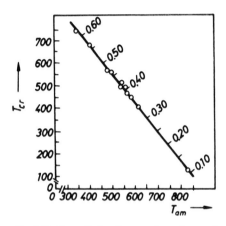

Fig. 7.95. Crystallinity diagram for polypropylene.

The most critical step of the method is the separation of amorphous and crystalline areas. In several cases, especially for cellulose derivatives, the amorphous area has been determined rather arbitrarily, the curve being close to a straight line. For synthetic polymers the measurement of the amorphous areas was consistent with the so-called absolute methods, thus carrying over errors associated with such techniques.

A further problem, which was pointed out earlier, is that if the total scattering of the samples with different crystallinities is constant in the angular range studied ($T_{cr} + T_{am} = $ const.), then Eq. (7.86) after some rearrangement reduces to Eq. (7.58) without correction factors. This method under such circumstances gives a simple area ratio. Under standardized conditions the constancy of the total area has been observed by Hermans and Weidinger in the case of cellulose, and a similar effect has been observed by Bodor et al. for polypropylene (1964).

7.13.4. Determination of crystallinity by density measurements

Determination of crystallinity by density measurements is based on the assumption that the two phases of polymers (that is, the crystalline and amorphous phases) can be characterized by well-defined, constant densities (or specific volumes); that there is no intermediate phase, and that the overall density can be calculated from those of the two phases by a simple linear rule.

The volumetric crystallinity is given by:

$$x_v = \frac{V_{am} - V}{V_{am} - V_{cr}} \cdot 100 . \tag{7.93}$$

and the mass crystallinity is given by:

$$x_m = \frac{\rho - \rho_{am}}{\rho_{cr} - \rho_{am}} \cdot 100 . \tag{7.94}$$

Volumetric crystallinity can also be determined from density data using:

$$x_v = \frac{\rho - \rho_{am}}{\rho_{cr} - \rho_{am}} \cdot 100 \frac{\rho_{cr}}{\rho} . \tag{7.95}$$

Besides the classical methods of density measurement, density gradient columns and electromagnetic methods have become increasingly popular.

The density gradient column is especially suitable for rapid density measurements on polymer samples. The essence of this method is that two liquids with different densities are poured into a glass column so that the lighter liquid is above. If the liquids mix with each other a density gradient results, and the density of the mixture increases continuously (practically linearly) from the top to the bottom. Such a density gradient is usually produced in a long glass tube with glass spheres at the ends, so that the density gradient is established between two large volume, constant density liquids (Fig. 7.96).

When a polymer sample is dropped into a density gradient column with the appropriate density range, it begins to sink as the density of the surrounding medium is lower than that of the polymer. On sinking further, however, the sample becomes surrounded by an increasingly dense liquid mixture, and it finally stops sinking when the density of the surrounding medium is equal to that of the polymer. The position of the sample can be measured exactly by a cathetometer, and the density of the liquid column is calibrated point-by-point using standardized glass spheres. The density

gradient column, must, of course be held at constant temperature, since the
density gradient is very sensitive even to minor changes in temperature.

The liquids to be used must meet a number of requirements. They must
be miscible, but must not act as swelling-agents for the polymer to be studied,
and their densities must be different (remaining within the density range to be
studied). Using this method densities can be determined to four significant

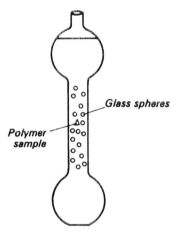

Fig. 7.96. Schematic of the density gradient column.

figures. Density gradient columns are extremely stable, and can frequently be
used for a year or more. The samples studied can be retrieved with a small net,
and the original density gradient is restored within a short time automatically
(Tang 1956; Wiley 1962).

The principle of electromagnetic density measurements is based on
Archimedes' principle. The weight of the sample is measured in air and in a
liquid. In the liquid the sample weight is reduced by the weight of the liquid
which is displaced by the volume of the sample, thus its volume can be
accurately measured. The ratio of weight and volume gives the density. The
name of the method stems from the fact that the weight of the polymer in the
liquid is determined by electromagnetic means. The sample to be studied is
placed in a small glass basket which is attached to a glass float. The glass rod
of the float contains a steel wire and is surrounded by a solenoid at the outer
surface of the vessel. The float, together with the rod and the basket exhibits
lower density than that of the liquid used in the measurement. If, however,
current flows through the solenoid, it attracts the steel wire in the rod, and the
float sinks. The current needed to move the steel wire depends on the weight of

the sample. This sensitive "balance" can be calibrated by plotting the weight versus the current, and so the weight of the sample placed into the glass basket can be determined. The scheme of the apparatus is shown in Fig. 7.97. This method determines densities up to five significant figures.

The density of the totally amorphous polymer, needed for crystallinity calculations, can be measured directly for some polymers, but many others

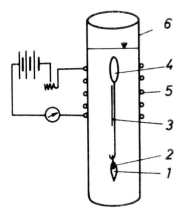

Fig. 7.97. Schematic of the electromagnetic density measurement. *1* sample, *2* sample holder basket, *3* steel wire, *4* float, *5* solenoid, *6* glass tube.

cannot be obtained in a totally amorphous form (examples are polyethylene and cellulose). In these cases the amorphous density is obtained by extrapolating the melt density, or by measuring the densities of model compounds. According to Kargin et al. (1957), fully amorphous polymers show a bundle-like structure, so that even these materials cannot be regarded as totally disordered.

The densities of fully crystalline polymers can be obtained from X-ray diffraction measurements, and are calculated from the the number and atomic mass of the atoms being present in the unit cell, and its dimensions:

$$\rho_{cr} = \frac{N M_{\text{atomic mass unit}} \times 1.660 \times 10^{-24}}{v \cdot 10^{-21}} \tag{7.96}$$

where ρ_{cr} is the density of the totally crystalline sample in g/cm^3 units, N the number of molecules present in the unit cell, M the molecular mass, and v the volume of the unit cell in nm^3 (Buerger 1960). For polymers M is taken as the molecular mass of the monomer (the repeat unit of the chain). Amorphous

Table 7.8

Crystalline and amorphous densities of some important polymers

Polymer	ϱ_{cr}	ϱ_{am}	Reference
Polyethylene	1.000	0.855	Matsuoka (1961)
Polypropylene	0.937	0.854	Miller (1961)
Polystyrene (isotactic)	1.111	0.947	Miller (1961)
Poly(vinyl alcohol)	1.345	1.269	Onogi (1962)
Poly(ethylene terephthalate)	1.455	1.335	Farrow and Ward (1960)
Polycarbonate (Bisphenol-A)	1.300	1.200	Prietschk (1958)
Polyamide-6	1.230	1.097	Bodor et. al. (1959)
Polyamide-6/6	1.220	1.069	Starkweather (1956)

and crystalline densities for the more important polymers are summarized in Table 7.8.

Crystallinity determination by density measurements knowing ρ_{am} and ρ_{cr}, is a relatively simple and rapid method. Its main disadvantages are that if more than two phases are present, and the densities of any crystalline modifications are different, it cannot be used; it does not give any information about crystalline size distribution, or orientation; and bubbles adhered to the surface or air inclusions, can easily lead to false density values. This latter effect can be avoided only by careful sample preparation and duplicate measurements.

When performing density measurements it is necessary to take into account the possibility of the formation of crystalline modifications with different densities, since this effect can substantially alter the calculated crystallinity values. If, for example, polyamide-6 is crystallized, a hexagonal, β, form can appear with $\rho_{cr} = 1.15$ g/cm^3. If, under the action of heat this transforms into the monoclinic α form, then the crystalline density changes to $\rho_{cr} = 1.23$ g/cm^3. As a result, annealing increases the density of polyamide-6. The question arises as to whether this change is simply a consequence of transcrystallization, or it is accompanied by an increase in the crystallinity.

Characteristic X-ray and density data for the hexagonal, monoclinic and amorphous phases are summarized in Table 7.9.

When measuring the density of such a three-phase system, the specific volume of the sample is the sum of three terms:

$$V = (1-x)V_{am} + x_h V_h + x_m V_m \qquad (7.97)$$

where V_h and V_m are specific volumes of the hexagonal and monoclinic crystalline phases respectively, whereas V_{am} refers to the specific volume of the

Table 7.9

Characteristic X-ray and density data for
polyamide-6

Modification	Characteristic X-ray interference lines	Density g/cm³
Hexagonal (β)	$d = 4.10$ Å	1.150
Monoclinic (α)	$d = 4.40$ Å and 3.70 Å	1.230
Amorphous	$d = 4.2$ Å (vague)	1.097

Bodor 1967.

amorphous phase, and x_h and x_m are the mass fractions of the hexagonal and monoclinic phases. By introducing $z = x_h/x_m$ and rearranging Eq (7.97):

$$x_m = \frac{V_{am} - V}{V_{am} - V_m + z(V_{am} - V_m)} \tag{7.98}$$

where z is unknown. Its value can be determined by spectrum analysis of the X-ray diffraction curve to be treated later. The X-ray diagram for the two crystalline forms can be easily distinguished (Fig. 7.98).

From the diffraction curve, and assuming that the area under the diffractogram is proportional to the intensity scattered by the the two crystalline types both z and the total crystallinity can be determined. The results of a typical series of measurements are shown in Table 7.10. From the

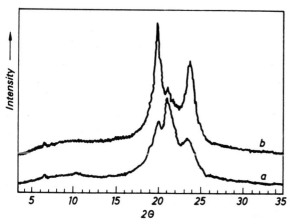

Fig. 7.98. X-ray diffractograms of polyamide samples. *a* Mainly hexagonal modification; *b* the same sample after annealing, mainly monoclinic modification. (After Bodor 1967.)

Table 7.10

Change of the hexagonal portion, specific volume and crystallinity in polyamide-6 on annealing

Annealing temperature	Hexagonal portion (from the total area) %	V	Z	Total crystallinity %
Quenched	100	0.8917		
90	65	0.8865	1.9	41.3
140	47	0.8818	0.89	42.2
170	36.4	0.8804	0.576	40.9
190	21.8	0.8782	0.280	39.8
210	3.4	0.8730	0.0356	40.2

data it can be concluded that, although the density increases on annealing, this change is mainly due to the hexagonal to monoclinic transformation process, and the total crystallinity remains effectively constant (Bodor 1967).

7.13.5. Survey of other methods for crystallinity determination

Of the other methods for crystallinity determination the study of infrared spectra is the most important. According to accumulated experience there are certain absorption bands on the infrared spectra of polymers which appear on crystallization and disappear on melting, these are known as crystalline bands. In some other cases amorphous bands appear, the intensity of these bands is greatest in the amorphous state and decreases with increasing crystallinity. The appearance of crystalline bands is probably due to the intermolecular interactions between atoms in the crystalline unit cell (Nikitin and Pokrovsky 1954). Crystalline bands can be used in relative, and amorphous bands in absolute crystallinity determinations. Crystalline peaks frequently show strong polarization under the effects of orientation, so that crystallinity determination in oriented samples using this method requires extreme care (Elliot 1959).

Crystallinity determination from amorphous bands is by simple proportionality. If the transmission in the amorphous sample is E_1, and that in the studied sample is E_2, then:

$$1 : (1-x) = E_1 : E_2. \tag{7.99}$$

Table 7.11

Absorption bands sensitive to changes in
crystallinity

Polymer	"Amorphous" band cm^{-1}	"Crystalline" band cm^{-1}
Polyethylene	1,300	731, 1,894
Poly(ethylene terephthalate)	790, 898	1,120, 1,350
	1,125	972, 848
Poly(tetrafluoro ethylene)	780	—
Polypropylene	—	1,170, 900
Polystyrene	—	1,980, 1,302
PVC	—	1,428, 1,290
Polyamide-6	—	930
Polyamide-6/6	1,136	935
Polyamide-6/10	1,136	853
Poly(vinyl alcohol)	—	1,146
Cellulose	910	1,430

After Kocskina 1968.

The crystallinity-sensitive absorption bands of some important polymers are
listed in Table 7.11. It has to be stressed, however, that for IR spectroscopic
crystallinity determinations calibration data are needed which have been
obtained by independent methods (such as X-ray diffraction or density measurements). Other problems may also arise, for example, the "crystalline"
bands can be due to the helical configuration of the chain. For these reasons
this is not a direct measure of crystallinity.

The calorimetry method of crystallinity determination has been referred
to earlier in connection with Eq. (5.9). The determination of the heat of
melting for the fully crystalline sample is a serious problem here, as such
polymers are always only partially crystalline. Such data can be obtained by
extrapolating the melting heats of low molecular mass crystalline homo-
logues for polyethylene (Dole 1952) and also for polyamides (Wilheit 1953).
The melting points and melting heats of the totally crystalline parts of several
polymers are now known and Table 7.12 lists these data for some important
polymers. Thermodynamically unequivocal evaluation of calorimetry
measurements is, however, complicated by the structural transformations
caused by heating.

Table 7.12

Melting points and corresponding melting heats of
some important polymers

Polymer	T_m °C	ΔH_{cr} J/g
Polyethylene	140	293
Polypropylene (isotactic)	180	138
Polypropylene (syndiotactic)	138	50
Poly(vinyl chloride) (syndiotactic)	273	180
Poly(vinylidene chloride)	190	—
Poly(ethylene oxide)	72	222
Poly(ethylene adipate)	65	121
Poly(decamethylene adipate)	79.5	151
Poly(ethylene sebacate)	83	138
Poly(decamethylene sebacate)	80	147
Poly(ethylene terephthalate)	270	126
Poly(tetramethylene terephthalate)	230	126
Poly(decamethylene terephthalate)	138	151
Polyamide-6	228	188
Polyamide-11	198	226
Polyamide-6/6	268	197
Polyamide-6/10	225	201
Polyamide-10/10	216	103
Poly(vinyl alcohol)	265	163
Poly(acrylonitrile)	317	96
Cellulose triacetate	315	—

After Fatou 1979.

Crystallinity can also be determined by NMR (nuclear magnetic resonance) spectroscopy, where atoms of the crystalline lattice exhibit broad, resonance peaks while those of the mobile ("liquid-like") phase are narrow.

This method will be treated in more detail in Chapter 8.

Another interesting method is based on chemical reactivity differences. In this method it is assumed that the amorphous areas will react with certain reagents, while the functional groups of the crystalline phase are resistant.

The results of such measurements depend, of course, on the amount of the reagent. During the reaction structural changes may occur in the crystalline phase (such as swelling) which can alter the crystallinity values. Such studies have been performed for example on cellulose (Goldfinger 1943; Nelson 1948).

There have been attempts to measure crystallinity from mechanical data. Wildschut (1946) determined the crystallinity of some rubbers from the temperature dependence of their elasticity.

Comparing all the different methods it can be concluded that those based on X-ray diffraction and density measurements are most widespread; in practice these (or their combination) prove to be most suitable for measurement of crystallinity in polymers.

7.13.6. Mathematical description of X-ray interference profiles

Study of crystallinity and crystalline particle size distribution is an important area for investigating polymer structure and properties. For example, Compostella et al. (1963) found that the properties of polypropylene fibres and films depend to a great extent on crystallinity, and on the relative amount of crystalline and smectic liquid crystalline phases. The smectic modification of

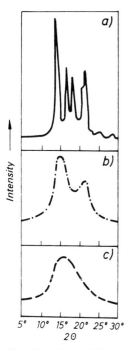

Fig. 7.99. X-ray diffractograms of monoclinic *a*, smectic *b* and atactic, amorphous *c* polypropylene samples. (After Natta 1959.)

polypropylene can be obtained from the melt by quenching. Fig. 7.99 shows the X-ray interference profiles for:
— isotactic crystalline;
— isotactic smectic liquid-crystalline;
— atactic polypropylene.

Better processing of X-ray interference profiles is made possible by the mathematical modelling of diffraction curves. For example, Bodor et al. approximated the diffraction lines by Gaussian curves and correlated the parameters of these with the properties studied. The basic notions of mathematical modelling of interference lines are discussed briefly, below.

As mentioned earlier in connection with Eq. (7.49), line broadening is characterized by the half-width of the bell-shaped curve (the line width relating to half of the maximum intensity), see Fig. 7.73.

Another possible width parameter is the integral linewidth. This parameter refers to the quotient of the area of the interference line and its height. This is shown in Fig. 7.74.

It is usual to write the integral linewidth as:

$$\beta = \frac{T}{I_0} \qquad (7.100)$$

where T is the area of the interference line, and I_0 the maximum intensity. The half-width parameter is written as $2w$ or B:

$$2w \equiv B. \qquad (7.101)$$

The difference between these two parameters is characteristic of the shape of the curve. Therefore, the division of the halfwidth by the integral line-width is called the shape factor.

The two curve types generally used are the Gauss and Cauchy (or Lorentz) functions. The analytical form of the Gauss function is:

$$G(x) = I_0 e^{\frac{-\pi(x-x_0)^2}{\beta^2}} = I_0 e^{-4(\ln 2)z^2} \qquad (7.102)$$

where $G(x)$ is the function value at point x; I_0 the maximum value of the function at $x = x_0$, so that:

$$z = \frac{x-x_0}{2w}. \qquad (7.103)$$

For Gauss curves the relationship between halfwidth and integral linewidth can be expressed as:

$$w_G^2 = \frac{\beta_G^2 \ln 2}{\pi}. \qquad (7.104)$$

The functional form of the Cauchy curve is:

$$C(x) = \frac{I_0}{1 + 4z^2} \qquad (7.105)$$

where z is given by Eq. (7.103), but in this case the halfwidth and integral width have the relationship:

$$\beta_C = \pi w_C. \qquad (7.106)$$

The value of the shape factor in the case of Gaussian curves is:

$$\frac{2w_G}{\beta_G} = \sqrt{\frac{4 \ln 2}{\pi}} = 0.939\,44, \qquad (7.107)$$

while for Cauchy curves it is given by:

$$\frac{2w_C}{\beta_C} = \frac{2}{\pi} = 0.636\,62. \qquad (7.108)$$

The area under the Gaussian curve can be given by:

$$T_G = I_0 \beta_G = I_0 \sqrt{\frac{\pi}{4 \ln 2}} \cdot 2w_G = 1.9788\, w_G I_0. \qquad (7.109)$$

whereas the same parameter for the Cauchy curve is:

$$T_C = \pi w_C I_0. \qquad (7.110)$$

In experimental work a convolution of the two types is often observed, which can be given by:

$$V(x) = \int_{-\infty}^{\infty} C(u) G(x - u)\, du. \qquad (7.111)$$

This can be solved quite easily using a computer, and is known as the Voigt function, which can be approximated by the analytical function (it will be referred to as the Voigt approximation function):

$$W(x) = \frac{I_0}{1 + \alpha 4z^2 + (1 - \alpha) \cdot 16z^2} \qquad (7.112)$$

where I_0 and z have been defined previously, α is the line-shape factor, its value varies between 0.743 and 1; the first value corresponds to the Gaussian limit, whereas the second refers to the Cauchy limit (Shirane and Cox 1962;

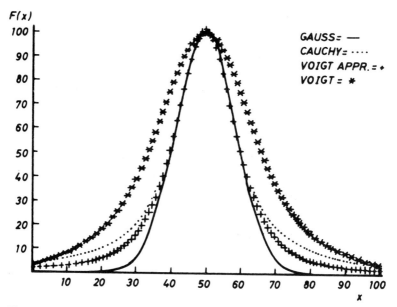

Fig. 7.100. Lineshapes of the Gauss, Gauchy, and Voigt approximation functions. Normalized to equal height.

Meisel 1971). The shape-factor of the Voigt approximation function is given by:

$$\frac{2w}{\beta} = 0.7792 \tag{7.113}$$

(Langford 1978).

Fig. 7.100 shows the functional form of these four curves calculated with normalized peak height ($I_0 = 100$) and the same linewidth parameter. It can be seen that the Voigt function is broader than the other three because of the convolution.

α, the line-shape factor of the Voigt approximation function was chosen to be 0.85. The effect of this parameter on the lineshape is demonstrated in Fig. 7.101. The upper curve refers to the pure Cauchy function, whereas the lower to the pure Gaussian function ($\alpha = 0.743$).

In practice the lineshape varies between wider limits than the Cauchy and Gaussian types. These limits can be described by the Pearson VII function which can be given by:

$$P(x) = \frac{I_0}{[1 + 4z^2(2^{1/m} - 1)]^m} \tag{7.114}$$

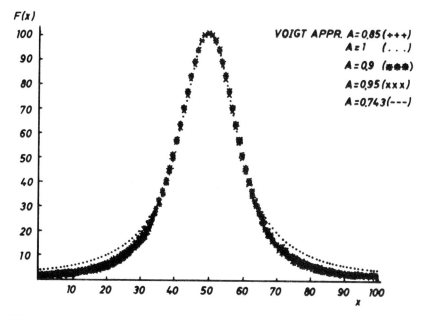

Fig. 7.101. The effect of the lineshape parameter on the form of the Voigt approximation function.

where m is a line-shape factor, $m=1$ refers to the Cauchy, and $m=\infty$ to the Gauss function (Fig. 7.102).

The mathematical modelling of distribution of scattered X-ray intensity has the advantage that if the intensity profile is approximated by an analytical function and its parameters are known, the crystalline particle size can be calculated quite easily.

This is especially important if, because of the superposition of the lines, only the envelope curve is known, so that the width of the component curves cannot be measured directly.

Under these circumstances the shape of individual curves can be calculated if the x_0 parameters are known. For example take the envelope curve of a sample exhibiting seven interference lines, so that:

$$G(x)= \sum_{i=1}^{7} G_i(x)= \sum_{i=1}^{7} I_{0,i}\exp\left[-4(\ln 2)z_i^2\right] \qquad (7.115)$$

where $G(x)$ is the intensity of the envelope curve at point x, $G_i(x)$ is that of the i-th component curve at the same point, and $I_{0,i}$ the intensity of the i-th component curve at point $x=x_{0,i}$:

$$z_i = \frac{x-x_{0,i}}{2w_i}, \qquad (7.116)$$

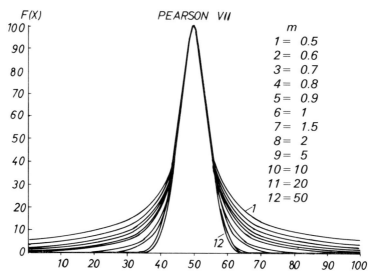

Fig. 7.102. Lineshape of the Pearson VII function.

where $x_{0,i}$ is the maximum position of the i-th component, and $2w_i$ the halfwidth of the corresponding line. Eq. (7.115) can frequently be used to describe X-ray interference profiles.

The mathematical description of interference lines is an interesting problem in itself. It has become particularly important since it became possible to simulate the diffraction profiles of samples with different crystallinities and varying crystalline particle size. Some examples of such simulation, and its application, are given in the subsequent sections.

7.13.7. The effect of crystalline particle size on X-ray interference profiles

This section considers the study of crystalline particle size by mathematical modelling. Other parameters (such as crystallinity, sample volume, and exposure time) must be kept constant. Two assumptions are made, these are:

1. The integrated intensity of the X-ray scattering for a given amount of material is independent of the phase composition, that is, independent of the amorphous or crystalline state. This assumption is usually accepted when evaluating X-ray crystallinity of polymers.

2. The line broadening effect of the crystalline particle size does not influence the area under the interference lines. If the interference line is described by the Gauss function, then the area under the line is given by:

$$T_G = I_0 \beta_G. \tag{7.109}$$

This area is constant, if line broadening occurs due to the decreased crystalline particle size, then:

$$I_0 \beta_G = \text{const} \tag{7.117}$$

and as a result, the height of the broadened line is lower.

Using these assumptions the X-ray diffraction profile of any sample with any particle size can be calculated. Fig. 7.103 shows the results of such

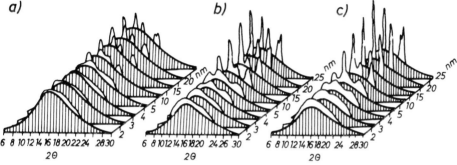

Fig. 7.103. Modelling the X-ray diffractogram of polypropylene. Dependence of the diffractogram as a function of crystalline particle size in three samples of different crystallinities.

calculations for polypropylene samples with different crystallinities (*a*, *b* and *c*) and crystalline particle sizes (2, 3, 4, 5, 10, 15, 20 and 25 nm respectively) assuming a Müller–Mikro 111 diffractometer (with a slit of 1°).

7.13.8. The effect of crystallinity on X-ray interference profiles

According to assumption 1. of the previous section, the integrated scattered intensity is a constant from a given amount of material. Thus changes in the scattered intensity originating from differing proportions of amorphous component are compensated by changes of the crystalline scattered intensity. Assuming identical exposure conditions, and constant mass of the sample, the total darkening of the film is constant. Using this condition, Bodor et al. have determined the shape of the interference curves assuming various crystallinity values.

Such calculations should be performed over the full range of scattering angles, as mentioned earlier. For polypropylene the $2\Theta = 6-30°$ range is, however, sufficient, since it appears that the total scattered intensity within this range remains constant on annealing.

This model takes into account the presence of α (monoclinic) modification only, but comparing Figs 7.103 and 7.99 it seems that there is no smectic liquid crystalline modification at all. Depending on the thermal history of the sample and on the resulting crystalline particle size the original 5 interference lines merge into 3, and finally 2 separately discernible lines. Below the 3 nm particle size an amorphous scattering curve is obtained. In practice, by quenching from the melt diffraction profiles for the 3 nm particle size can be obtained.

Based on the present model it seems probable that no distinct smectic phase exists, and that it is the crystalline particle size that plays a pivotal role in the polymer's properties. This interpretation is corroborated by the fact that the infrared spectra of the two "modifications" are identical, and thermal analysis shows no phase transition. It does not mean, however, that polypropylene has no other crystalline forms. It is well known from the work of Turner-Jones that when polypropylene is cooled from 190–230°C to 100–120°C, a β-phase appears.

A hotter melt quenched to below 90°C or above 130°C, results predominantly in crystals of the α-phase. Cooling slightly crystalline samples other than the α-form allows development of the γ-modification.

Using this model, it is clear that, for polypropylene, it is impossible to determine the height of the amorphous curve from the minima between the crystalline interference lines, and the error introduced by this procedure is inversely proportional to the crystalline particle size.

The essence of these conclusions is that the amorphous curve cannot be drawn, and the X-ray crystallinity analysis cannot be performed, if the crystalline particle size is not known.

For a realistic estimation of the amorphous scattering it is necessary to know the extent of the smearing of interference lines. These principles can be demonstrated on crystallinity measurements performed by other authors. For example, the data published by Kuribayashi and Nakai (1962) include measured densities and X-ray crystallinities, as well as crystalline particle sizes, determined from well-separated interference lines. If crystallinities are calculated from density data, and compared to those determined from X-ray scattering, a systematic difference is observed, which increases with decreasing crystalline particle size. The data calculated from the measurements of these Japanese authors are shown in Fig. 7.104.

If the diffraction profiles are analysed taking into account the effect of particle size, much better agreement is found between the crystallinities calculated from density values and X-ray data. Results from Bodor et al. on polypropylene samples with different crystallinities are plotted in Fig. 7.105.

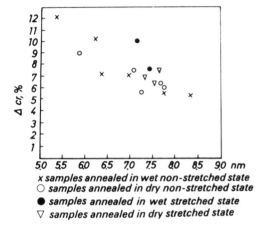

x samples annealed in wet non-stretched state
○ samples annealed in dry non-stretched state
● samples annealed in wet stretched state
∇ samples annealed in dry stretched state

Fig. 7.104. Difference between crystallinity values calculated from density and X-ray data as a function of crystalline particle size. (After Kuribayashi and Nakai 1962.)

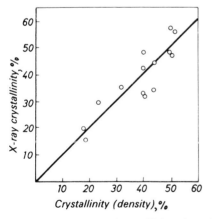

Fig. 7.105. Correlation of crystallinity values of polypropylene obtained from density and X-ray measurements. (After Bodor et al. 1964.)

Comparing these results with those shown in Fig. 7.87 it can be concluded that the differences observed by Farrow are probably due to the line-broadening effect, and to the superposition of interference lines, caused by the decreasing crystalline particle size.

7.13.9. Structural changes during the annealing of polypropylene

The analysis of experimentally observed X-ray diffraction profiles has led to some very interesting results. Probably the most important is that the crystallinity of unoriented polypropylene samples is not altered by annealing. Detailed analysis shows that total scattered intensity remains constant on annealing. The true amorphous content does not change, and the effect of the annealing process manifests itself in a continuous increase of the crystalline particle size. This is shown in Fig. 7.106 for an actual annealing process. It can

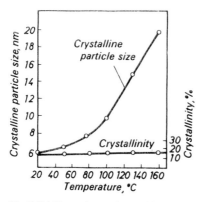

Fig. 7.106. Dependence of crystallinity and crystalline particle size on the annealing temperature. The sample was originally "smectic". (After Bodor et al. 1964.)

be seen from the Figure that while the crystallinity changes are only very slight, the crystalline particle size increases from 5.8 nm to 20 nm. During the annealing of polypropylene it is predominantly the crystalline particle size that changes, and this is the factor responsible for the changes observed in properties.

7.14. SMALL ANGLE X-RAY SCATTERING (SAXS)

There has already been discussion in the section on optical characterization of supermolecular structure, indicating that small angle X-ray scattering is a powerful tool in the study of such structures. At this stage, having become acquainted with X-ray methods, it is possible to take a closer look at the small angle X-ray diffraction technique.

According to the Bragg equation, the product of $\sin \Theta$, the Bragg angle and d, the interplanar spacing, is a constant, which is equal to half the wavelength of the radiation, so that:

$$\frac{\lambda}{2} = d \sin \Theta . \tag{7.106}$$

It follows that, if d is large, the interference appears at small angles. So that for d equal to 10 nm the diffraction appears at 0.5°, while for $d = 100$ nm, the diffraction angle is about 0.03°.

7.14.1. Various types of small angle X-ray scattering

Polymers produce two kinds of small angle X-ray scattering, known as diffuse and discrete.

Diffuse scattering appears in both solid and liquid phase polymers, including solutions and melts. This kind of scattering shows a maximum intensity at 0° and decreases practically to zero at between 1–2°.

Discrete scattering, however, can be obtained only from crystalline polymers. It is characterized by the presence of one or more peaks in the small angle scattering range. Usually one or more maxima appear at angles relevant to the 7.5–20 nm interplanar spacings.

Both kinds of scattering vary with any mechanical deformation of the polymer sample. Discrete scattering transforms from a ring into interference lines on the meridian which refer to a periodic structuring along the molecular chain axis.

7.14.2. Interpretation and evaluation of diffuse small angle scattering data

Diffuse scattering is due to inhomogeneities in electron distribution, the phenomenon being essentially similar to light scattering, which is also a consequence of inhomogeneity in the sample. Small angle scattering is proportional to the electron density differences of inhomogeneities.

The spatial distribution of the inhomogeneities is reflected by the scattering. Diffuse small angle scattering depends on the shape of the inhomogeneity of electron distribution, and also on its size and distribution. A homogeneous sample shows no small angle scattering.

The inhomogeneity of electron distribution can be detected and measured by small angle X-ray scattering, but the method cannot determine whether the scattering is due to systems with higher or lower electron densities than the average. This is the Babinet reciprocity relationship which states that, as far as the scattering pattern is concerned, it is not important whether it is the result of a particular structure, or of a complementary one (complementary from the point of view of electron densities). The nature of inhomogeneities cannot therefore be decided simply from the scattering diagrams. For this information from other sources is required.

Small angle scattering patterns are not so characteristic as the wide angle variety. Often very similar curves can be obtained from non-identical samples.

The angular range of small angle diffuse scattering is inversely proportional to the size of the inhomogeneities (such as inclusions or holes).

With respect to the quantitative evaluation of diffuse scattering it is necessary to distinguish between dilute and concentrated systems.

Inhomogeneities in dilute systems are distant from one another, and spatially disordered. Their interference therefore normally cancels out, and the resulting scattered intensity is the sum of that scattered by discrete individual particles.

Guinier (1927) has shown that the scattered intensity in this case varies with the scattering angle according to a Gaussian function:

$$I(\Theta) = I_0 \exp[-K\bar{S}^2\Theta^2] \tag{7.118}$$

where $K = 16\,\pi^2/3\lambda^2$, I_0 the intensity at $\Theta = 0°$, $\bar{S} = (\bar{s}^2)^{0.5}$, the scattering mass radius. This value for a given body can be obtained if the distances of the mass elements from the centre of gravity are raised to the second power and averaged. The square root of this average is the scattering mass radius. (This expression is equivalent to the radius of gyration discussed earlier with respect to the conformational properties of polymer chains.) If the logarithmic form of Eq. (7.118) is taken, then:

$$\ln I(\Theta) = \ln I_0 - K\bar{S}^2\Theta^2 \tag{7.119}$$

and it is clear that from the slope of the $\ln I(\Theta)$ versus Θ^2 plot (the Guinier plot) KS^{-2} can be obtained.

Fig. 7.107 shows the angular dependence of diffuse scattering, while Fig. 7.108 shows the Guinier plot. In this way the scattering mass radius, $\bar{S}$, of

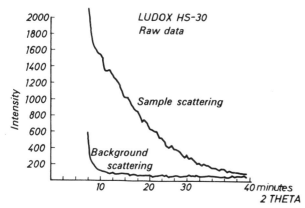

Fig. 7.107. Small angle diffuse pure particle scattering as a function of scattering angle. The sample is an aqueous solution containing glass spheres of 14.6 nm (Ludox HS 30). (Sample: courtesy of Du Pont de Nemours.)

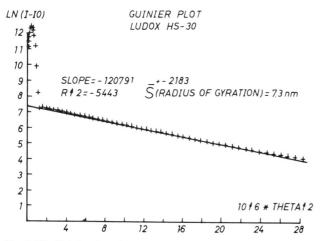

Fig. 7.108. Guinier plot of scattering data obtained from the Ludox HS 30 sample.

the scattering centres can be determined. The scattering function is relatively insensitive to the shape of the scattering centres, for example, there are only slight differences for a sphere and cube.

The intensity of diffuse scattering cannot be measured at $\Theta = 0°$, but it can be calculated from the extrapolation of the Guinier plot. The I_0 intensity is proportional to the volume of the scattering particles, or, if the scattering centres are molecules, to the volume of the molecules. This method gives the possibility of molecular mass determination if absolute intensities can be measured. Standard polyethylene samples can be obtained from several for absolute intensity measurements.

If the system is polydisperse, that is, it contains particles of identical shape but different size, then the scattering curve will be the superposition of the component curves. It is normal to process the Guinier plot by drawing successive straight tangential lines, and thus to determine the size distribution function by successive approximation. This approach is shown in Fig. 7.109.

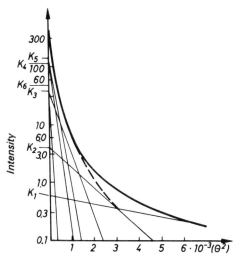

Fig. 7.109. Analysis of small angle X-ray scattering data obtained from a polymeric solution. (After Jellinek et al. 1946.)

Firstly the tangent with the smallest slope is determined with intercept K_1 on axis I_1, this is subtracted from the envelope curve. A new tangent is then drawn to the resulting curve with intercept K_2, and so on. The intercepts K_1, K_2, ... K_n are proportional to the mass fraction and the scattering radii, such that:

$$K_n \sim m_n(s)\bar{S}_n^2. \tag{7.120}$$

Dividing the K_n intercepts by $\bar{S}_n^2$ gives an approximate value for m_n, which reflects the polydispersity of the sample inhomogeneities.

Several attempts have been made in the past to determine the distribution of inhomogeneities. One example is the work of Brill et al. (1968). They derived an expression for spherical colloidal particles with diameters smaller than 100 nm:

$$\rho(a) = Aa^{-2} \int_0^\infty [h^4 I(h) - C_4] \alpha(ha) dh \qquad (7.121)$$

where $\rho(a)$ is the distribution function of the diameter, a, A is a normalization constant, and h the scattering vector of the reciprocal space:

$$h = \frac{4\pi \sin \Theta/2}{\lambda} = 2\pi s; \qquad (7.122)$$

$I(h)$ being the measured intensity, and:

$$C_4 = \lim_{h \to \infty} [h^4 I(h)], \qquad (7.123)$$

$$\alpha(x) = [(1 - 8x^{-2}) \cos x] - \left[4 - 8x^{-2} \frac{\sin x}{x} \right]$$

where $x = ha$. \qquad (7.124)

This equation was originally derived for spherical particles exhibiting no interactions. It can, however, with certain limitations be applied to inhomogeneous polymer systems as well.

Inhomogeneities in polymer systems include impact modifying elastomers, crazes in polymers exposed to mechanical fatigue, and polymer blocks in block-copolymers. Fig. 7.110 shows the Guinier plot of an impact

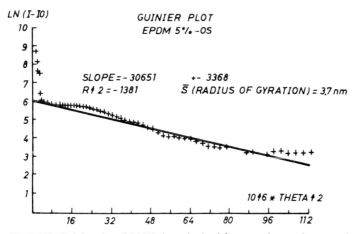

Fig. 7.110. Guinier plot of SAXS data obtained from a polypropylene sample containing 5% EPDM rubber.

polypropylene sample containing 5% elastomer. In this case the elastomer was EPDM rubber (ethylene-propylene-diene monomer copolymer). From the Guinier plot the radius of gyration is determined to be 3.7 nm. This, however, is only the lower limit of the particle size. According to the distribution analysis, inhomogeneities appear within a very narrow size range. The distribution function exhibits a sharp maximum around 7.5 to 8.0 nm (Fig. 7.111) (Bodor and Füzes 1979).

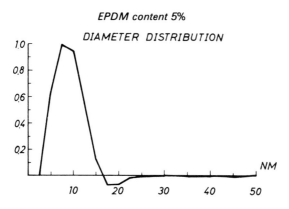

Fig. 7.111. Inhomogeneity distribution calculated from the SAXS curve of the PP sample containing 5% EPDM rubber.

For the quantitative evaluation of small angle X-ray scattering in lamellar semicrystalline polymers Vonk and Kortleve (1967) and Strobl (1970) have introduced new methods. They calculated the scattering intensity function of stratified amorphous–crystalline structures. The autocorrelation of the electron density function has been determined directly from the intensity data. This function is the correlation of small angle scattering which can be calculated from the one dimensional intensity by Fourier transformation:

$$\gamma_1(x) = \int_0^\infty I_1(s) \cos (2\pi x s)\, ds \bigg/ \int_0^\infty I_1(s)\, ds. \tag{7.125}$$

Fig. 7.112 shows a calculated correlation function of this kind. Curve a shows the correlation function of an ideal periodic lamellar structure with crystalline fraction $x = 0.2$ and two other correlation functions belonging to non-ideal models assuming a non-perfect periodicity (curves b and c). Curve d is for the case when transition layers appear between the two phases (Vonk 1982).

It is important to note that between points U and Y the function shows a negative value, which is equal to $-\left(\dfrac{x}{1-x}\right)$ when $x < 1/2$ and $-\left(\dfrac{1-x}{x}\right)$ if $x > 1/2$. This value can only be obtained from the correlation function if the structure is close to ideal lamellar, which is reflected by the $\overline{UY}$ section being a horizontal straight line. If this is not the case, and the experimental curve is similar to curve c, then the base line is equal to the minimum value below $\gamma_{\min}$. Vonk wrote a computer program to calculate the correlation function from experimental data and the software is freely available.

Strobl developed the method further, and calculated several other molecular parameters from experimental data. His method uses the correlation function to calculate the invariant, Q, the thicknesses of

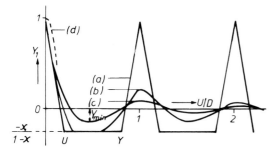

Fig. 7.112. Electron density correlation function of a lamellar model. (After Vonk 1982.)

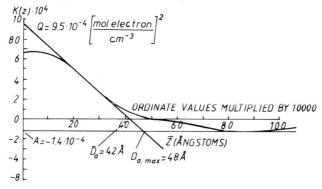

Fig. 7.113. Electron density correlation function of a PE.

amorphous and crystalline layers, D_a and D_{cr} respectively, the long period, L, and the specific interfacial area S/V. Fig. 7.113 gives a short overview of these calculated parameters.

Vonk (1987) proposed a method for the determination of the lamellar thickness in semicrystalline polymers by SAXS, from S/V, the specific surface area:

$$S/V = 2x/D_{cr} \tag{7.126}$$

where D_{cr} is the lamellar thickness of the polymer.

7.14.2.1. Scattering in concentrated systems

Diffuse scattering in solid polymers is due to the presence of microvoids. As mentioned earlier, this cannot be directly proven by scattering measurements. The presence of voids rather than high electron density inhomogeneities has nevertheless been demonstrated by gradually filling the void with a substance of higher electron density, such as iodine. In the course of such experiments it has been observed that the small angle diffuse scattering weakens initially, but at higher levels of iodine concentration it increases again. The period of weakening demonstrates that the electron density difference between the scattering area and the bulk at first decreases, arising the conclusion that inhomogeneities really are voids. When the iodine concentration is increased further, the electron-density difference increases, but at this stage the voids are filled with a material of higher electron density, and the small angle scattering is the result of an inverted inhomogeneous structure.

Voids are empty areas in the polymer (usually smaller than 10 nm) which are not filled by the polymer molecules. The volume fraction of these microvoids in viscose fibre for example is about 0.75%, and similar values have also been measured for other polymers. No connection has been found between crystallinity and the presence of microvoids.

The diffuse scattering pattern for drawn fibres shows a typical "diamond" shape, the apices appear on the meridian and the equator. Further stretching leads to increased scattered intensity on the equator, the scattering angular range extending in this direction. The scattering area in the direction of the meridian and the equator changes in parallel with the elongation of the microvoids under the effect of drawing. The extension ratio of the fibre is directly related to the elongation of the microvoids. A typical diamond shaped small angle X-ray diffractogram is shown in Fig. 7.114. The left hand diagram was obtained from a poly (acrylo nitrile) fibre, drawn in water at

Fig. 7.114. Small angle X-ray scattering of poly(acrylonitrile) fibre. *a* 8 fold stretching in water at 100°C; *b* 8 fold stretching on an iron edge at 121°C.

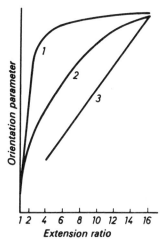

Fig. 7.115. Dependence of WAXS X-ray orientation *1*, IR-orientation *2* SAXS orientation (*S* ratio, see Eq. (7.127)) on the drawing ratio.

100°C, while the right hand diagram is from another poly (acrylo nitrile) sample drawn on a hot iron edge. The draw ratio was 8 in both cases. The intensity ratio, *S*, in the direction of the equator and the meridian is:

$$S = \frac{I_{equator}}{I_{meridian}} \tag{7.127}$$

and *S* changes with draw ratio as shown in Fig. 7.115. The Figure shows the orientation parameters obtained from wide angle X-ray scattering (WAXS) and infrared spectroscopic, IR measurements for comparison.

7.14.3. Discrete maxima in small angle X-ray scattering

Marked interference spots are frequently observed in the small angle scattering range for both natural and synthetic polymers. Fig. 7.116 shows as an example, the small angle X-ray diffraction photograph of collagen, where d = 62.7 nm. Synthetic polymers do not usually exhibit so numerous and well-defined interference maxima, but one or more discrete reflections frequently appear on diffraction photographs of semicrystalline polymers. These maxima demonstrate the presence of what is termed the "long period" in these polymers. Such reflections were noted on small angle X-ray scattering photographs for the meridian as early as 1944. According to current knowledge all fibrous polymers exhibit discrete small angle X-ray reflections, with three exceptions. These exceptions are poly(tetrafluoroethylene), isotactic polystyrene and poly(acrylo-nitrile).

Owing to second order distortions of the crystalline lattice in polymers, these small angle interference maxima are not very sharp, higher order reflections broaden and merge into the background scattering. If the statistical deviation is higher than 25%, only the first maximum can be detected. The line broadening effect of finite crystallite size also plays an important role in the resulting line broadening.

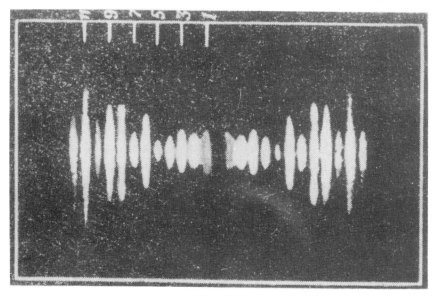

Fig. 7.116. SAXS diffraction pattern of a collagen fibre.

Many theories have been proposed to explain the presence of discrete interference lines at small angles.

It is now clear that the long period is not related to the chemical identity period. This is shown by the fact that the long period can be changed by annealing. The long period is also independent of the molecular mass, and cannot therefore be due to chain-end imperfections. The long period must be related to the crystalline state, since it has never been observed in amorphous polymers. Processing conditions, such as temperature, pressure and time, influence the long period substantially. These parameters influence the long period itself, the shape of the reflection, and its sharpness.

Hess and Kiessig (1944) assumed that long periods are the result of the regular periodic sequence of ordered and disordered areas, the long period corresponding to one crystalline + amorphous structural unit (Fig. 7.117).

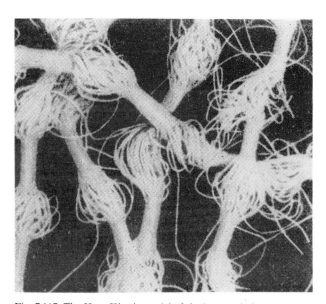

Fig. 7.117. The Hess–Kiessig model of the long period.

Statton (1967) found a direct relationship between the crystallite size determined by WAXS measurements and the long period for polyethylene fibres (Fig. 7.118) thus giving support to the Hess–Kiessig model. Subtraction of the crystallite size from the long period is proposed to give the thickness of the amorphous zone.

In the case of polymeric single crystals chain folding occurs, and the long period here is equal to the chain length between the folds. In fact, the thickness of these crystals is equal to the value calculated from the line broadening of the WAXS interference lines and to the long period. The other two dimensions of the single crystals, cannot, however, be detected by either of these methods.

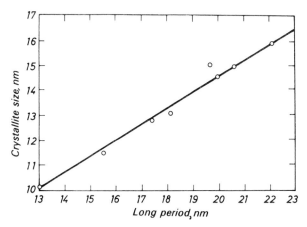

Fig. 7.118. Relationship between the crystalline particle size obtained from WAXS data and the long period in polyethylene. (After Statton 1967.)

Chain folding also occurs in bulk crystallized samples and fibres. In these cases small angle interference lines can be explained not only by the Hess–Kiessig model, but also by the presence of chain folding. The long period increases on annealing, as does the lamellar thickness of single crystals. An excessively long anneal produces a predominantly lamellar, chain folded structure leading to a lower strength, and brittle polymer.

7.14.4. Experimental technique for small angle X-ray scattering

Small angle X-ray scattering measurements are technically very complicated, since the intensity of the direct radiation is orders of magnitude higher than that of the scattered one. A very finely adjusted slit system is needed to obtain scattering curves of acceptable quality (Fig. 7.119). For collimation, elongated slit or circular hole systems are used. Fig. 7.120 shows the Kiessig system, while Fig. 7.121 gives that applied in the Kratky camera.

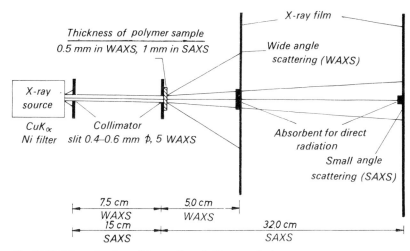

Fig. 7.119. Schematic view of the small angle X-ray scattering camera.

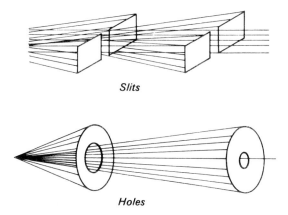

Fig. 7.120. Slit and hole systems used in small angle X-ray scattering measurements.

A specific feature of the Kratky camera is that, instead of slits, prisms of equal size are positioned close to one another determining the path of the radiation beam. Since these prisms can be machined with high precision, cameras of high resolution can be produced by this method.

The intensity of the scattered radiation at small angles is orders of magnitude weaker than that of the interference spots in the WAXS technique. Therefore the use of films for detection is relatively slow, with exposure times in the order of 100 hours. Cameras with proportional counters are, however, available, and these make a great contribution to faster data collection and processing. Scattering measurements are strongly disturbed by scattering caused by air, cameras are therefore built so that the path between the sample and the detector, at least, can be evacuated. Figs 7.107, 7.108, 7.110 and 7.111 were obtained using the step scanning method. This

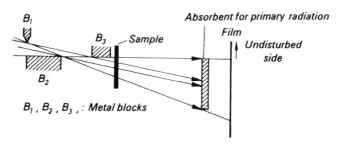

Fig. 7.121. Schematic view of the Kratky camera.

means that the detector measures the scattered intensity at a given position for a certain time (say, for 100 seconds) it is then stepped by a predetermined angular value (say, 1 minute), the measurement is afterwards repeated. Intensity data are stored in this case on a punched tape, and processed by a programmable calculator, or computer.

Statton (1964) presented a very good review of the SAXS method, and a detailed review of the WAXS and SAXS investigations, and light diffraction studies on polymer mixtures was published (Bodor 1981).

CHAPTER 8

SPECTROSCOPIC STUDIES ON POLYMERS

Spectroscopy is the study of the interaction of electromagnetic radiation with matter. The most important types of interaction are absorption and emission of energy quanta. If the frequency of absorbed or emitted radiation is measured, then energy changes can be determined. These energy changes can be used to reconstruct the discrete energy levels of the particular substance, these transitions yield several pieces of information about, for example, chemical composition, branching, crystallinity, orientation and crosslinking. Polymer spectroscopy is a branch of general spectroscopy, which will not be dealt with here, but some basic notions are discussed to make further treatment easier.

It is well known that electromagnetic radiation can be characterized by its frequency and velocity. The unit of frequency, v, in the SI system is the Hertz (Hz), its dimension is (s^{-1}). Frequency does not usually change even when the radiation, for example, light passes from one substance to another.

The velocity of light in a vacuum is 3×10^{10} cm/s, and if the refractive index of the medium is n, then:

$$c_n = \frac{c}{n}. \tag{8.1}$$

The division of velocity by frequency gives the wavelength:

$$\lambda = \frac{c}{v}. \tag{8.2}$$

The wavelength is the minimum distance between wave fronts vibrating in the same phase. In spectroscopy electromagnetic radiation, or light, is frequently characterized by the wavenumber, v^*, which is the reciprocal of the wavelength:

$$v^* = \frac{1}{\lambda} = \frac{v}{c} \, cm^{-1}. \tag{8.3}$$

The frequency range used in practical spectroscopic measurements is limited to between 10^8 and 10^{14} Hz.

In polymer spectroscopy absorption methods are normally used, as emission spectroscopy requires such high temperatures that most polymers would decompose.

When light is transmitted through the material, those components are absorbed which correspond to the energy difference between two states, E_1 and E_2:

$$hv = \Delta E = E_2 - E_1 \tag{8.4}$$

where h is Planck's constant ($h = 6.6254 \times 10^{-34}$ Js).

Energy scales. Various spectroscopic techniques are related to such energy differences. These energy differences are usually expressed in erg/molecule or J/mole units. Another energy unit is the electron volt (eV), which is the energy of a single electron accelerated by 1 V potential difference ($1 \, eV = 1.602 \times 10^{-19}$ J, 1 eV/molecule = 96.4 kJ/mole). Different types of radiation used in spectroscopy, together with their characteristic data, are listed in Table 8.1.

Spectrum lines for the generated polymer molecules are due to transitions between electronic vibrational or rotational states. The highest energy difference corresponds to changes in the electronic state, these frequencies appear in the visible or ultraviolet (UV) spectrum range. Energy

Table 8.1

Various kinds of electromagnetic radiation used in different branches of spectroscopy

Type of radiation	Mean wavelength	Mean wavenumber	Mean frequency	Energy $E = hv$	
Unit	cm	$v^* = \frac{1}{\lambda} cm^{-1}$	Hertz, s^{-1}	J	eV
Infrared	10^{-3}	10^3	$3 \cdot 10^{13}$	$2 \cdot 10^{-20}$	$1.24 \cdot 10^{-1}$
Visible	$5 \cdot 10^{-5}$	$2 \cdot 10^4$	$6 \cdot 10^{14}$	$4 \cdot 10^{-19}$	2.48
Ultraviolet	10^{-5}	10^5	$3 \cdot 10^{15}$	$2 \cdot 10^{-18}$	$1.24 \cdot 10$
X-ray	10^{-7}	10^7	$3 \cdot 10^{17}$	$2 \cdot 10^{-16}$	$1.24 \cdot 10^3$

differences between vibrational states are much lower. Transition frequencies in this case can be found in the near infrared (IR) spectrum range. Energy differences between the rotational changes are even lower, these transitions appear in the far infrared and microwave range.

If I_0 is the intensity of the incident light, and I that of the transmitted light, then the optical density (absorbance) is expressed as:

$$D = \log \frac{I_0}{I}. \tag{7.15}$$

Plotting absorption versus the wavelength or wavenumber, should identify absorption maxima so long as Eq. (8.4) is fulfilled. The degree of absorption (or extinction) is characterized by the optical density, D, which, according to the Lambert–Beer law, is proportional to the concentration of the absorbing units and sample thickness:

$$D = acd \tag{8.5}$$

where c is the concentration ($g/100$ cm^3), d the thickness of the sample (cm), and a is the specific absorbance (an earlier term was extinction coefficient).

The frequency positions of the absorption maxima yield information about the chromophore, that is, the absorbing group. For example, aromatic unsaturated double bonds exhibit absorption maxima in the UV frequency range. The concentration of these chromophore groups in the polymer can be determined using Eq. 8.5. UV spectroscopy can be used in some polymers which react with UV-absorbing substituents. In this case the degree of substitution can be inferred from the UV spectrum. UV spectroscopy is useful in the qualitative and quantitative analysis of aromatic stabilizers in polymers.

8.1. INFRARED SPECTROSCOPY OF POLYMERS

Rapid developments in IR absorption spectroscopy began after World War II, when reliable electronic amplifiers and radiation detectors became available. Nowadays IR spectroscopy is one of the most important and widespread structural characterization methods used in polymer science, mainly because of the simplicity of polymer identification. If the spectrum (or the absorbed intensity) is plotted versus the wavelength or wave number, the functional groups of the molecule give characteristic absorption bands with frequencies lying within a relatively narrow range, almost independent of the composition of the rest of the molecule. The relative constancy of these group frequencies makes possible the determination of characteristic functional groups in the molecule. The exact frequency values give further information about the neighbourhood of these functional groups. Thus details of the probable structure can be determined.

In most molecules several independent vibrational modes exist, and in practice, there are no two polymers which give identical vibrational–rotational spectra, the IR spectrum can then be regarded as kind of "fingerprint" of the molecule. In a similar way, for crystalline materials, the X-ray diffraction pattern can be regarded as a fingerprint.

Besides qualitative and quantitative analysis, IR spectroscopy can be used to monitor chemical changes in polymer molecules, for crystallinity and orientation measurements.

Absorption bands. It has been mentioned with regard to energy scales (see Table 8.1) that IR absorption bands belong to rotational–vibrational transitions. These transitions are analysed by quantum mechanics.

A molecule consisting of N atoms has 3N degrees of freedom and 3N wavenumbers ascribable from the equations of motion for the individual atoms. Three of these belong to the translation of the molecule, 3 to the rotation, whereas the corresponding wave numbers are zero. The number of independent normal vibrations is 3N-6. As an example, Table 8.2 shows the normal vibrations of the methylene group. Here the motion of 3 atoms is studied, and there are 9 degrees of freedom. The translational modes are not independent degrees of freedom with respect to the rest of the molecule, the rotational modes must, however, be taken into account as vibrational modes, as the C atom of the CH_2 group is regarded as fixed. These modes are methylene wagging, rocking and twisting. The total number of vibrational degrees of freedom is $3N - 3 = 6$, so that the methylene group can be vibrationally excited at 6 frequencies.

Not all of the possible vibrations appear in the infrared spectrum, and one cannot be sure that all vibrations are non-degenerate. With increasing symmetry the number of observable bands decreases because of degeneracies. In simpler cases the vibrational frequencies can be calculated by quantum mechanics methods.

Polymer molecules contain a very high number of atoms, but their IR spectra are relatively simple. This can be explained by the fact that the polymer (with the exception of end-groups) consists of identical monomeric units. To a first approximation it is assumed that the characteristic frequencies of the repeat unit are independent of the rest of the macromolecule. These group frequencies can be conveniently used in the evaluation of infrared spectra. The main absorption lines originate from the vibrations of

Table 8.2

Stretching and bending type vibration modes of the methylene
($-CH_2-$) group

Vibration mode	Direction of the transition dipole moment	Description
	↑	Symmetric stretching
	→	Asymmetric stretching
	↑	Scissoring
	⊙	Wagging
	None	Twisting
	→	Rocking

$+$ and $-$ denote the axis
perpendicular to the plane of the paper

the atoms and groups of the repeat unit. The unit cell of polyethylene, for example, contains 4 $-CH_2-$ groups. The maximum number of normal vibrations of the atoms in the unit cell is 36, of these, only 15 appear in practice, in the infrared spectral range.

Vibrational frequencies are not normally calculated for each group of atoms. Empirical rules and tables are used instead for the identification of

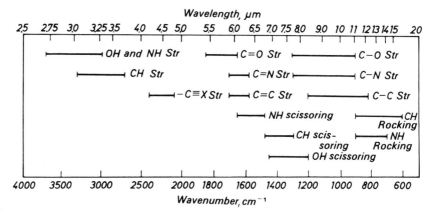

Fig. 8.1. Main vibration frequency regions of polymers as a function of wavelength and wave number (*str* = valence stretching).

absorption bands (Bellamy 1958; Nakanishi 1964; Szymansky 1963; Henniker 1967; Hummel 1978, 1984). As an example Fig. 8.1 shows the frequency ranges for some recurring vibrations in polymers as a function of wavelength (upper scale) or wavenumber (lower scale). Figures of this type can be used for a rough, first interpretation of absorption bands.

Before a more detailed discussion of spectra, a short review is given below on sample preparation and spectrometers.

8.1.1. Sample preparation for IR measurement

Both soluble and insoluble materials can be studied by this method. Soluble polymers are frequently studied in solution, the absorption of the solvent is compensated for. (The reference beam is passed through pure solvent.) Soluble or insoluble samples can be in film, powder, thin section, fibre or bulk forms.

Films can be transilluminated directly. The simplest method is to fix a film of suitable thickness using a magnetic plate in the path of the infrared beam.

The illuminated area is about 2 cm². The optimum film thickness depends on the chemical composition of the film and on the wavenumber used, usually ranging between 5 and 20 μm.

Powders are generally studied after mixing with an IR transparent powder and pelletization. Pellets are formed by pressing. Matrix material is usually K Br. The amount of sample needed for the measurement is 5–20 mg, the pellet is normally referred to as a K Br disc. Powders can also be studied in the form of a suspension, in, for example, paraffin oil.

Thin sections can be prepared by standard techniques used in microscopy. These can be studied in a similar way to powders.

Special microtechniques have been developed for studying fibres, but parallel winding is also used. Short cut fibres can be studied as with powders.

Bulk samples can be investigated using reflection or the attenuated total reflectance (ATR) technique. Fig. 8.2 shows schematic diagrams for the transillumination (*a*), reflection (*b*) and the ATR (*c*) methods.

In the case of the ATR technique the radiation penetrates a few microns into the sample at the interface of the sample and the optical element, it is then

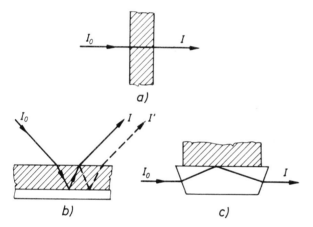

Fig. 8.2. Main methods for infrared investigation.

reflected back into the optical element. If the incident angle and the refractive indices are chosen correctly, the same beam is repeatedly reflected from the surface of the sample, and the sample absorbs some of the radiation at the corresponding wavelengths. Spectra obtained this way are, as a rule, weaker than absorption spectra, but they can be used successfully in the study of coatings. The optical material is usually silver chloride or K RS-5 (which is 42 wt% thallium bromide and 58 wt% thallium iodide).

8.1.2. Equipment

Various IR spectrometers are commercially available. The usual wavenumber range is 3000 to 250 cm^{-1} (3–40 μm). Better spectrometers can be used from 4000 to 200 cm^{-1}.

Double-beam spectrometers are normal. The main elements being the radiation source, monochromator and radiation detector, together with the recording unit. A schematic of such a spectrometer is shown in Fig. 8.3.

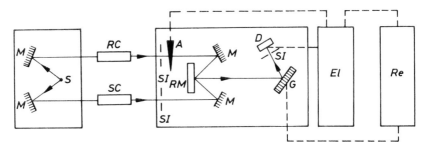

Fig. 8.3. Schematic view of a double beam infrared spectrometer. *S*—radiation source, *M*—mirror, *SC*—beam passing through the sample, *RC*—reference beam, *SI*—slit, *A*—optical wedge, *RM*—monochromator, *G*—optical grating, *D*—detector, *EI*—electronic amplifier, *Re*—recorder.

The radiation source is a Globar lamp, which is a silicone–carbide rod heated to 1200–2000 °C. Another popular radiation source is the Nernst lamp, which consists of rare earth metal oxides, the operating temperature being around 2000 °C.

The radiation is split into two beams, one passes through the sample, the other is used as a reference. In the case of solution studies the reference beam is passed through the pure solvent. The monochromator selects the correct wavelength. Earlier spectrometers used prisms for this purpose made of lithium fluoride, sodium chloride, potassium bromide, caesium bromide or caesium iodide.

Nowadays diffraction gratings are used more and more frequently. In both cases several prisms or gratings are needed to cover the total frequency range. The measurement and reference beams reach the monochromator and detector alternately.

The detector can be a thermocouple, bolometer or Golay detector. The latter is a metal box filled with gas, heated by the radiation. The gas changes volume under the effect of radiation which is recorded optically, using a mirror and photodetector. The signal from the photodetector is amplified electronically. The recorder unit can record the relative intensity of the two component beams, I/I_0, which is proportional to the transmission:

$$T\% = I/I_0 \cdot 100, \tag{8.6}$$

or the absorption percentage:

$$A\% = (100 - T\%) \tag{8.7}$$

or the optical density, D, as given by Eq. (7.15), also known as the absorbance, as a function of the wavenumber (v^*, cm^{-1}) or the wavelength (λ, μm):

$$D = \log I_0/I = \log \frac{1}{T}. \tag{7.15}$$

It is important to note that comparisons can be made only between optical densities, since $T\%$ or $A\%$ values are unsuited for this purpose.

Normal coordinate scales are compared in Fig. 8.4. In scientific publications some authors denote total absorption (zero transmission) at the bottom of the scale (Fig. 8.4a), where absorption bands appear as minima. Others place total absorption at the top of the diagram, so that absorption bands appear as maxima.

Absorption is the opposite of transmission, both expressed in percentages. Absorption is not identical with absorbance which is equal to log $1/T$. In order to avoid confusion absorbance is better known as optical density. The optical density relating to total absorption is infinite. Owing to practical, numerical reasons this absorbance is taken to be 3.2.

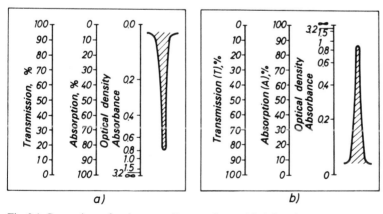

Fig. 8.4. Comparison of various coordinate scales used in infrared spectroscopy.

In modern publications abscissae are usually given in cm^{-1} (wavenumber) units. If, however, wavelengths are used, expressed in μm units, the data can be converted by using the formula:

$$v^* = \frac{10,000}{\lambda}.$$

(8.8)

For orientation studies polarized infrared radiation is used. Polarized beams are produced using transmission polarizers made of silver chloride or metallic selenium.

8.1.3. Qualitative structural analysis with the aid of IR absorption spectroscopy

Several atomic groups (such as —CH, —CO, —C$\equiv$N or —CH$_3$) can be identified relatively easily based on a single absorption line. The presence of other groups can be detected by the simultaneous appearance of several absorption bands. For example, the presence of vynilidene (H$_2$C$=$C$\diagdown$) groups is indicated by the simultaneous appearance of C—H, C$=$C and $=$CH$_2$ frequencies.

Identification of unknown polymers is facilitated by IR handbooks. The type of the polymer can be identified using small compilations (Nygist 1960; Haslam and Willis 1965), for closer identification, spectrum data banks consisting of 1900 spectra are available (Hummel 1984). The qualitative identification of polymers from IR spectra is discussed in detail in the book by Henniker (Henniker 1967).

Small but important differences between similar, but non-identical polymer samples can be studied by differential IR spectroscopy.

Additives (such as stabilizers and lubricants) are usually added in small amounts to the polymers. They have therefore to be concentrated (by extraction or selective dissolution) prior to spectroscopic investigation.

8.1.3.1. Polyethylene

Polyethylene is one of the most frequently studied polymers (Fig. 8.5). Assignment of the individual absorption bands is given in books and articles (Krimm 1960; Wood and Luongo 1961).

The concentration of methyl groups in polyethylene has been studied by several authors. Methyl groups can appear as end-groups of chain molecules or branchings. Determination of methyl group concentration thus gives

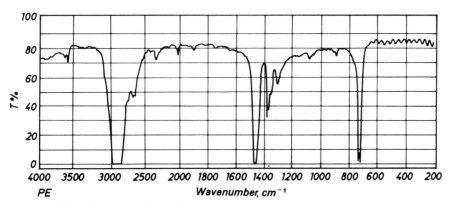

Fig. 8.5. Infrared spectrum of polyethylene.

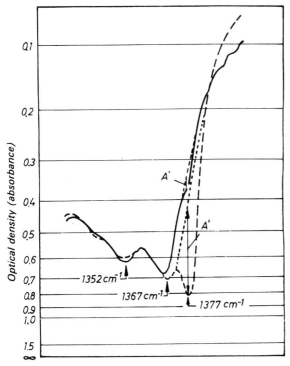

Fig. 8.6. Determination of chain branching from the analysis of infrared absorption spectrum in the neighbourhood of 1370 cm^{-1}.

information about the degree of branching, especially if the number average molecular mass is known. Physical properties of polyethylene highly depend on the linearity of the molecular chain.

The scissoring vibration of the methyl group in polyethylene appears at $1377\ cm^{-1}$ (7.25 μm). (It is disturbed by the $1367\ cm^{-1}$ peak of the methylene group, which has to be compensated for.)

Fig. 8.6 shows the IR spectrum of two polyethylene samples in the neighbourhood of $1370\ cm^{-1}$. The continuous line denotes 1.6, and the dashed line 34 methyl groups/1000 C atoms respectively. D is the optical density of the $1377\ cm^{-1}$ bands with respect to the base line which is the dotted line on the Figure (Reding and Lovell 1956). High pressure, low density polyethylene (LDPE) contains 20–30 methyl groups/1000 C atoms, while low pressure, high density polyethylene (HDPE) contains 3–10 CH_3 groups//1000 C atoms. The linear (or Phillips) polyethylene contains only 1—3 CH_3 groups/1000 C atoms (Fox and Martin 1940; Wilbourn 1959).

In polyethylene thermal or oxidative degradation can be monitored by measuring the absorption intensity in the 1300–2200 cm^{-1} range. At 1718

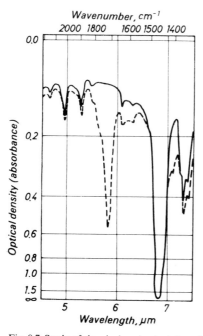

Fig. 8.7. Study of chemical ageing by infrared spectroscopy ———— original sample, — — — sample aged for two weeks.

cm^{-1} (5.82 μm) a strong carbonyl ($>$C$=$O) band appears under the action of oxidizing agents. Fig. 8.7 shows the IR spectrum of virgin and aged poly-ethylene samples in the relevant frequency range.

Crystalline bands appear at 731 and 1894 cm^{-1}, the amorphous band at 1300 cm^{-1} (see also Table 7.11).

8.1.3.2. Polypropylene

The IR spectrum of polypropylene is shown in Fig. 8.8. The infrared spectrum of isotactic polypropylene was studied, among others, by Luongo (1960) and McDonald and Ward (1961). Atactic polypropylene was investigated by

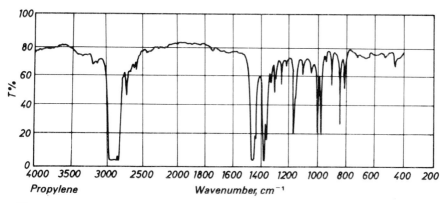

Fig. 8.8. Infrared spectrum of polypropylene.

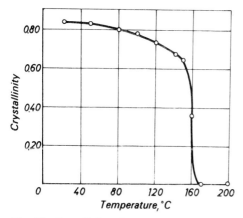

Fig. 8.9. Crystallinity of polypropylene as a function of temperature calculated from the D_{846}/D_{1171} absorbance ratio.

Immergut (1961). The characteristic of the syndiotactic modification is the appearance of the 867 cm⁻¹ band (Natta 1959).

On melting polypropylene, bands disappear at 1329, 1305, 1223, 1106, 1048, 1003, 944, 904 and 846 cm⁻¹ wavenumbers. Bands appearing at 1450, 1370, 1171, 978 and 890 cm⁻¹ do not depend on temperature. The first group can be used in crystallinity measurements, the latter as an internal standard. As an example the temperature dependence of IR-crystallinity is shown according to Heinen (1959) (Fig. 8.9).

The 977 cm⁻¹ band is useful in measuring the isotactic helix content, while the 867 cm⁻¹ band is characteristic of the syndiotactic helical details (Woodbrey and Trementozzi 1965).

Intensities of the bands appearing at 1256 and 1220 cm⁻¹ wavenumbers have been used in orientation measurements (Samuels 1965).

8.1.3.3. Polypropylene–polyethylene copolymers

Fig. 8.10 shows the infrared spectrum of a polyethylene–polypropylene block copolymer. Mixtures, blends and copolymers cannot usually be distinguished by IR spectra. This is possible only if the spectrum of stereo-regular polymers are studied, where the frequencies of *trans* and *gauche* isomers are different. This kind of isomery is discussed later with reference to Fig. 8.37. In these cases crystalline structure influences intermolecular forces, and frequency differences become more marked (Haslam and Willis 1965).

The similarity of the IR spectra of mixtures and copolymers can also be exploited, artifical mixtures can help in calibrating the method, and copolymer compositions can be determined.

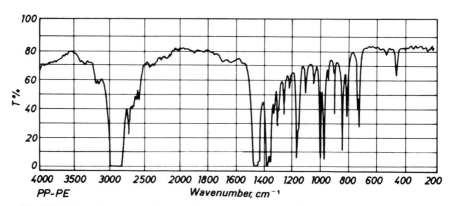

Fig. 8.10. Infrared spectrum of a polyethylene–polypropylene block copolymer.

8.1.3.4. Polystyrene

Fig. 8.11 shows the IR spectrum for polystyrene. This compound has also been studied by several authors, by Liang and Krimm (1958) and by Kämmerer et al. (1962), amongst others.

Isotactic polystyrene was first studied by Natta and Corradini (1955). The D_{560}/D_{543} optical density ratio was suggested by Takeda (1959, 1960) to be a measure of isotacticity.

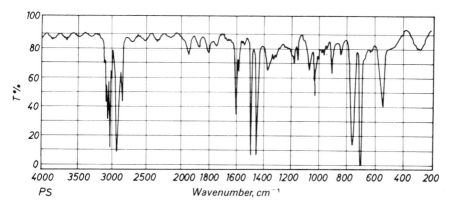

Fig. 8.11. Infrared spectrum of polystyrene.

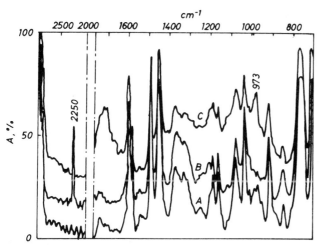

Fig. 8.12. Infrared spectra of polystyrene and styrene copolymers.

High impact polystyrene is produced by copolymerization or by mixing with elastomeric additives. Acrylonitrile gives the characteristic nitrile peak at 2250 cm^{-1}, while polybutadiene gives a characteristic peak at 973 cm^{-1}. Fig. 8.12 shows an example for normal and toughened polystyrenes. The acrylonitrile content can be determined easily, as its absorption is well separated from the other bands. Nitrile content as low as 1% can be detected. The measurement of the polybutadiene content is not so simple, as the absorption bands of polystyrene are also present in the same neighbourhood.

8.1.3.5. Poly(tetrafluoro ethylene)

The infrared spectrum of poly(tetrafluoro ethylene) (PTFE, Teflon) is shown in Fig. 8.13. In this Figure the ordinate is $T\%$ (transmission percentage). For comparison, the spectrum of the same material, in the same frequency range is shown in Fig. 8.14, but in this case the ordinate is optical density.

The chemical structure of the —CF$_2$— group is in certain aspects similar to that of the methylene group, so that PTFE is similar to polyethylene. Vibrational analysis was performed by Liang and Krimm (1956). Intermolecular forces between the (—CF$_2$—) chains are stronger than between the (—CH$_2$—) chains of polyethylene. As a result, —CF$_2$— wagging vibration appears at 638 cm^{-1} as opposed to 625 cm^{-1} in polyethylene.

Crystallinity determination methods for PTFE were developed by Pokrovskii and Kotova (1956) and by Bro and Sperati (1959).

Infrared spectroscopy, as we have seen, is a very powerful tool in polymer identification, for qualitative and quantitative group analysis. Crystallinity,

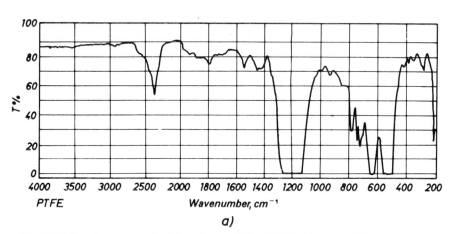

Fig. 8.13. Infrared spectrum of poly(tetrafluoro ethylene) (Teflon), in transmission.

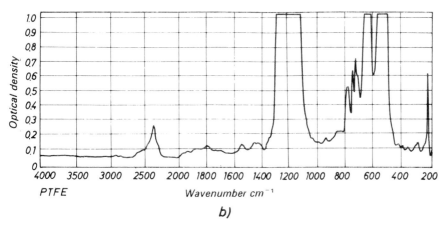

Fig. 8.14. Infrared spectrum of Teflon in optical density representation.

thermal and oxidative degradation can be monitored by the method. With polarized radiation orientation measurements can be performed. Polymer additives can also be identified. In well-equipped IR laboratories various data-bases (such as spectral atlas, punched cards, and magnetic disk systems) are available.

8.2. ESR SPECTROSCOPY OF POLYMERS

Atoms, atomic groups or molecules containing unpaired electrons (for example, paramagnetic materials and radicals) exhibit a nonzero electron–spin moment. When placed into a magnetic field of constant strength, H_0, the spin moments, under the effect of radiofrequency electromagnetic radiation, become aligned with the field. Since the relative direction of the field and the magnetic moment are restricted by quantum mechanics, rules (spatial quantization), the spin transitions are excited only by well-defined frequencies. These studies are referred to as electron-spin resonance (ESR) spectroscopic measurements. Resonance frequency and energy level differences can be calculated by the formula:

$$\Delta E = h\gamma = g\beta\mu_0 H_0 \tag{8.9}$$

where g is the spectroscopic splitting factor, β the magnetic moment of the electron, μ_0 the magnetic permeability of free space, and γ the frequency.

The main application of ESR spectroscopy is in the detection of free radicals. It has been shown that mechanical fracture produces free radicals, and chemists have succeeded in coupling low molecular mass paramagnetic molecules with otherwise ESR-inactive groups, this is the essence of the spin-labelling technique. The application of spin labels in studing the structure of polymers was discussed by Törmala and Lindberg (1976), and a review of the ESR study of synthetic spin-labelled polymers was published by Bullock and Cameron (1976).

8.3. NMR SPECTROSCOPY OF POLYMERS

8.3.1. The NMR phenomenon

Nuclear magnetic resonance (NMR) spectroscopy is a relatively new and powerful tool in the structural research of polymers.

The fact that certain nuclei exhibit different net magnetic moments (and, consequently spin) has been known for a long time (1924). The spin quantum number, I, of the atoms is related to their mass number, A, and atomic number, Z:

— if both A and Z are even, then $I = 0$;
— if A is odd and Z is odd or even, then I is equal to 1/2, 3/2, 5/2 etc.;
— if both A and Z are odd, I is equal to 1, 2, 3 etc.

Some frequently occurring isotopes, such as ^{12}C, ^{16}O or ^{32}S have no magnetic moment and cannot therefore be studied by NMR. Other nuclei, for example ^1H, ^{13}C, ^{17}O, ^{19}F do have magnetic moments, and are NMR active.

Table 8.3

Some important characteristics of nuclei studied by NMR spectroscopy

Isotope	NMR frequency		Natural abundance %	Relative sensitivity if equal number of nuclei are present	Magnetic moment μ	Spin (multiples of $h/2\pi$) (I)
	MHz at 14.092 kG	at 23.487 kG				
^1H	60.00	100.00	99.985	1.00	2.7927	1/2
^2H (D)	9.210	15.351	0.015	$9.65 \cdot 10^{-3}$	0.8574	1
^{11}B	19,250	32.084	80.42	0.165	1.8006	3/2
^{13}C	15.087	25.144	1.108	$1.59 \cdot 10^{-2}$	0.7022	1/2
^{14}N	4.334	7.2238	99.63	$1.01 \cdot 10^{-3}$	0.4036	1
^{17}O	8.134	13.56	0.037	$2.91 \cdot 10^{-2}$	—	5/2
^{19}F	56.446	94.077	100.0	0.833	2.6273	1/2
^{31}P	24.288	40.481	100.0	0.066	1.1305	1/2

After Bovey, 1972.

Some important data of nuclei frequently studied by NMR are summarized in Table 8.3.

If nuclei having magnetic moments are placed in a strong magnetic field, the magnetic moment, μ, of the nucleus tends to align with the field. This ordering is, however, accompanied by energy changes, so that orientation is limited to certain directions determined by energy quanta. This phenomenon is known as spatial quantization.

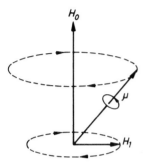

Fig. 8.15. Nuclear magnetic moment in an outer magnetic field. After Bovey 1972.

The nuclei, as small elementary magnets, do not become totally parallel with the field, but perform a precessing motion (like tops in a gravitational field) (Fig. 8.15). This is Larmor precession, where the radial frequency is ω_0 (radian/sec), frequency is ν_0 (Hertz), and the two quantities are related:

$$\omega_0 = 2\pi\nu_0. \tag{8.10}$$

The interaction energy between the external field, H_0, and the magnetic moment is described by:

$$E_m = -\frac{h}{2\pi}\gamma H_0 m \tag{8.11}$$

where γ is the gyromagnetic ratio of the nucleus, which can be calculated from:

$$\gamma = \frac{2\pi\mu}{Ih} \tag{8.12}$$

where h is Planck's constant; m the magnetic quantum number with possible values $m = I, I-1, I-2, \ldots, -I$, (total possibilities are $2I+1$), I the spin quantum number. If, for example, I is 1/2, the possible quantum numbers are 1/2 and $-1/2$. The spin quantum number of the proton is e.g. 1/2, i.e.

protons can occupy two positions in the magnetic field, either parallel with or perpendicular to it.

This is shown in Fig. 8.16, where a schematic description of the orientation process due to an external field is given. If H_0 is increased in an attempt to influence the degree of orientation, the real effect is to accelerate the precession of the nuclei. Orientational excitation can be facilitated by a

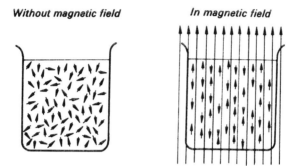

Without magnetic field   **In magnetic field**

Fig. 8.16. Ordering of protons in a magnetic field.

secondary magnetic field, which is perpendicular to the H_0 field, and rotates at a frequency of ν_0. This second field is marked as H_1 in Fig. 8.15. Such a rotating magnetic field can be produced by electromagnetic radiation.

If the frequency of the H_1 field is close to the precession frequency, ν_0, but is not equal to it, μ, the magnetic moment, remains unchanged. If, however, the rotational frequency of H_1 is exactly equal to ν_0, a strong oscillation is observed in the angle between H_0 and μ. A resonance phenomenon is observed, and energy is absorbed. The condition necessary for this quantum transition is, in radial frequency terms:

$$\omega = \gamma H_0, \tag{8.13}$$

or, expressing it using rotational frequency and Eq. 8.12:

$$h\nu_0 = \frac{\mu H_0}{I}. \tag{8.14}$$

If the frequency of the electromagnetic radiation is swept, transitions appear at well-defined frequencies as resonances. Such resonances can also be obtained by changing H_0. The energy difference between the two states can be expressed by the quantum conditions:

$$\Delta E = h\nu_0 = 2\mu H_0. \tag{8.15}$$

Table 8.4

Relation between H_0 (kG) and v_0 (MHz) values for various magnetic nuclei

Isotope	H_0					
	10	14.1	23.4	52	70.5	100
	v_0 value					
^1H	42.6	60	100	220	300	426
^2H (D)	6.5	9	15	34	46	65
^{13}C	10.7	15	25	56	75	107
^{19}F	40.0	56	94	208	282	400

As one can see from Eq. 8.15, v_0 is proportional to H_0 and μ. μ, the magnetic moment for the proton is 2.79, and for ^{13}C it is 0.702. The interrelation of H_0 and v_0 values for some important nuclei are given in Table 8.4. It can be seen that if the magnetic field strength is 14,100 G, v_0 for protons is 60 MHz. If the field strength is 23,400 G, the frequency is 100 MHz.

These are the most widespread field strength values. Newer spectrometers equipped with superconducting magnets can produce 52 kG field strength with a corresponding resonance frequency for protons of 220 MHz. Table 8.4 also includes 100 kG field strength, but such spectrometers are quite rare.

The frequency dependence of energy absorption can be detected as a spectrum line. Its position, intensity and shape depends on the intermolecular interactions, it can therefore be used in the study of such interactions.

8.3.2. The importance of NMR spectroscopy in the structural investigation of polymers

Nuclear magnetic resonance was first observed in solids in 1945. The first observation in polymers dates back to 1946, and in 1958 a comprehensive review was published on NMR studies of polymers (Slichter 1958). NMR has been used for stereochemical studies in vinyl polymers since 1960.

NMR studies on polymers can be roughly divided into two classes:

i) Nuclear relaxation measurements use the broad line NMR technique (usually proton relaxation). Solid samples are studied to clarify morphological details or mobilities. Amorphous content and chain orientation can also be determined in semicrystalline polymers.

In solutions temperature dependent relaxation times can be determined from the time dependence of the magnetization, which gives information about molecular motions in the polymer.

ii) By studying high resolution NMR spectra the sequential distribution of polymers can be determined. This is the only method which gives information about the sequence of consecutive structural units, and is the most important method in tacticity analysis and in submolecular structure determination of copolymers. NMR techniques have developed rapidly. The resolving power of the instrument has increased dramatically. New methods include cross polarization, and high power resonant proton decoupling techniques. Solid state studies by the magic angle spinning method make possible the solution of stereochemical problems, even if the polymer is insoluble.

Areas most important from the point of view of structural characterization are summarized in Table 8.5.

Table 8.5

Applications of NMR spectroscopy in the structural analysis of polymers

Structure

Structure of the repetition unit;
Structure and analysis of end groups and branchings, polymer tacticity (configuration);
Copolymer composition;
Order and sequence of monomeric units in the polymer;
Head-to-head and head-to-tail structures in polymers;
Crystallinity;
Orientation

Chemical information

Details of chain intitiation and propagation, including stereochemical mechanism;
Direct estimation of activity ratios for monomers;
Conditional probabilities of chain growth;
Identification of chemically active sites;
Kinetics

Physical information

Chain conformation;
Conformation transitions in solutions;
Equilibria;
Transition temperatures (T_g, T_m) in solid polymers (linewidth measurements)

8.3.3. The NMR spectrometer. Detection of resonance

Fig. 8.17 shows a schematic view of an NMR spectrometer. The sample is placed into tube A (diameter is about 5 mm). Tube A is placed into the magnetic field of electromagnet E (the field strength being between 14.1 and 52 kG). The Larmor, or precession frequency, is 60 MHz when the field strength is 14.1 kG, but a 52 kG it shifts to 220 MHz. The same frequency is produced by a radio frequency generator in the accelerator coil B. The axis of this coil is perpendicular to the direction of the static magnetic field. The

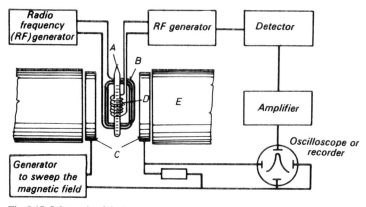

Fig. 8.17. Schematic of the NMR spectrometer. A—sample holder tube, B—radio-frequency coil, C—adjustable magnetic field, D—sensor coil.
After Bovey 1972.

magnetic field vector of the H_1 radio frequency field oscillates parallel to the sample tube, and perpendicular to the static field. To detect the resonance signal the magnetic field of smaller, ancillary electromagnets, C, is slowly varied until the resonance condition (Eq. 8.15)) is reached and resonance occurs. In this case the magnetic nuclear dipoles change from the lower energy state to a higher energy, and induce currents in coil D. This coil is perpendicular to field H_0, and to the direction of coil B. The induced current is amplified and recorded.

8.3.4. Spin–lattice relaxation

The energy difference between the magnetic levels is very small. In the case of protons in a field of 14.1 kG it is just 0.08 Joule, the value of $2\mu H_0/kT$ being about 10^{-5}. Expressing the distribution of spins between the states the ratio for protons can be expressed as:

$$\frac{N_+}{N_-} = 1 + \frac{2\mu H_0}{kT} \tag{8.16}$$

where N_+ and N_- denote the relative probability of the spins in the higher and lower states respectively, T is the Boltzmann spin temperature.

At a temperature of 300 K, and 14.1 kG field strength, the ratio of parallel to perpendicular protons is 1,000,010 to 1,000,000, so out of 2 million nuclei, only 10 excess parallel spins lead to a weak magnetic polarization of the sample. If the spin of the nuclei exceeds 1/2, there will be $2I + 1$ energy levels with equal energy spacings of $\mu H_0/I$. Relative populations can be described by equations similar to Eq. (8.16).

Under the influence of the H_1 field, spins are excited from a lower to a higher energy state corresponding to a high Boltzmann temperature. If there is a relaxation in this higher energy state, this corresponds to a cooling of the spin system. Similarly, when the sample is first placed into the magnetic field, spins are randomly oriented, and a relaxation phenomenon occurs until the equilibrium distribution corresponding to the H_0 field is established.

Since the energy differences involved are very small, the spin temperature can be much higher than the actual, observable sample temperature. The heat capacity attributed to the spins is very small.

The relaxation process is always strongly influenced by the environment of the spin. This environment is generally termed the "lattice", which can be

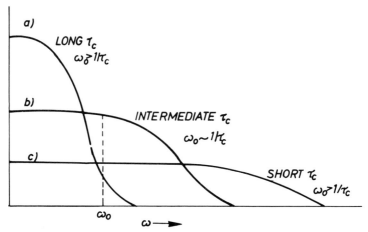

Fig. 8.18. Frequency spectrum in materials of *a* high, *b* medium and *c* low viscosity. (Courtesy of F. A. Bovey.)

liquid or solid condensed material. Each spin is sensitive to the presence and position of several other spins in its neighbourhood. If the fluctuating local field also has an H_1 component with v_0 precession frequency, the same transitions can occur between the magnetic states as under the effect of the original H_1 field. In solid materials and liquids of high viscosity, molecular motions are slower, and the frequency spectrum can be described by curve *a* in Fig. 8.18. If, on the other hand, relaxation is studied in a low viscosity fluid the frequency spectrum becomes flattened owing to the enhanced molecular mobility, as shown by curve *c* in Fig. 8.18. The thermal relaxation in liquids of medium viscosity can be described by curve *b*.

If the actual population difference between the states at time t is n, and n_{equ} denotes the equilibrium population difference, then:

$$n = n_{equ}(1 - e^{-t/t_1}) \tag{8.17}$$

where t_1 is the spin–lattice relaxation time. This is the time in which the equilibrium spin-population difference decreases to n_{equ}/e, where e is the base of the natural logarithm.

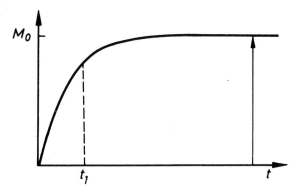

Fig. 8.19. Time dependence of spin polarization. t_1 denotes the spin–lattice relaxation time.

Fig. 8.19 shows the exponential–asymptotic increase of the polarization to the equilibrium value, and the position of t_1 on the time scale. t_1 is frequently referred to as the longitudinal relaxation time, since the process is accompanied by energy changes, which modify the magnitude of the nuclear magnetic momentum component parallel to the applied static magnetic field. Later t_2, the transverse relaxation time, will be introduced, and Fig. 8.27 shows the orientation dependence of the magnetic moment change related to the spin–lattice relaxation.

Nowadays experimental linewidths are at least 0.3 Hz, which corresponds to about 0.5 sec spin lifetime. t_1 usually falls in the order of 1 and 10 sec in liquids.

8.3.5. Chemical shift

In the case of a single nucleus (such as a proton), and a fixed v_0 frequency, not all protons show the resonance phenomenon at the same H_0 magnetic field strength. This fact renders NMR spectroscopy an important tool for structural investigation. The H_0 value where resonance appears is different for every type of proton, depending on the chemical bond and intramolecular position. In organic compounds this variation of H_0 falls in the order of 10 ppm, that is, it is 600 Hz in the case of a 14.1 kG magnetic field. This change in the resonance field strength can be attributed to the electronic shell around the nuclei. When a molecule is placed into a magnetic field, it induces currents in the molecule. According to Lenz's law this current results in a small local field of opposite sign to H_0, the inducing field, even when the number of electrons is even, and there is no net magnetic dipole moment of the molecule.

This phenomenon explains the diamagnetic behaviour observed in all types of molecule. All nuclei are partially shielded by the electrons and, in practice, the H_0 field needed for resonance is somewhat higher than it would be expected for a "bare" nucleus.

Electronegative groups such as —OR, —OH, —OCOR, or halogens, attract the shielding electrons and the resonance appears at lower H_0 fields. This is the reason for the fact that, in NMR spectroscopy, there is no natural zero-point or scale unit. The following two conventions are usually accepted:
   — the scale unit is the relative change of the H_0 field in ppm (parts per million) units;
   — an arbitary reference material is chosen to fix the zero point and all resonance shifts are relative to this standard. These resonance shifts are also referred to as chemical shifts. The advantage of this dimensionless scale is that chemical shifts are made independent of the H_0 value used in the actual spectrometer, which would not be true for gauss or Hz scales.

As a reference material tetramethyl silane (TMS) has been used since 1958. This is a volatile liquid, which is chemically inert, boils at 20 °C, and can therefore be easily removed. It contains 12 equivalent protons resulting in a single, strong resonance line, and there are usually no other absorption lines in its vicinity.

On the so called τ-scale the τ-value assigned to tetramethyl-silane is 10.00, as its protons are usually more strongly shielded than the protons of other organic compounds. The chemical shift values of protons on this scale usually lie between 0 and 10. On another arbitary scale the TMS value is the zero point. This is the δ-scale, which is related to the τ-scale, such that $\delta = 10 - \tau$.

Fig. 8.20 shows the chemical shift values of some organic protons in carbon–tetrachloride solvent.

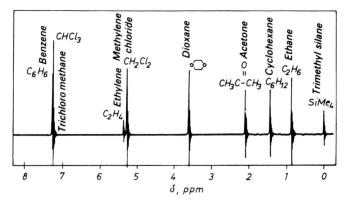

Fig. 8.20. NMR signals of some compounds which exhibit only one line with different chemical shifts.

8.3.6. Intramolecular spin–spin coupling

Nuclei in the molecule interact through chemical bonds, resulting in partial spin-polarization and a deformation of the valence electron orbitals. If a particular nucleus has n identically bound neighbours, then the resonance line splits into $n+1$ lines, according to the $n+1$ possible spin states.

One neighbouring spin therefore causes a doublet, two spins a triplet, three spins a quadruplet, and so on, where the relative intensities are 1:1, 1:2:1 and 1:3:3:1 respectively. The strength of the coupling is characterized by the distance between the split peaks. J, the spin coupling constant is usually given in Hertz units. It is important that the J value should be independent of the applied H_0 magnetic field strength, so that it characterizes the intramolecular nuclear interactions, a molecular constant. Protons of saturated hydrocarbons usually exhibit 5 to 8 Hz spin–spin coupling constants. Coupling of more than three spins results in such weak lines that in polymer spectra they are usually not observable, the lines being smeared out.

Fig. 8.21 shows a schematic diagram of spin–spin coupling. In the case of protons there are two roughly equally populated energy states. About half of the A spins are parallel to the H_0 field, the other half being perpendicular. Similarly, half of the B spins are parallel, and half perpendicular to the static field. These two states interact differently with the electrons; in half of the molecules the local magnetic field increases at nucleus A, in the other half of the molecules the local magnetic field decreases, because of the different position of the B spins. Thus two resonance signals are observed, and the peak splits into a doublet. Exactly the same splitting phenomenon appears at nucleus B because of the two different positions of the A spin. The spectrum therefore consists of a doublet pair, with splitting constant J. If another magnetic nucleus is joined to the neighbouring atom, these two spins can interact with

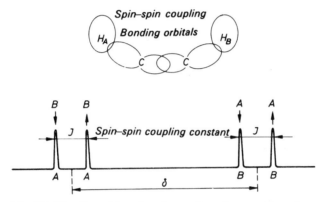

Fig. 8.21. The effect of the spin–spin coupling of two nuclei on the spectrum.

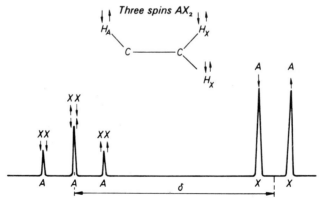

Fig. 8.22. NMR spectrum of three interacting nuclei.

the A nucleus in 3 different ways. In half of the molecules the two spins are perpendicular, and have no effect on the position of the A line, thus one line of the triplet coincides with the chemical shift of the unperturbed A nucleus. This is shown in Fig. 8.22. If further neighbouring magnetic nuclei are present in the molecule, the number of lines increases according to Fig. 8.23. As an

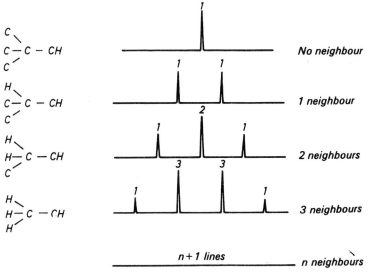

Fig. 8.23. The effect of coupling neighbouring spins on the NMR spectrum.

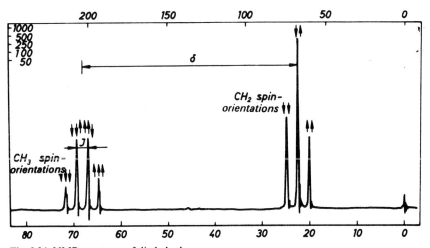

Fig. 8.24. NMR spectrum of diethyl ether.

example the NMR spectrum for the diethyl ether molecule is shown in Fig. 8.24, where the two ethyl groups are symmetrical. The methylene ($-CH_2-$) group gives four lines, while the methyl ($-CH_3$) group appears as a triplet. The chemical shift depends on the H_0 field, J does not. These two effects influence the appearance of the spectra. If the chemical shift difference, Δv (Hz), between two different nuclei is relatively high and J is relatively low, the spectrum consists of two well-separated doublets of almost equal intensity. This is known as a first order spectrum. The chemical shift of the nuclei is equal to the midpoint between the corresponding doublets.

If $J/\Delta v$ exceeds 0.1–0.2, in the case of two-spin systems the intensities become perturbed, the external lines become weaker, and the internal ones more intense. Here the midpoints between the doublets are no longer equal to the chemical shifts, but the difference between the doublets remains equal to J.

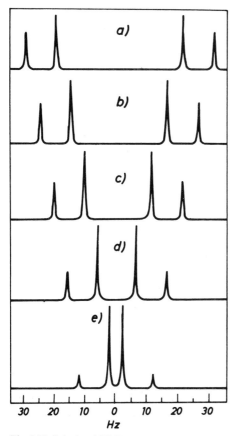

Fig. 8.25. Calculated NMR spectra with $J = 10$ and $J/\Delta r = a$ 0.20, b 0.25, c 0.33, d 0.50 and e 1.00.

These spectra are known as second order, or strongly perturbed. The computer simulated spectra of such a two-spin system at $J = 10$ Hz and $J/\Delta v$ = 0.20, 0.25, 0.33, 0.50 and 1.0 are shown in Fig. 8.25. Both types of spectra are observed in polymers.

Second order spectra are best analyzed by means of computer. Such spectra can be simplified if H_0 is increased, which reduces the $J/\Delta v$ value. Under the effect of stronger fields second order spectra can transform into first order ones. Besides increasing the field strength other alternative methods of spectrum analysis can be found in the literature, such as selective deuteration

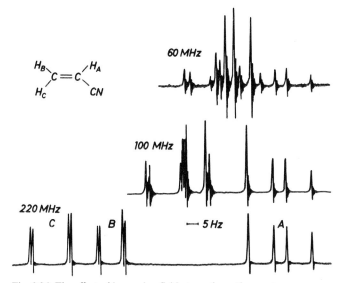

Fig. 8.26. The effect of increasing field strength on the spectrum.

or double resonance. Fig. 8.26 shows the effect of field strength on the spectra (greater field strengths are indicated by the increasing base frequencies). From this diagram it is quite clear why high-field spectrometers are so popular among scientists.

8.3.7. Dipolar line broadening. Broad line spectra

The local magnetic field acting on protons distant from one other is essentially equal to H_0, in fact it is somewhat lower, owing to electronic shielding. If H_0 is homogeneous in the sample the linewidth of the absorption lines is approximately 10^{-4} G, 10^{-2} ppm or 0.3–0.5 Hz.

In most materials protons have an appreciable influence on the local magnetic fields of one another. Let us assume that isolated proton pairs are present in the system, each proton is influenced by the H_0 static field and by an H_{loc} local field. The magnitude and sign of this H_{loc} field depends on the internuclear distance, r, the relative orientation of the nuclei and the static field, H_0. The latter is characterized by the angle, Θ, between H_0 and the internuclear vector. The energy of these local pairs can be expressed as:

$$\Delta E = 2\mu(H_0 \pm H_{loc}) = 2\mu\left[H_0 \pm \frac{3}{2}\mu r^{-3}(\cos^2\Theta - 1)\right].\tag{8.18}$$

The $\pm$ sign depends on the orientation of the neighbouring spin relative to the H_0 field, it can either decrease or increase the local field.

The spectrum of such protons consists of two lines, the separation of which is proportional to r^3 if Θ is constant. This is shown in Fig. 8.27a. If $\cos^2\Theta = 1/3$ ($\Theta = 54.7°$), the lines coincide, and a single line is observed.

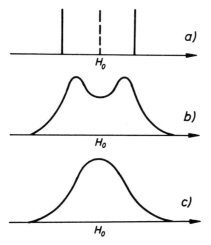

Fig. 8.27. Schematic NMR spectra for the demonstration of spin interactions. *a* Isolated proton pair, *b* partially isolated proton pair, *c* randomly oriented proton pairs.

If the protons are slightly oriented (for example, in compounds containing water of hydration), the spectrum lines broaden. Such lines are shown in Fig. 8.27b. In polymers, protons are usually not limited to the nearest neighbours, so both r and Θ vary within wide limits. This results in a multiplication of the lines, and the component peaks cannot be separated. In this case a single broad line is observed, as shown in Fig. 8.27c. In practice, under the effect of local magnetic fields, the half-width of the absorption line in solid polymers can reach 10 G, or 10^5 Hz.

Nuclei can take part in molecular motions, for example, certain groups of the polymer can rotate, and Eq. (8.18) becomes time dependent. If it is assumed that r, the internuclear distance, is constant and only Θ is time dependent (this is the case with protons of the methyl or methylene groups, and of the benzene ring), then the average time of the local field is:

$$H_{\text{loc}} = \mu r^3 t_2^{-1} \int_0^{t_2} (3 \cos^2 \Theta - 1) dt, \tag{8.19}$$

where averaging is performed for a time interval of t_2. t_2 is the spin–spin or transverse relaxation time. It is called transverse because the spin–spin relaxation influences the field component perpendicular to the H_0 field. If the direction of the field is parallel to the z axis of the xyz coordinate system, then the spin–spin relaxation acts in the xy plane (Fig. 8.28). Spin–lattice relaxation, on the other hand, influences the magnetic moment parallel to the H_0 field (z direction, longitudinal relaxation).

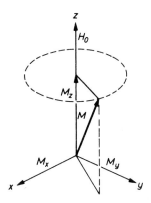

Fig. 8.28. Macroscopic nuclear moment and its x, y and z components.

In mobile systems $t_2 = t_1$. If, however, mobility is decreased, t_2 decreases monotonically while t_1 exhibits a minimum. If solid lattices are formed, t_2 becomes a constant. This is shown in Fig. 8.29, which plots t_1 and t_2 versus molecular mobility. Molecular mobility is characterized by t_{corr}, the correlation time. This is the time needed for the molecule to rotate 1 radian around its axis.

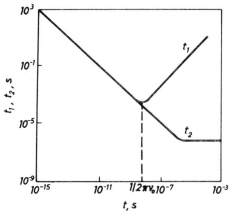

Fig. 8.29. Theoretical relations between the t_1 and t_2 relaxation times on correlation time, t_{corr}.

If Θ changes rapidly, the result of averaging can be a virtually zero H_{loc} local field. As a consequence, the linewidth decreases dramatically, it can be 10^{-4}–10^{-5} times less than in the case of slow angular change.

8.3.8. Magic angle spinning (MAS)

In certain spectrometers there is the possibility of rotating the sample at an angle of 54.7° relative to the static field. Rotation frequency is higher than 2 kHz. In these experiments the linewidth of the solid specimen approximates that of liquids, and is close to 1 ppm. The technique is referred to as magic angle spinning (MAS).

The structure of crosslinked polyimides has been studied by the magic angle spinning technique, and it has been shown that the acetylene end groups undergo an aromatization reaction during heat treatment (Sefcik et al. 1979).

The development of the carbon fibre structure has also been studied by this technique. Polyacrylonitrile (PAN) fibres are transformed into carbon fibres in two steps. First the fibres are pyrolized at lower temperatures in air, which leads to partial aromatization and oxidation. In the next step a high temperature carbonization is performed in an inert atmosphere, which results in the carbon fibre product. The first stage has a profound effect on fibre performance, but it cannot be studied by conventional methods. Aromatization proceeds in two steps:

A ^{13}C-MAS study of the oxidation process of PAN fibre has shown that, under the effect of oxidation, the number of aliphatic carbon atoms decreases, while that of aromatic carbons increases. This method is used in both laboratory investigations and in the quality control of the carbonization process (Stjeskal et al. 1982).

A comprehensive review of solid state NMR spectroscopy was given by O'Donnel (1984).

Möller et al. (1986) studied surface phenomena of lamellae in polyethylene and n-alkanes using the ^{13}C-MAS technique. They succeeded in demonstrating that in the premelting range, a phase transition occurs; the density of the methylene groups decreases, and chain ends become mobile.

8.3.9. Spin–spin interactions

Spin–spin interactions and dipolar line broadening are closely related phenomena, but not identical. Spin–spin coupling is an intramolecular phenomenon, where neighbouring molecules are not involved. Line broadening is the result of local fields of both intra- and intermolecular origin. For example, in a lattice containing two types of spin, direct spin–spin coupling is impossible owing to the widely differing precession frequencies. Dipolar interactions, however, are quite possible. Such an effect is frequently observed in ^{13}C polymer spectra.

Lineshape analysis is complicated by overlapping. ΔH_2^2, the second central moment, plays an important rôle in data processing. Its value can be calculated using quantum mechanics methods so long as the polymer structure is relatively simple. It can also be obtained from experimental data using the formula:

$$\Delta H_2^2 = \frac{\int\limits_{-\infty}^{\infty} (H-H_0)^2 f(H) dH}{\int\limits_{-\infty}^{\infty} f(H) dH} , \tag{8.20}$$

where $f(H)$ is the lineshape-function, that is, the signal value as a function of field strength. $H-H_0$ is the deviation from the central H_0 resonance value.

Broad line spectra are frequently recorded as a derivative and not as absorption lines. Fig. 8.30 shows three different absorption lines and their derivatives. In the literature the recording of the absorption signal is referred to as an absorption mode, while that of the derivative is the dispersion mode.

Line half-width refers to the width of half the maximum absorption. The distance between the extremes of the derivative corresponds to that between the inflexion points of the original absorption spectrum. Narrow lines of high resolution spectra can be characterized by the Lorentz function. In this case, the distance between the extremes of the derivative is very close to the

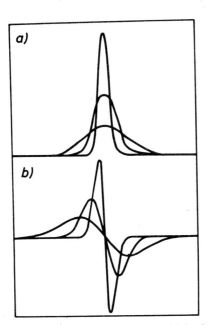

Fig. 8.30. NMR lines of various width in absorption and dispersion mode.

halfwidth, and is equal to $1/\pi t_2$ Hz, where t_2 is the spin-spin relaxation time. Broad line spectra lines cannot be described by the Lorentz function, the relative position of the extremes in the derivative spectrum yields only approximate t_2 values.

8.3.10. Quantitative analysis, integrated spectra

An important feature of NMR spectroscopy is that the relative intensities of the absorption signals are proportional to the relative concentrations of the corresponding nuclei.

This is a great advantage over optical spectra where integral absorption intensities are proportional not only to the concentration but also to the absorption coefficient, they must therefore be calibrated using solutions of known concentration. Peak intensities themselves are not sufficient for intensity determination, since peak widths are proportional to t_2^{-1}, and are not usually the same for different protons. Absorption intensities can be measured by several methods, and most spectrometers are equipped with electronic integrators, which plot the integral spectra as a series of steps. The

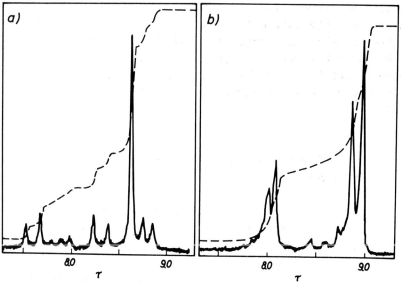

Fig. 8.31. 100 MHz NMR spectra of two poly(methyl methacrylate) samples together with the integrated spectra (dashed lines). *a* Mainly isotactic, *b* mainly syndiotactic polymer. A more detailed analysis of the configurational sequences is given in Fig. 8.38. After Bovey 1972.

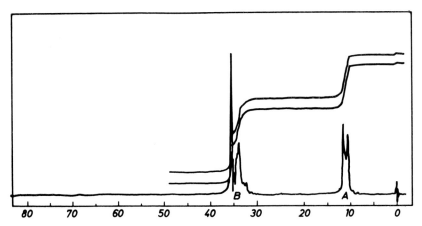

Fig. 8.32. An example of the quantitative analytical use of NMR spectra. Ethylene oxide–propylene oxide copolymer. Signals *A* and *B* are the methyl and methylene absorption respectively. The copolymer composition can be found from the integrated ratios. (After Mathas and Meller 1966.)

heights of these steps are proportional to the number of protons generating the signal. As an example Fig. 8.31 shows the spectra of two poly(methyl methacrylate) samples, the dashed lines representing the integrated spectra.

Fig. 8.32 shows the analysis of an ethylene oxide/propylene oxide copolymer. The resonance signal appearing at 1 ppm belongs to the methyl protons of propylene oxide. The height of the step related to this proton compared to the total integrated height yields the composition after a simple calculation.

8.3.11. Analysis of head-to-head and head-to-tail structures in polymers

Fig. 8.33 shows the structure of the poly(vinylene fluoride) chain, while Fig. 8.34 shows the ^{19}F NMR spectrum. The ^{19}F nucleus of the head-to-tail structure, *A*, appears at 91.6 ppm, while *B*, *C* and *D* structures appear at 94.8, 113.6 and 115.6 ppm values respectively.

The intensity of the *D* line related to the reversed *D* structure is roughly equal to the intensities of lines *B* and *C*, and so the reversed structure shows no tendency for accumulation. Based on the relative intensity, the proportion of head-to-head structure in the sample is 7%.

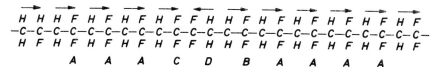

Fig. 8.33. Poly(vinylidene fluoride) molecule. A, B, C and D denote various possible sequential structures.

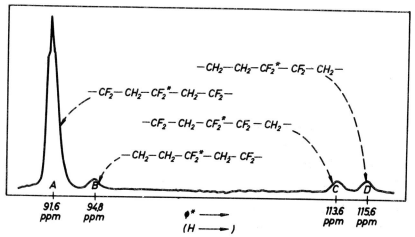

Fig. 8.34. Analysis of head-to-tail and head-to-head linkages by NMR. The curve shows the ^{19}F NMR spectrum of poly(vinylidene fluoride). (After Wilson-Santee 1965.)

8.3.12. Analysis of chain branching

The anticipated ^{13}C NMR signals due to methyl-, ethyl-, propyl- and butyl-branchings in polyethylene have been calculated. These theoretical spectra are shown on the right hand side of Fig. 8.35. The presence of these bands has been confirmed in polymers of known structure which contain such side groups. These real spectra are shown of the left hand side of Fig. 8.35. Spectrum a is from hydrogenated polybutadiene, and spectrum b from ethylene/1-hexene copolymer. Curves c and d correspond to high- and low molecular mass branched polyethylenes respectively. It can be seen that polyethylenes contain mainly butyl branches, but sample d also has some ethyl branches (Dorman et al. 1972).

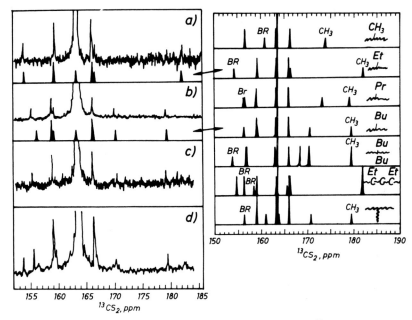

Fig. 8.35. Examples for the study of chain branching in polymers. ^{13}C NMR spectra of the following compounds. Left hand side, *a* hydrogenated polybutadiene, *b* ethylene-1-hexene copolymer, *c* high molecular mass, branched polyethylene, *d* low molecular mass, branched polyethylene. Right hand side. Expected spectra for various branches. The presence of ethyl groups is clearly evident in spectrum *a*, that of butyl groups in *b*. In polyethylene mainly butyl groups are present, but in sample *d* a small amount of ethyl groups can also be found.

8.3.13. Analysis of stereoregularity

NMR spectroscopy has proven to be extremely successful in the analysis of stereoregularity in polymers. In order to understand the process, a brief discussion of the stereochemistry of polymers and related NMR results is necessary.

8.3.13.1. Geminal and vicinal protons

The proton–proton interactions most frequently encountered in polymers originate either from geminal protons (bound to the same carbon atom) or from vicinal ones (situated on two, neighbouring atoms). This first type of interaction is frequently referred to as ^2J, and the latter as ^3J, the superscript

Table 8.6

Dependence of proton–proton coupling constants on the molecular geometry and chemical structure

---

Geminal protons

$J_{gem} = {}^2J = +5 \ldots -30$ c/s   (Hz)

Vicinal protons
a) nonsaturated bond

cis

cc. half of the *trans* coupling

trans

$0-14$ c/s   (Hz)

b) saturated bond

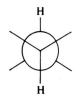

Trans

$8-13$ c/s   (Hz)

Gauche

Lower than *trans* coupling
$2-4$ c/s   (Hz)

---

gives the number of valence bonds between the coupled spins. ^1J type, or direct coupling also exists in polymer NMR spectroscopy, for example, between ^{13}C and attached protons or the ^{19}F nucleus.

Geminal proton–proton coupling constants range usually between -12 and -15 Hz, for example, in methane it is -12.4 Hz, and in the methylene group of poly(methyl methacrylate) -14.9 Hz. The value increases with increasing H–C–H valence angles. An electronegative substitution on the vinyl group decreases the value, for example, in vinyl chloride to $^2J = -1.3$ Hz.

In the case of vicinal coupling, geometrical parameters and substitution influence the strength of coupling. The vicinal coupling constants in unsaturated compounds fall between 0 and $+20$ Hz. In ethylene $J_{cis} = 11.5$ Hz, $J_{trans} = 19.0$ Hz, while in vinyl chloride $J_{cis} = 7.4$ Hz, $J_{trans} = 14.8$ Hz. The strength of *trans*-coupling is usually double that of *cis*-coupling. In saturated bonds the strength of *trans*-coupling is also much higher than that of *gauche*-coupling. Typical values are $J_{trans} = 8$ to 13 Hz, $J_{gauche} = 2$ to 4 Hz. Table 8.6 summarizes these findings.

8.3.13.2. Isomerism of polymer chains

Polymer chains from vinyl monomers can form in different ways. Head-to-head and head-to-tail structures have already been mentioned, together with branching and their analysis by NMR spectroscopy. Heterotactic (iso-, syndio- and atactic) polymers formed from α-substituted monomers were discussed in the first introductory chapter. A simplified view of these structures is listed in Table 8.7. CH_2 groups are denoted as a vertical line.

Asymmetric polymeric structures exhibit neither rotational nor translational symmetry. Dissymmetric structures, on the other hand, have no mirror plane, but they can have rotation axes of glide planes. The isotactic structure is neither asymmetric, nor dissymmetric, as it has a mirror plane and a translation symmetry but no axis of symmetry.

Syndiotactic structure, on the other hand, has a translational symmetry but no mirror plane, it is therefore dissymmetric.

α, β disubstituted vinyl monomers can lead to further steric structures, the most important of which are disyndiotactic (4), erythrodiisotactic (5) and threodiisotactic (see Table 8.7).

Table 8.7

Isomery of polymer chains exhibiting head-to-tail structure obtained from vinyl monomers

a — substituted vinyl monomers

      (A or B may also be H)

1.   isotactic

2.   syndiotactic

3.   heterotactic or atactic

a, β — disubstituted vinyl monomers

    or

4.   disyndiotactic

5.   erytrodiisotactic

6.   treodiisotactic

8.3.13.3. Model compounds for the identification of isotacticity

The structures referred to above can be distinguished from their NMR spectra. The basis of this distinction is the thorough study of model compounds which contain such structures in a well-defined way. The 2, 4, 6 trisubstituted heptanes are good model compounds for the study of isotactic, syndiotactic and atactic vinyl polymers. Several studies have been made which include trichloro, tricarboxy, tricarbmetoxy, and tricyano heptanes. Their general structure is shown in Fig. 8.36. These compounds have two

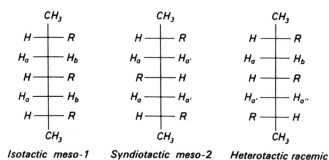

Isotactic meso-1   Syndiotactic meso-2   Heterotactic racemic

Fig. 8.36. 2, 4, 6 trisubstituted heptanes as model compounds for isotactic, syndiotactic and atactic polymer sequences.

meso- and one racemic steric isomers. As they are closely related to the isotactic, syndiotactic and atactic triadic sequences in polymeric chains, these names are also applied to the isomers. In the meso-isomers there are two identical methylene positions, in the racemic isomer the two methylenes are in different positions, so all four methylene protons behave differently.

8.3.13.4. Configurational sequences

The configuration of these model compounds can be generalized for polymers. Assigning meso, m, or racemic, r, configuration to the diadic sequences, all further configurations can be described. In triads mm corresponds to isotactic, mr to the heterotactic, rr to the syndiotactic structure. Table 8.8 covers tetrads and pentads as well. Further description of NMR spectra refers to this notation.

Methylene ($-CH_2-$) is described as heterosteric if the two protons are non-equivalent. The heterosteric methylene group is also referred to as a

Table 8.8

Polymeric chains built up from α-substituted vinyl monomers

Sequence	Schematic view	Notation	Sequence	Schematic view	Notation
Diad:		Meso, *m* (protons of the β-methyl group are not equivalent)	Triad:		isotactic (*i*), mm
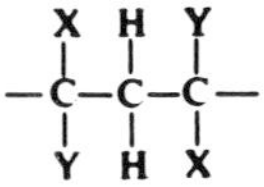	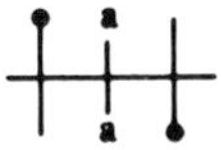	racemic, *r* (the protons of the β-methyl group are equivalent)		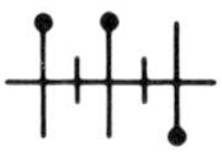	hetero-tactic (*h*), mr
				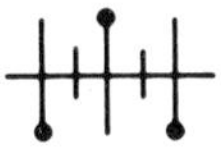	syndio-tactic (*s*), rr

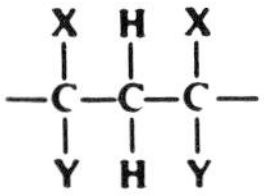

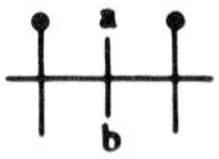

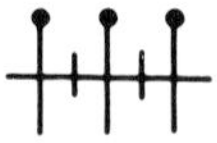

Tetrad:

mmm

mmr

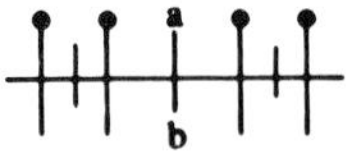

rmr

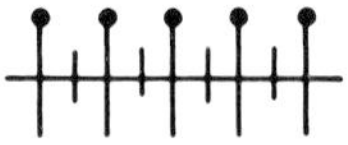

mrm

rrm

rrr

Pentad:

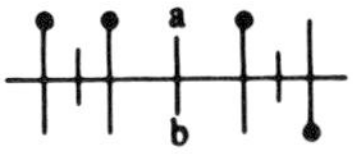

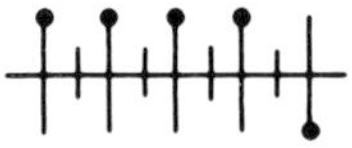

mmm (isotactic)

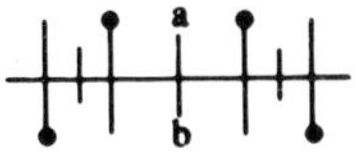

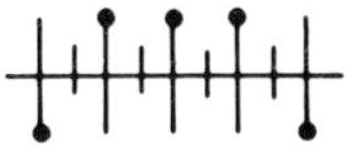

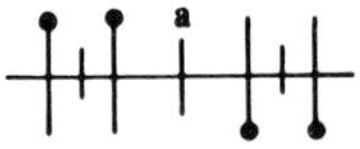

mmmr

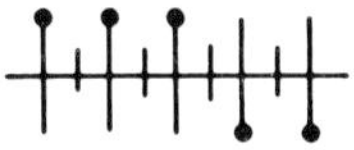

rmmr

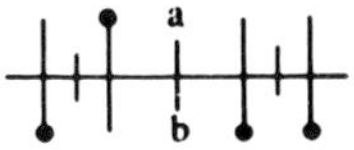

mmrm

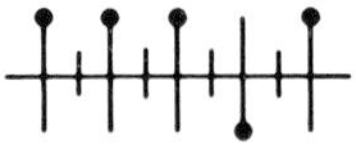

mmrr

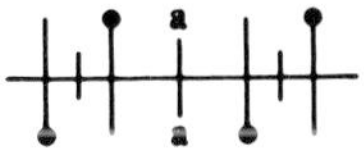

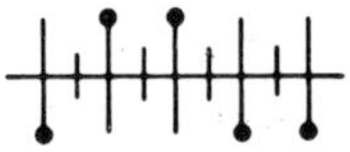

rmrm (heterotactic)

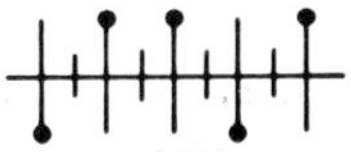

rmrr

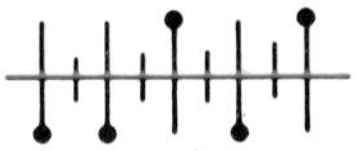

mrrm

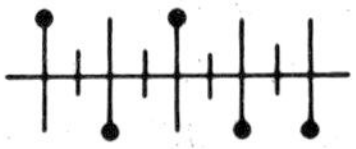

rrrm

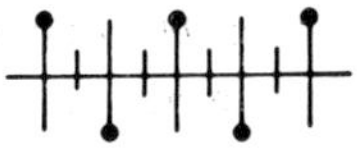

rrrr (syndio-tactic)

meso, *m*, group. The methylene group is homosteric if the two protons are equivalent. These groups are also referred to as racemic, *r*, groups.

Another common description of polymer chains is based on the generalization of the *trans* and *gauche* notation. According to the convention introduced by Yoshino (1966), the configuration is known as *trans* if the chain C-atoms are in the *trans* position. The structure is called *G* if the *R* substituent is in the *trans* position with respect to the chain, and it is denoted as *G'* if both the *R* substituent and the chain C-atoms are in the *gauche* position with respect to the chain (Fig. 8.37).

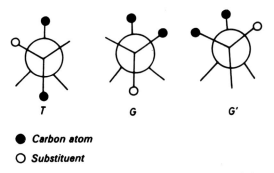

● **Carbon atom**

○ **Substituent**

Fig. 8.37. Various conformations of vinyl polymer chains.

According to this convention the meso position is given as TG, and the racemic as TT or GG. Isotactic chains are either in TGTG or GTTG conformations, whereas atactic chains can be described as TTTG, TTGT or GGTG, and syndiotactic sequences as TTTT or GGTT. Since in the case of 2, 4, 6 trisubstituted heptanes (trimeric model compounds) the number of possible TG conformations is 81, this notation is preferentially used in description of conformations rather than configurations.

8.3.13.5. Observation of configurational sequences in polymers. Analysis of poly(methyl methacrylate)

Sequences observed in model compounds, and predicted by calculations, can be identified in polymers. As an example, the NMR spectra of poly-(methyl methacrylate) is examined below at different frequencies, and the sequences observed at different levels of resolution are discussed.

Fig. 8.38 shows the 60 MHz spectra of two PMMA samples prepared by two different catalyst systems. The peaks can be identified quite easily using

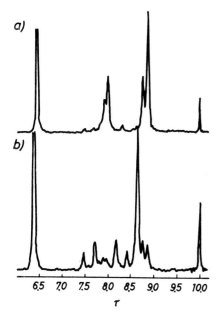

Fig. 8.38. 60 MHz spectrum of poly(methyl methacrylate) in a 15% chlorobenzene solution. *a* Polymerization by free radical initiator; *b* polymerization by anionic initiator. Ester methyl proton resonance appears at 6.5 τ; that of the β-methylene proton at 8.0 τ, the resonance of the α-methyl group with 3 peaks appears between 8.5 and 9.0 τ.

the experience gained on low molecular model compounds. The peak appearing at 6.42 τ can be ascribed to the ester methyl group. The α-methyl resonance appears between 8.7 and 8.8 τ, while the methylene resonance of the main chain appears in the 7.3 to 8.4 τ-range. The reference peak of tetramethyl silane appears at 10.00 τ.

Three α-methyl peaks appear at 8.67 τ, 8.79 τ and 8.9 τ respectively. The relative heights strongly depend on the preparation conditions. The peak appearing at 8.67 τ is characteristic of the *i* units (mm), the 8.90 τ peak corresponds to the syndiotactic s(rr) configuration while the 8.79 τ peak can be assigned to methyl groups, where the central monomeric unit is in heterotactic, h(mr) configuration. The relative intensities of the peaks appearing at 8.67, 8.79 and 8.90 τ are proportional to the number of the *i*, *h* and *s* units respectively.

The chain, β, methylene resonance should give a single resonance in the syndiotactic polymer, since the two protons are in equivalent environments:

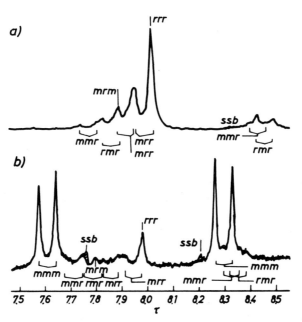

—·—·— α-methyl groups

———— β-methylene groups

In the isotactic polymer, however, the two protons of the β-methylene group are in different environments (different shielding), and owing to the spin–spin interaction four resonance peaks are expected. These four peaks correspond to two doublets, with roughly equal intensities so long as the shielding difference is much higher than the J coupling. If the shielding

Fig. 8.39. Details of the 220 MHz NMR spectra of the compound shown in Figs 8.31 and 8.38. This Figure illustrates the fine structure in the β-methylene proton region. Spectrum lines split according to tetradic structures. It can be seen that sample *a* consists mainly of rrr syndiotactic, sample *b* predominantly of mmm isotactic sequences. The spectrum of the methylene group can be used in the analysis of dyadic and tetradic stereosequences.

difference is close to the coupling value, the intensity of the outer peaks decreases.

In Fig. 8.38*a*, where the sample consists mainly of isotactic units, the latter situation can be observed, the centre being around 7.90 τ. The sample shown in Fig. 8.38*b* contains mainly syndiotactic units. Here the methylene signal is essentially a single line showing some fine structure. This can be explained by the appearance of tetrads, which can be studied more successfully on high resolution spectra.

100 MHz spectrometers offer a much better resolution as shown in Fig. 8.31. Further resolution can be achieved at even higher frequencies. Fig. 8.39 shows the splitting of the β-methylene proton resonance of the same

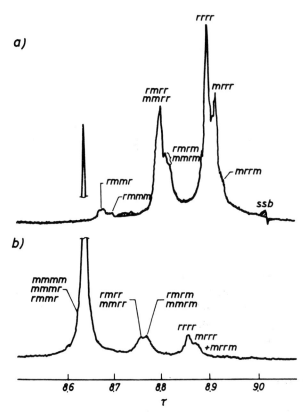

Fig. 8.40. Details of the 220 MHz NMR spectra of the compounds shown in Figs 8.31 and 8.38. This Figure illustrates the fine structure in the α-methyl proton resonance region. Spectrum lines split according to pentadic structures. The spectrum of the α-methyl group can be used in the analysis of triadic and pentadic stereosequences.

PMMA samples at 220 MHz. The upper spectrum corresponds to a mainly syndiotactic polymer, while in the lower spectrum isotactic units are predominant. (The ssb abbreviation in the Figure corresponds to spin side band.)

Fig. 8.40 shows the α-methyl resonance region of the same polymer samples at 220 MHz. In this Figure the splitting of the mm, mr and rr peaks is due to the appearance of pentad sequences (Frisch et al. 1968).

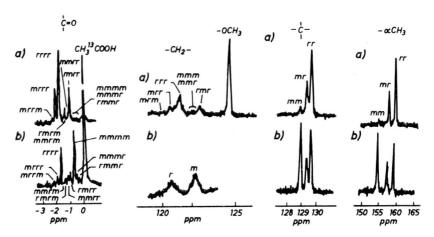

Fig. 8.41. ^{13}C NMR spectra of poly(methyl methacrylate) (PMMA) samples. *a* Mainly syndiotactic PMMA; *b* stereoblock PMMA. Pentadic stereosequences can be measured by the carbonyl carbon resonance. (After Inoue et al. 1971.)

Configurational sequences can also be observed on ^{13}C NMR spectra. Fig. 8.41 shows the spectra of a predominantly syndiotactic PMMA (*a*) and of a stereo-block (*b*) PMMA polymer, in which the pentad sequence distribution can be clearly observed. Sequence distribution can be measured on the resonance signals of the carbonyl C-atom.

8.3.13.6. Isotacticity in polypropylene

Deciphering the sequence distribution in PP is a fairly complex task mainly because of the strong coupling of the main chain and side chain protons which strongly influences the spectra. Useful results have been obtained on samples prepared from a deuterated monomer. The structure of this deuterated monomer can be given by:

$$\begin{array}{c} CD_3 \\ \underset{H}{|} \\ C = CH_2. \end{array}$$

(8.21)

Using the Ziegler–Natta polymerization technique three fractions were prepared by solvents. Fig. 8.42 shows the spectra of these fractions together with the spectrum of the unfractionated sample (Stehlin 1964). The *a* crystallizing fraction gives four methylene resonances, and is almost totally isotactic. The *c* fraction, which is amorphous, contains mainly syndiotactic units. The *b* block contained both isotactic and syndiotactic units, but its composition cannot be determined from the NMR spectrum. A good quality spectrum of the nondeuterated polymer could be obtained by using a 220 MHz spectrometer with 52 kG field strength and superconductive solenoid magnets. This is shown in Fig. 8.43. Assignment of the main lines of this spectrum is given in Table 8.9.

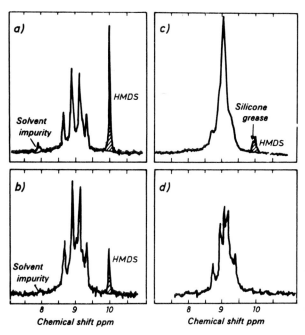

Fig. 8.42. NMR spectrum of poly(2,3,3,3 d_4-propylene). Measuring frequency is 60 MHz, temperature is 110 °C, solvent is 2-chloro thiophene. *a* Isotactic fraction; *b* stereo-block fraction; *c* amorphous fraction; *d* unfractionated sample. Internal standard is hexamethyl disiloxane (HMDS), which gives an absorption line at 9.94. τ.

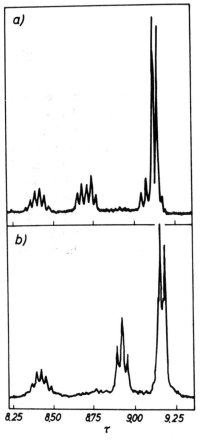

Fig. 8.43. 220 MHz NMR spectrum of polypropylene. Solvent is o-dichloro benzene. *a* Isotactic; *b* syndiotactic polymer. (After Ferguson 1967.)

Fig. 8.44 shows the ^{13}C NMR spectrum of polypropylene. On this spectrum the methylene group appears at 146 ppm, the methylene carbon at 166 ppm, and the methyl carbon at 173 ppm. Spectrum *a* corresponds to the isotactic, and spectrum *b* to the atactic polymer. The band of the methyl carbon in the atactic polymer splits into a triplet, where the presence of mm, mr and rr triads can be observed.

The NMR spectra of other vinyl monomers have been analysed from the point of view of tacticity by several other authors using the general principles discussed here. Detailed information for these studies can be found in the references given at the end of this book.

Table 8.9

Typical bands of the NMR spectrum for polypropylene

CH₃ H_B CH₃	CH₃ H_A H_B
—C—C—C—	—C—C—C—
H_C H_A H_C	H_B' H_A' CH₃
Isotactic	Syndiotactic

Proton	τ	Proton	τ
H_A (Anti)	8.73	H_A and $H_{A'}$	8.97
H_B (Syn)	9.13	H_B and $H_{B'}$	8.46
H_C	8.42		
CH_3	9.14	CH_3	9.19

After Ferguson 1967.

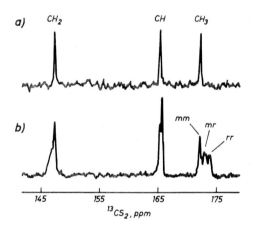

Fig. 8.44. ^{13}C NMR spectrum of polypropylene (25 MHz). *a* Isotactic PP, 5% solution in o-dichloro-benzene at 60°C, average of 100 measurements; *b* atactic PP. 30% solution in o-dichloro-benzene at 60°C, average of 100 measurements. (After Johnson et al. 1970.)

8.3.14. Studies on vinyl copolymers

High resolution NMR is a very powerful tool in the structural analysis of vinyl copolymers.

First let us review the structural units to be distinguished. If the polymer consists of CH_2=CX_2 type monomers, the configurational sequences given

Table 8.10

Configurational sequences of co-monomers of the $CH_2 = CX_2$ type (A and B)

in Table 8.10 can be distinguished. If one of the monomers is of the $CXY=CH_2$ type, whereas the other is of the $CX_2=CH_2$ type, the sequences are limited to those given in Table 8.11. NMR can be used to measure the average composition of copolymers. It is faster and probably more exact than conventional methods, but since this book is devoted to structural problems, studies of this kind do not belong here.

Various copolymers of monomers A and B, as block copolymers (...AAAAABBBB...), alternating copolymers (...ABABAB...) or random copolymers (...AABABBBAA...), can be described by NMR spectroscopy provided that A–B type bonds can be distinguished from A–A and B–B type bonds, either from the chemical shifts or coupling interactions. If no A–B type bond can be detected, the system is a simple mixture of two

Table 8.11

Configurational sequences of comonomers of the CXY = CH$_2$ (A) and
CX$_2$ = CH$_2$ (B) type

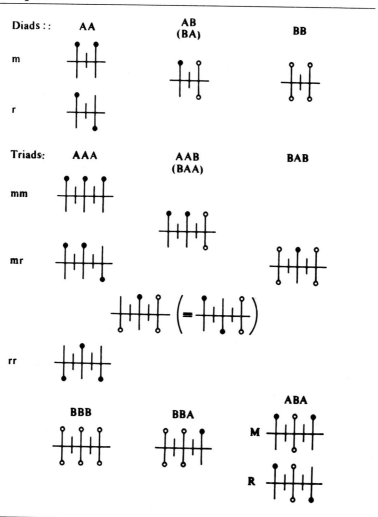

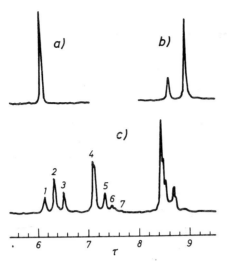

Fig. 8.45. 60 MHz NMR spectra. *a* Poly(vinylidene chloride), *b* polyisobutylene, *c* vinylidene chloride–isobutylene copolymer, 70 mole% vinylidene chloride content. Peaks marked by numbers are tetradic sequences in the following order. 1 AAAA; 2 AAAB; 3 BAAB; 4 AABA; 5 BABA; 6 AABB; 7 BABB. (After Kinsinger et al. 1966.)

homopolymers, which might otherwise be demonstrated by some far more cumbersome method, such as selective dissolution.

As an example the study of a vinylidene chloride copolymer is given below. One of its important copolymers is prepared with isobutylene. The NMR spectra of the homopolymers and of a copolymer containing 70 mole% vinylidene chloride is shown in Fig. 8.45. On this spectrum seven different methylene peaks can be distinguished. This means that the resolution gives information up to tetrads. If only diads and triads were distinguishable, three methylene groups would be observed. Interpretation of the lines is given in the Figure caption, *a* is the spectrum of vinylidene chloride, and *b* that of isobutylene. The other three peaks belong to the methyl group absorption range, around $8.6\,\tau$.

8.3.15. Crystallinity determination from NMR spectra

Crystallinity measurements by NMR are based on the fact that should the frequency of the group motion exceed about 10^4 Hz, the resonance curve becomes very sharp. The polymer to be studied has to be held at a temperature where the correlation time of the motions relating to the crystalline parts is $t_{corr, cr} \gg 10^{-4}$ s and the time for amorphous areas $t_{corr, am} \ll 10^{-4}$ s.

Under these circumstances the NMR absorption is the superposition of a narrow (amorphous) and a broad (crystalline) resonance. These can be resolved by graphical means, and from the intensity ratio the crystallinity can be calculated.

Slichter and McCall (1957) studied the mobilities of two polyethylene samples, one being a linear PE (Marlex 50), the other a crosslinked PE.

They found that in the case of Marlex 50 the $t_{corr, cr} \gg 10^{-4}$ s and $t_{corr, am} \ll 10^{-4}$ s condition is valid over a wide temperature range. The measured crystallinity was 90%. In the case of the crosslinked PE samples the authors were unable to find a temperature where the correlation time condition was fulfilled.

Farrow and Ward (1960) studied the crystallinity of poly(ethylene terephthalate) in both oriented and unoriented states. They could only obtain a curve for the resolution of mobile and immobile parts above the secondary transition point, at 110°C. Fig. 8.46 shows the derivative curve of such a

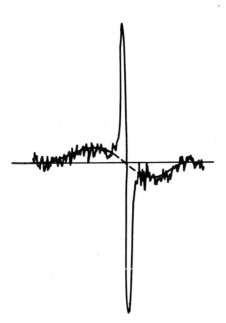

Fig. 8.46. Broad line NMR spectrum of poly(ethylene terephthalate) above the second order transition temperature in dispersion mode. The superposition of broad and narrow components can be observed.

Table 8.12

X-ray and NMR crystallinity of poly(ethylene terephthalate) samples

Sample	Crystallinity, %		
	NMR	X-ray	
	180 °C	25 °C	180 °C
Flake	72	52	46
Flake obtained by slow cooling	72	62	44
Oriented film	80	53	—

superposed NMR spectrum. The result strongly differs from that obtained by X-ray methods, as shown in Table 8.12.

It is generally true that the correlation between crystallinities obtained by X-ray and NMR techniques is not very satisfactory. The probable reason for this discrepancy may lie in the fact that, while X-ray diffraction studies the regularity of the scattering centres, NMR detects the mobility of the structural units, irrespective of the regularity of the interstructural distances.

8.3.16. Temperature dependence of molecular motions

In the study of the temperature dependence of molecular motions, we are concerned in the main with two characteristic quantities:
    — the spin-lattice relaxation time, t_1;
    — the linewidth and closely related second moment, or the inversely proportional spin–spin relaxation time, t_2.
As mentioned earlier in Fig. 8.18, there is a slight, but important, exchange process between the spins and the lattice. Protons excited to high spin temperatures transfer some heat energy to the lattice in order to reach the equilibrium spin population which appears in the molecular motion of the lattice. If the temperature of the sample is decreased, and the viscosity increases, the spectral intensity of the local magnetic noise increases, and passes through a maximum at the Larmor-frequency, then subsequently decreases.

As shown in Fig. 8.27 the protons "feel" not only the external H_0 field, but also the field of the neighbouring proton, which can add to, or subtract from, the H_0 value, so that the resonance line splits into a doublet. If the interactions are not restricted to individual pairs, but are a superposition of fields from protons at different r distances and ϑ angles relative to the H_0 field, the magnetic interactions exhibit a Gaussian distribution. In this case the NMR spectrum transforms into a broad line as shown in Fig. 8.27c.

If molecular motion changes as a result of increasing or decreasing temperatures, the field strength of the magnetic interactions becomes time dependent. t_{corr}, the correlation time (the time interval under which the spin rotates one radian, so that the rotation time is $2\pi t_{corr}$) may be so short that H_{loc} can be regarded as a constant. Under these circumstances the resonance lines become 10^4–10^5 times narrower, compared to the time dependent H_{loc} value. Starting from a rigid lattice, and increasing mobility by moving to higher temperatures, the resonance line becomes narrower when:

$$\frac{1}{t_{corr}} \geq 2\pi\delta v \tag{8.22}$$

where δv is the halfwidth of the line obtained from the rigid lattice (in Hz units).

Magnetic nuclei are not only magnetic dipoles, they also exhibit a precession motion around the magnetic field, even in the solid state. This has a component which is parallel to the H_0 field, but also has a rotating component. This component can excite the neighbouring spin if their Larmor frequencies are equal. During these magnetic spin-exchange processes (also known as a flip-flop process) the energy is constant, but the phase coherence of the precession motion decreases. The loss of phase coherence can be characterized by t_2, the spin–spin relaxation time.

If the coupling between the spins increases, the spin–spin relaxation time decreases, and t_2 is inversely proportional to the linewidth:

$$t_2 = \frac{1}{2\pi\delta v}. \tag{8.23}$$

In glassy or crystalline polymers the conformation of the polymeric chains (either in the ordered or non-ordered state) is frozen, and the structure is rigid. The side groups are, however, mobile, even in this state. In rubbers, chain conformations change rapidly via rotations around the σ-bonds of the main chain, and side groups are also mobile. On cooling, the mobile rubbery phase transforms into glass, or into a partially crystalline matrix. The linewidth increases from 0.1–1 Gauss to 2–8 Gauss when the frequency of the motion of the polymeric main chain becomes comparable with the linewidth

(10^4–10^5 Hz). At even lower temperatures a second line-widening phenomenon occurs due to the freezing of side group activity.

The rate of thermodynamic equilibration between the spin system and
the lattice can be characterized by the t_1 relaxation time. The equilibration is
faster if the frequency of the molecular motion approaches the NMR
frequency, which is 5–100 MHz. So t_1 can show several minima, since on
cooling various molecular motions pass the critical frequency range. t_1 can be
measured in high resolution spectrometers in the liquid state by monitoring
the signal strength when switching the H_1 field on and off. When switched off
the spin system is already saturated, that is, the relative population
approximates the equilibrium value. Time dependence is studied by rapid
consecutive measurements. This method can be used only if t_1 is longer than 2
to 3 sec. t_2 can be deduced from linewidth data only if it is shorter than 1 sec,
otherwise field inhomogeneity has a greater effect on the linewidth, and t_2 is
underestimated.

In liquids t_2 approximates to t_1 and this method cannot be applied. In
polymer solutions the linewidths are of the order of 5–10 Hz or higher, which
means 1/60 sec for t_2. In this case t_2 can be measured relatively accurately if
other sources of error (such as overlapping multiplicity) are not present.

If t_1 is shorter than 2 to 3 sec, or if t_2 is longer than 1 sec, the pulse
technique is used. The essence of this method is as follows: for a short time
period, τ, a strong radio frequency field is applied. During this time interval
practically no relaxation occurs. Under the effect of this pulse, M, the macroscopic magnetic moment, turns in the direction of the xy plane, and the
ratio of rotation is γH_1. If $\gamma H_1 \tau$ is chosen so that M rotates exactly to the xy
plane, it is called a 90° pulse. The resulting signal will decrease with t_2
remaining constant, but t_2 is still influenced by the field inhomogeneity. True
t_2 values can be measured by the "spin echo" technique which compensates
for the effect of field inhomogeneity.

When measuring t_1, a second 90° pulse is applied to turn the net
moment in the direction of the original axis. If the time interval between the
pulses is shorter than t_1, the z-component of the magnetization cannot reach
the equilibrium M_0 value. By measuring M_z at different pulse sequence times,
t_1 can be determined.

Fig. 8.47 shows the temperature dependence of the t_1 relaxation time for
polyisobutylene at 30 MHz and 50 MHz, and for natural rubber at 30 MHz
(Slichter and Davis 1964).

It can be seen that, at a given temperature, t_1 increases with the applied
frequency. On both curves two minima appear. In the case of natural rubber
the lower temperature minimum (at $-150°C$) is related to the methyl

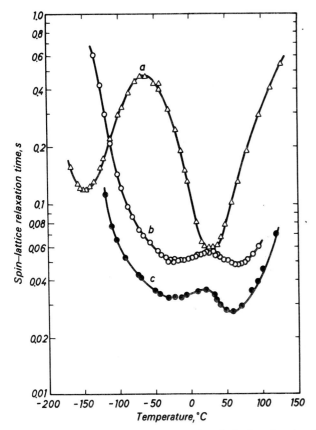

Fig. 8.47. Temperature dependence of t_1, the spin–lattice relaxation time of polyisobutylene and natural rubber. *a* Hevea-rubber, 30 MHz, *b* polyisobutylene, 50 MHz, *c* polyisobutylene, 30 MHz.

rotation, while the higher temperature minimum (at about 30°C) can be attributed to the translation and rotation of the chain segments. In the case of polyisobutylene the lower temperature peak is much broader and appears at a temperature about 100°C higher. This is probably due to the fact that the methyl group rotation in this polymer is hindered, and becomes free only if segmental motion is enhanced.

    This interpretation is borne out by the temperature dependence of the t_1 relaxation time in *cis*-1,4-polybutadiene (Fig. 8.48). Here there is no methyl group, thus the low temperature minimum is absent. The single minimum, however, appears at lower temperatures (at about −45°C), which corre-

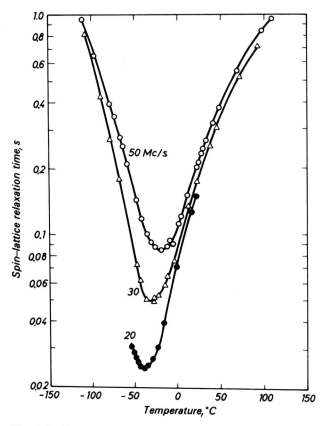

Fig. 8.48. Temperature dependence of t_1, the spin–lattice relaxation time of cis-1,4-polybutadiene.

lates well with the fact that the glass temperature of cis-polybutadiene ($-110\,°C$) is much lower than that of natural rubber or polyisobutylene ($-70\,°C$).

From studies on rigid polymers Fig. 8.49 shows the results obtained on poly(methyl methacrylate) (Powles et al. 1964). Both syndiotactic and isotactic polymers exhibit two t_1 minima. The lower temperature minimum is due to methyl group rotation, whereas the higher one to segmental motion. Both transitions appear at temperatures about $50\,°C$ lower in the isotactic polymer as compared to the commercially known (mainly syndiotactic) version, that is, stereochemical configuration influences both motions.

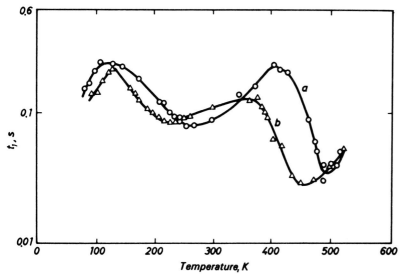

Fig. 8.49. Temperature dependence of the spin–lattice relaxation time of poly(methyl methacrylate) at 21.5 MHz. *a* syndiotactic, *b* isotactic polymer.

Fig. 8.50 shows the temperature dependence of the second moment of the NMR spectra of polyisobutylene, natural rubber and atactic polypropylene. Moments calculated for non-moving polymer chains are shown on the left hand side of the Figure. The lowest Figure relates to natural rubber. This level is reached by the natural rubber sample at about 50 K. At about 100 K the second moment suddenly decreases because of the onset of methyl rotation. This plateau is observed up to 220 K, where again the second moment decreases, now because of onset of segmental motion. This second temperature value is close to the T_g of natural rubber shown on the abscissa. In the case of atactic polypropylene the measured value does not reach the calculated one even at 77 K, and the methyl groups are free to move even at this temperature. It is expected that the "freezing in" of this motion should begin at around 50 K. Similarly to natural rubber, a plateau is observed in polypropylene up to 250 K, corresponding to the much higher T_g value.

Second moments calculated from rotating methyl groups are shown on the right hand side of the Figure. It can be seen that in the case of natural rubber, the agreement between calculated and experimental values is quite satisfactory, while for polypropylene the calculated value is somewhat higher than experiment indicates.

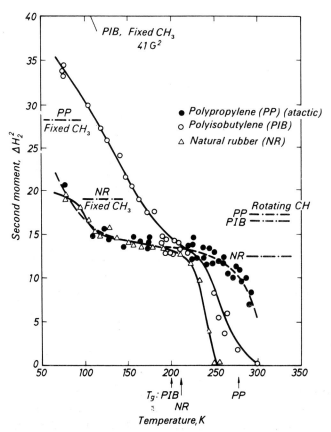

Fig. 8.50. Temperature dependence of the second moment of the NMR spectra in polyisobutylene (PIB), natural rubber (NR) and atactic polypropylene (PP). (After Slichter 1961.)

In the case of polyisobutylene the temperature dependence of the second moment is somewhat different. It reaches the value calculated for rotating methyl groups only at about 200 K, and there is no pronounced transition due to the onset of methyl rotation, only an inflexion is observed at 225 K. This agrees well with the earlier observation on the temperature dependence of t_1, that the rotation of the methyl group in polyisobutylene is hindered, and becomes free only when segmental motion begins.

The behaviour of polymers which are rigid at room temperature is similar to that of elastomers, but the curves shift towards higher temperatures.

Below 150 K the spectra are the superposition of narrow and broad components, which can be explained by the presence of amorphous and

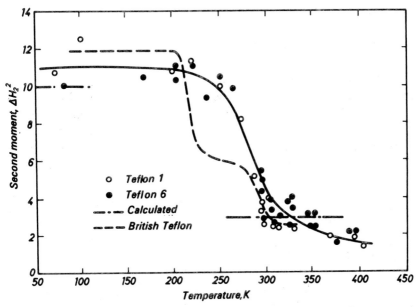

Fig. 8.51. Temperature dependence of the second moment of the NMR spectrum in poly(tetrafluoro ethylene).

crystalline regions. As an example for a highly crystalline, rigid polymer, Fig. 8.51 shows the temperature dependence of the second moment of poly(tetrafluoro ethylene) (Teflon). The calculated static second moment (10.0 G^2) is the superposition of a dominant contribution due to the helical structure determined by X-ray techniques (7.5 G^2), and of an estimated inter-chain dipole–dipole interaction contribution (2.5 G^2). The experimentally observed 10.9 G^2 value at low temperatures is in resonable agreement with the theoretical value. The Figure shows data for three different Teflon samples (Schmidt 1955). Teflon 1 and Teflon 6 differ in molecular mass, the curve shapes being practically the same, whereas in the 240–310 K interval the second moment drastically decreases. In this temperature range it is generally known that the crystal structure, heat capacity and density of Teflon change. These effects can be explained by the onset of chain rotation. The sample denoted as British Teflon shows the onset of chain motion at lower temperatures which indicates the presence of low molecular impurities (acting as plasticizers) or low molecular mass fractions.

8.4. MECHANICAL SPECTROSCOPY OF POLYMERS

8.4.1. Thermomechanical curves

Mechanical and dielectric properties of polymers change as a function of temperature in a similar fashion to the second moments of the NMR spectra. If the tensile, E, or shear, G, moduli of polymers are plotted versus temperature, the analogy is quite clear. Moduli can be determined in deformational or dynamic, for example, torsional experiments. Such a thermomechanical curve is shown in Fig. 8.52, which plots the shear modulus, G, versus temperature.

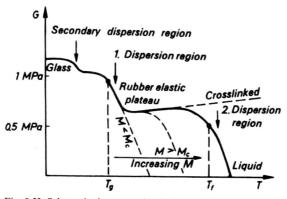

Fig. 8.52. Schematic thermomechanical curve of an amorphous polymer.

The analogy is not by chance, the effects of various modes of motion manifest themselves in the mechanical properties. If a rigid amorphous polymer is heated from its glassy state, a secondary dispersion region is observed first, which is related to the onset of methyl, or other side group, rotation. Following a plateau region, the primary transition region is reached and segmental motion of the chains begins. This is the T_g or glass–rubber transition zone, there follows a second plateau, which is known as the rubber–elastic region. The upper limit of this region is T_f, the flow point in linear polymers. In crosslinked systems there is no flowpoint, these remain at the plateau-level up to high temperatures, where they gradually decompose. The second dispersion at the flowpoint results in a viscous liquid, and elasticity disappears.

It should be noted that, if the molecular mass is lower than the minimum value needed for rubbery properties, meaning rubber elasticity, then the thermomechanical curve shows a very short rubbery plateau, or none at all. This critical molecular mass is noted as M_c in the Figure.

The thermomechanical curve of partially crystalline polymers is somewhat more complicated. The general shape is shown in Fig. 8.53. Above the glass transition temperature the crystalline parts are still intact, which

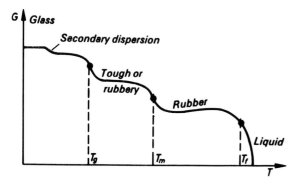

Fig. 8.53. Schematic thermomechanical curve of a partially crystalline polymer.

reduces the segmental mobility similarly to crosslinking. Segmental motion becomes possible only above the melting point, T_m. The rubbery elastic state can appear both below or above the melting point, depending on the crystallinity level. If T_f, the flow point of the amorphous parts, is higher than T_m, the melting point of the crystalline part, then above T_m another, rubber–elastic plateau appears which ends at T_f, the flow zone. If T_f is lower than T_m, then the material flows just after melting at T_m. The latter case is observed if a high melting temperature is accompanied by high crystallinity.

8.4.2. Mechanical spectroscopy

Mechanical spectroscopy does not belong to the classical branches of spectroscopy, as it does not investigate the interaction of electromagnetic radiation with matter. Thermomechanical curves presented before, however, clearly show that the temperature dependence of the mechanical properties in polymers are closely related to molecular mobilites at a given temperature.

Since polymers under the action of mechanical stimuli display both elastic and viscous properties their behaviour is termed viscoelastic. If the elastic behaviour can be described by Hooke's law, and the viscosity relationship is Newtonian, then the material is termed linear viscoelasticity. The relationship between stress strain and shear stress and strain rate are:

$$\sigma = \varepsilon E \quad \text{Hooke's law for extension of a solid} \tag{8.24}$$

$$\tau = \gamma G \quad \text{for shear of a solid} \tag{8.25}$$

$$\tau = \eta \frac{d\gamma}{dt} \quad \text{for a Newtonian fluid} \tag{8.26}$$

where E is the tensile, and G the shear modulus; σ is the tensile, whereas τ the shear stress; ε is the tensile, and γ the shear strain, η is the viscosity coefficient (or simply viscosity), whereas $d\gamma/dt$ is the shear strain rate. Linear viscoelasticity in polymers is a first approximation, being valid only at small deformation levels.

After various external stimuli the polymer can recover its original equilibrium state so long as strains remain small. This is known as the relaxation process. The time and temperature dependence of this process are very important characteristics of the polymer material.

In stress relaxation experiments the stress decreases, that is, the stress level decreases to zero during the relaxation process. Elastic energy transforms irreversibly into viscous energy. If the strain level is constant, the process can be described by the differential equation:

$$\dot{\gamma} = \frac{1}{G}\frac{d\tau}{dt} + \frac{\tau}{\eta} = 0 \tag{8.27}$$

where γ is the time derivative of the shear deformation $d\gamma/dt$; G is the shear modulus, τ is the actual shear stress value, and η the viscosity.

Solving the differential equation with $\tau = \tau_0$ at $t = 0$ as initial conditions gives:

$$\tau = \tau_0 e^{-\frac{Gt}{\eta}}. \tag{8.28}$$

The η/G value in the exponent combines elastic and viscous effects, and is referred to as the relaxation time, Θ.

Mechanical spectroscopy is the study of the distribution of relaxation times under the action of various static and dynamic stimuli.

If the time dependence of the stress is studied, then the stress relaxation modulus, $E(t)$, can be calculated from equation:

$$E(t) = \frac{\sigma_{str}(t)}{\varepsilon_0} \tag{8.29}$$

where σ_{str} is the stress component parallel to the strain direction, and ε_0 the initial strain.

Similarly for the shear stress relaxation modulus:

$$G(t) = \frac{\tau(t)}{\gamma_0} \qquad (8.30)$$

where $\tau(t)$ is the time dependent shear stress, and γ_0 the initial shear strain.

The relaxation process can be described by the stress relaxation modulus itself. Such that:

$$\tau = \gamma G, \qquad (8.25)$$

from Eq. (8.28):

$$G(t) = Ge^{-t/\Theta}. \qquad (8.31)$$

If several relaxation processes are present, then the shear relaxation moduli superpose, so that for a relaxation consisting of z components:

$$G(t) = \sum_{i=1}^{z} G_i \exp(-t/\Theta_i). \qquad (8.32)$$

In the case of dynamic measurements the stress stimulus is applied periodically, for example sinusoidally, and the corresponding deformation is measured. Because of the viscous properties the amplitude decreases, and a phase lag appears between stress and strain. From such measurements a dynamic modulus can be obtained. Here it is assumed to be a shear process.

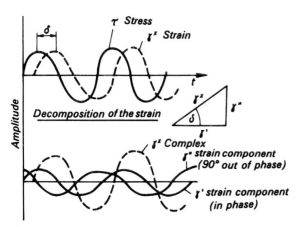

Fig. 8.54. Stress and stain curves in dynamic mechanical experiments.

This can be resolved into two components, one being in phase with the original stress wave, G', the other shows a 90° phase lag, G'' (Fig. 8.54). The components can be presented in complex form, such that:

$$G^x = G' + iG'', \tag{8.33}$$

or

$$|G^x| = \sqrt{G'^2 + G''^2}. \tag{8.34}$$

The ratio of the moduli components is known as tan δ or the loss factor:

$$\tan \delta = \frac{G''}{G'} \tag{8.35}$$

where δ is the phase angle between the stress and strain components. The term "loss factor" refers to the corresponding energy dissipation process.

Data obtained in dynamic measurements cannot be directly compared to quasistatic thermomechanical data, the results agree only if the dynamic measurements are extrapolated to very low frequencies. The exact relationship between the complex and relaxation moduli can be obtained by Fourier tranformation.

In dynamic measurements the realization of the time–temperature superposition principle has been of considerable importance. The principle suggests that in dynamic measurements an increase in the measurement frequency has a similar effect as a decrease in temperature. In quasistatic measurements an increase in temperature corresponds to a decrease in time. The principle is shown in Fig. 8.55. In experimental work the time interval can be varied by a factor of for example 10^3. If the temperature of the measurements is varied, a series of curves can be obtained, as shown on the left hand side of the Figure. If the curves are shifted horizontally a single master curve can be obtained from these component curves. The shift factor related

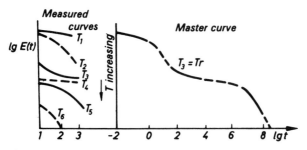

Fig. 8.55. Representation of the time–temperature superposition principle.

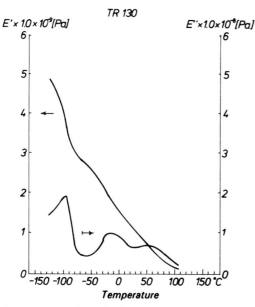

Fig. 8.56. Dynamic mechanical relaxation spectrum (DMA curve) of a polyethylene sample.

to temperature T, is measured on a logarithmic abscissa scale. If, for example, the temperature is increased from T_1 to T_2, the measurement time changes from t to t/a_T. By this method the modulus/time curve can be extended to long times which were not measurable by conventional techniques.

In practice a reference temperature, T_r, is chosen and higher and lower temperature curves are compared to this reference value.

Various types of equipment have been developed for the measurement of the mechanical relaxation spectrum. As an example Fig. 8.56 shows the DMA (dynamic mechanical analysis) curve of a polyethylene sample. The curves identify characteristic changes in G' and G'' as a function of temperature.

CHAPTER 9

Trends of development

The structural investigation of polymers is developing rapidly. Continuously new information is available on conventional and new materials. It is difficult to forecast the train of development, but it should be expected of an author to refer to foreseeable tendencies.

Firstly, concerning new materials. There is at least one high-volume industrial product which requires new structural concepts, and this is the ethylene-α-olefin copolymer, which is frequently referred to as LLDPE (linear low density polyethylene). It is also used in the medium and high density ranges, and in these cases it is called simply LPE (linear polyethylene).

Recently, homogeneous and heterogeneous ethylene copolymers of even lower density have been introduced to the market. They are not only industrially important but are also interesting from a theoretical point of view, on account of intriguing relationships between their molecular structure and crystallization behaviour.

These materials are Very Low Density Polyethylenes (VLDPE) with densities falling well below 900 kg/m^3 (Mathot and Pijpers 1987). Fig. 9.1 shows the thermal behaviour of such a VLDPE. The most remarkable features of these polymers are the wide crystallization and melting ranges, especially compared with an ethylene–propylene copolymer with virtually the same density, whose heat capacity curve has been included in the Figure. Such ethylene–propylene copolymers can be classified as homogeneous, since the distribution of the propylene units is homogeneous both intra- and intermolecularly. By contrast, fractionation experiments have revealed that the heterogeneity of comonomer incorporation in VLDPE is of an intermolecular nature.

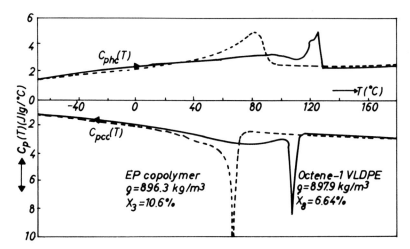

Fig. 9.1. Thermal behaviour of a very low density polyethylene, compared with an ethylene–propylene copolymer of the same density.

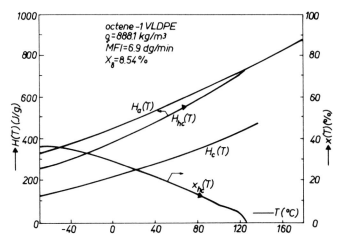

Fig. 9.2. Enthalpy and crystallinity of VLDPE versus temperature.

The relative width of the temperature ranges for the VLDPE in comparison to the homogeneous copolymer is also illustrated in Fig. 9.2. For a VLDPE of even lower density the $C_{p, hc}(T)$ curve (*hc* stands for the heating curve) was integrated, yielding a specific enthalpy curve $H_{hc}(T)$, which was compared to reference curves of 100% amorphous and 100% crystalline linear polyethylene ($H_a(T)$ *and* $H_c(T)$ respectively). The quantity $x_{hc}(T) = [H_a(T)$

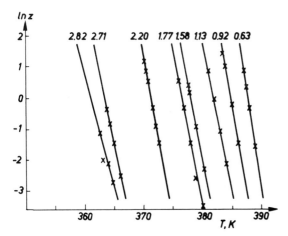

Fig. 9.3. ln k values of different branching fractions from the Avrami plot of LLDPE versus crystallization temperature.

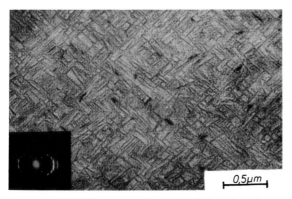

Fig. 9.4. Bright field electron micrograph of the epitaxial array of PE on PP.

$-H(T)]/[H_a(T)-H_c(T)]$ is an accurate measure of the degree to which these extreme conditions are realized and is very sensitive to experimental differences.

The $x_{hc}(T)$ curve shows a continuous decrease from $-60\,°$C onwards. Its value at $23\,°$C is 25%, lower than its peak value at $-60\,°$C. This illustrates that the melting process is in progress from $-60\,°$C onwards, while for this specific sample it is completed at the remarkably high temperature of about $130\,°$C.

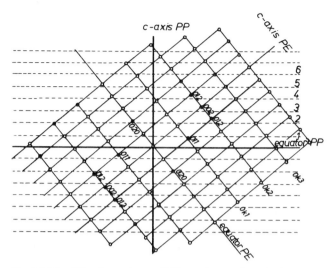

Fig. 9.5. The diffraction pattern of PE superimposed on the fibre texture of PP.

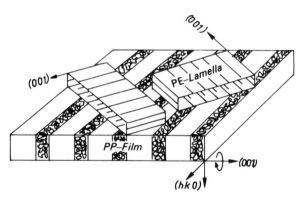

Fig. 9.6. Oriented PP structure reinforced by PE lamellae.

The HDPE and LDPE types crystallize at a very high rate. The ethylene-α-olefin copolymers are very sensitive to branching in their crystallization, and their structure shows a strong dependence on thermal history. As shown in Fig. 9.3 samples fractionated with respect to branching exhibit very different ln k values on the Avrami plot dependent on crystallization temperature (Bodor and Dalcolmo 1987). Such polymers must be characterized not just by fractionation with respect to molecular mass, but also with respect to chain branching.

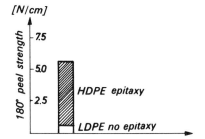

Fig. 9.7. Young's modulus of PP and PE blends, before and after the epitaxial structure is achieved.

Melting properties depend on the crystalline lamellar distribution. Two or more melting ranges are present in the LLDPE types. Methods for describing such distributions are of technical importance.

The mechanical properties of polymers are also improving. It seems that a new way to achieve higher tenacity is to use the epitaxy in some polymer pairs; Petermann et al. (1987) used the epitaxial crystallization of polyethylene on oriented polypropylene. Fig. 9.4 shows the bright field electron micrograph of the epitaxial array of PE on PP. Such structures are obtained when PP/PE blends are cold stretched and subsequently annealed above the melting point of the PE. The molecular order of this structure can be seen in Fig. 9.5. The [100] diffraction patterns of PE are superimposed on the fibre texture of PP. The physical effect is that the amorphous part of the oriented PP structure is reinforced by the PE lamellae, as shown in Fig. 9.6. The Young's modulus of such polymer blends is shown in Fig. 9.7 and is about double that of the normal blend when the epitaxial structure has been achieved.

Recommended Literature

Books of General Interest

Billmeyer, F. W. Jr. (1984): *Textbook of Polymer Science*. 3rd ed. New York–London–Sydney–Toronto, Wiley-Intersience.

Elias, H. G. (1977): *Macromolecules*. Vol. I–II. London–New York–Sydney–Toronto, Wiley.

Hoffmann, M, Krömer, H. and Kuhn, R. (1977): *Polymeranalitik*. I–II. Stuttgart, Georg Thieme Verlag.

Holzmüller, W. and Altenburg, K. (1961): *Physik der Kunststoffe*. Berlin, Akademie-Verlag.

Kämpf, G. (1982): *Charakterisierung von Kunststoffen mit physikalischen Methoden*. München, Wien, Carl Hauser.

Ke, B. (1964): *Newer Methods of Polymer Characterization*. New York–London–Sydney, Interscience.

Lipatov, Yu. S. (1984–1985) (ed.): *Spravotsnik po fizicheskoi khimii polimerov*. I–III. Kiev, Naukova dumka.

Lipatov, Yu. S. (1986) (ed.): *Fizikokhimiya mnogokomponentnikh polimernikh sistem*. I–II. Kiev, Naukova dumka.

Margerison, D. and East, G. C. (1967): *An Introduction to Polymer Chemistry*. Oxford, Pergamon Press.

Marihin, Z. A. and Myasnikova, L. P. (1978): *Nadmolekularnaya Struktura Polimerov*. Leningrad, Khimiya.

Perepechko, Y. Y. (1978): *Vivedenie v Fiziku Polimerov*. Moscow, Khimiya.

Rabek, J. F. (1980): *Experimental Methods in Polymer Chemistry*. Chichester–New York–Brisbane, Wiley.

Rogovin, Z. A. and Malkiy, A. Ya(ed) (1969): *Fizika Polimerov*. Moscow, Mir.

Schurz, J. (1974): *Physikalische Chemie der Hochpolymeren*. Berlin–Heidelberg–New York, Springer Verlag.

Sperling, L. H. (1986): *Introduction to Physical Polymer Science*. New York–Chichester–Brisbane–Toronto–Singapore, J. Wiley and Sons.

Stuart, H. A. (1952–1956): *Die Physik der Hochpolymeren*. Band I–IV. Berlin, Springer.

Tager, A. A. (1978): *Fizikokhimiya Polimerov*. 3. ed. Moscow, Khimiya.

Tanford, D. (1963): *Ch. in Physical Chemistry of Macromolecules.* New York–London, Wiley.

Tüdze, P. and Kavai, T. (1977): *Fizicheskaya Khimiya Polimerov. per. s. yaponsk.* Moscow, Khimiya.

Vollmert, B. (1962): *Grundriss der Makromolekularen Chemie.* Berlin–Göttingen–Heidelberg, Springer.

Chapter 1

Flory, P. J. (1953): *Principles of Polymer Chemistry.* Ithaca, N. Y., Cornell Univ. Press.

Huggins, M. L., Natta, G., Desreux, V. and Mark, H. (1966): Report on Nomenclature Dealing with Steric Regularity in High Polymers. *Pure Appl. Chem.* **12**, 645.

Natta, G. and Danusso, F. (1967): *Stereoregular Polymers and Stereospecific Polymerisations.* Oxford, Pergamon.

Chapter 2

Erdey-Gruz, T. and Schay, G. (1952): *Elméleti fizikai kémia.* (Theory of Physicochemistry) I. Budapest, Tankönyvkiadó.

Flory, P. J. (1942): Thermodynamics of High Polymer Solutions. *J. Chem. Phys.* **10**, 51.

Kerker, M. (1969): *The Scattering of Light.* London, Academic Press.

Huggins, M. L. (1942): Theory of Solutions of High Polymers. *J. Am. Chem. Soc.* **64**, 1712.

Huglin, M. B. (1972): *Light Scattering from Polymer Solutions.* London, Academic Press.

Máté János (1979): Az anyag szerkezete. (Structure of Matter) Budapest, Műszaki Könyvkiadó.

McIntyre, D. and Gornick, E. (1964): *Light Scattering from Dilute Polymer Solutions.* London, Gordon–Beach.

Schurz, J. (1972): *Viskositatmessungen an Hochpolymeren.* Stuttgart, Berliner Union.

Springer, J. (1970): *Einführung in die Theorie der Lichtstreuung verdünnter Lösungen grosser Moleküle.* Berlin–Dahlem, Applied Research Laboratories.

Stacey, K. A. (1956): *Light Scattering in Physical Chemistry.* London, Butterworths.

Vándor József (1952): Kémiai termodinamika I. (Chemical Thermodynamics) Budapest, Akadémiai Kiadó.

Chapter 3

Altgelt, K. H. and Segal, L. (1971): *Gel Permeation Chromatography.* New York, Dekker.

Beyer, G. L. (1959): Determination of Particle Size and Molecular Weight. In Weissberger (ed.): *Physical Methods in Organic Chemistry.* Part I. New York, Interscience, pp. 191–257.

Cantow, M. J. R. (1966): *Polymer Fractionation.* New York, Academic Press.

Determann, H. (1967): *Gelchromatographie.* Berlin, Springer.

Johnson, J. F. and Porter, R. S. (1970): Gel Permeation Chromatography. *Prog. Polymer Sci.* **2**, 201.

Yau, W. W., Kirkland, J. J. and Bly, D. D. (1979): *Modern Size-Exclusion Chromatography.* New York, Wiley–Interscience.

Chapter 4

Collier, J. R. (1969): Polymer Structure. *Industrial and Engineering Chemistry. 61*, 50.

Geil, P. H. (1963): *Polymer Single Crystals.* New York, Interscience.

Lindenmayer, P. H. (1974): Why Do Polymer Molecules Fold? *Polymer Engineering and Science* **14**, 456.

Wunderlich, B. (1973): *Macromolecular Physics.* Vol. 1. Crystal Structure, Morphology, Defects. New York, Academic Press.

Zachmann, H. G. (1974): Der Kristalline Zustand Makromolekularer Stoffe. *Angew. Chem.* **86**, 283.

Chapter 5

Fatou, J. G. (1979): Differential Thermal Analysis and Thermogravimetry of Fibres. In: *Applied Fibre Science.* ed by Happey, F. Vol. 3. London, Academic Press.

Kalló Dénesné (1966): Műanyagok szerkezetvizsgálata elektronmikroszkóppal. (Structural Analysis of Polymers by Electron Microscopy) *Kémiai Közlemények* **26**, 31.

Kalló Dénesné (1979): A pásztázó elektronmikroszkópia alkalmazása a műanyagok morfológiai kutatásában. (Application of Scanning Electron Microscopy in the Morphological Analysis of Polymers). *Műanyag és Gumi* **16**, 285.

McNaughton, J. L. and Mortimer, C. T. (1975): *Differential Scanning Calorimetry* in *IRS Physical Chemistry Series.* Vol. 10. London, Butterworths.

Miller, G. W. (1969): *The Thermal Characterization of Polymers.* Applied Polymer Symposia. No. 10, 35.

Turi, E. A. (ed.) (1981): *Thermal Characterization of Polymeric Materials.* New York, Academic Press.

Chapter 6

Schnuur, G. (1955): *Some Aspects of the Crystallization of High Polymers.* Delft, Rubber-Stiching.

Sharples, A. (1966): *Introduction to Polymer Crystallization.* London, Edward Arnold Publ.

Wunderlich, B. (1976): *Macromolecular Physics.* Vol. 2. Crystal Nucleation, Growth, Annealing. New York, Academic Press.

Wlochowitz, A. and Eder, M. (1984): *Polymer* **25**, 1268.

Zachmann, H. G. (1964): Das Kristallisations und Schmelzverfahren Hochpolymerer Stoffe. *Fortschr. Hochpolym. Forsch.* **3**, 581.

Chapter 7

Alexander, L. E. (1969): *X-Ray Diffraction Methods in Polymer Science.* New York, Wiley–Interscience.

Bodor G. (ed.) (1977): *Orientation Effects in Solid Polymers.* J. Polymer Sci.: Polymer Symposium. No. 58. New York, Wiley–Interscience.

Glatter, O. and Kratky, O. (ed.) (1982): *Small angle X-ray Scattering.* London, Academic Press.

Kakudo, M. and Kasai, N. (1972): *X-Ray Diffraction by Polymers.* Tokyo–Amsterdam–London–New York, Kodansha–Elsevier.

Kitaigorodsky, A. I. (1952): *Rentgenostrukturniy Analiz Melkokristallicheskikh i Amorfnich Tel.* Moscow, Gosteksudat.

Martinov, M. A. and Viletskhanina, K. A. (1972): *Rentgenografia Polimerov.* Leningrad, Khimiya.

Perepetshko, I. I. (1973): *Akusticheskie Metody Isledovanya Polimerov.* Moscow, Khimiya.

Samuels, R. J. (1974): *Structured Polymer Properties.* New York, Wiley–Interscience.

Urbanczyk, G. W. (1986): *Mikrostruktura Wlokna.* Warszawa Wydawnictwa Naukowo–Techniczne.

Chapter 8

Bovey, F. A. (1971): *Nuclear Magnetic Resonance* in *Encyclopedia of Polymer Science and Technology.* Vol. 9. New York, Wiley–Interscience. p. 356.

Bovey, F. A. (1969): *Nuclear Magnetic Resonance Spectroscopy.* New York–London, Academic Press.

Bovey, F. A. (1972): *High Resolution NMR of Macromolecules.* New York–London, Academic Press.

Bovey, F. A. (1984): *NMR and Macromolecules.* In NMR and Macromolecules. ACS Symposium Series No. 247. American Chem. Soc.

Dechant, J. (1972): *Ultrarotspektroskopische Untersuchungen an Polymeren.* Berlin, Akademie Verlag.

Füzes, L., Bodor, G., Tóth, G., Szőllősy, Á. and Almássy, A. (1986): The Determination of PE Branching with the ^{13}C-NMR Method. *Műanyag és Gumi* **22.** 145–147.

Holly, S. and Sohár, P. (1968): *Infravörös spektroszkópia.* (Infrared Spectroscopy) Budapest, Műszaki Könyvkiadó.

Hummel, D. O. (1978): *Atlas of Polymer and Plastics Analysis.* 2. ed. Münich, Carl Hanser Verlag.

Iving, K. J. (ed.) (1976): *Structural Studies of Macromolecules by Spectroscopic Methods.* London, Wiley.

Jelinski, L. W. (1984): Modern NMR Spectroscopy. *C and EN* Nov. 26–47.

Kössler, I. (1971): *Infrared- Absorption Spectroscopy* in *Encyclopedia of Polymer Science and Technology.* Vol. 7. New York, Wiley–Interscience. p. 620.

Krimm, S. (1960): Infrared Spectra of High Polymers. *Fortschr. Hochpolym. Forschung* **2,** 51.

Miller, R. G. J. and Stace, B. C. (1972): *Laboratory Methods in Infared Spectroscopy.* 2nd. ed. London–New York–Rheine, Heyden Ltd.

O'Donnel, D. J. (1984): *An Introduction to NMR Spectroscopy of Solid Samples* in *NMR and Macromolecules.* ACS Symposium Series No. 247, American Chem. Soc.

Siesler, H. W., Holland and Moritz, K. (1980): *Infrared and Raman Spectroscopy of Polymers.* New York, Basel, Marcel Dekker.

Sohár Pál (1976): *Mágneses magrezonancia-spektroszkópia.* (NMR Spectroscopy) Budapest, Akadémiai Kiadó.

Whiffen, D. H. (1966): *Spectroscopy.* London, Longman Green Co.

Zbinden, R. (1964): *Infrared Spectroscopy of High Polymers.* New York–London, Academic Press.

References

Alexander, L. E. (1969): *X-Ray Diffraction Methods in Polymer Science*. New York–London, Wiley–Interscience.

Alvang, F. and Samuelson, O. (1957): *J. Polym. Sci.* **23,** 57.

Anderson, F. R. (1964): *J. Appl. Phys.* **35,** 64.

Andrews, P. (1964): *Biochem. J.* **91,** 222.

Baker, W. O. (1945): *Advancing Fronts in Chemistry*. Vol. I. New York, Reinhold Publ.

Banks, G., Hoy, I. N. and Sharpless, A. (1964): *J. Polymer Sci.* **A–2.** 4059.

Bown, C. E. H., Freeman, R. F. J. and Kamaliddin, A. R. (1950): *Trans. Faraday Soc.* **46.** 677–684.

Bellamy, L. J. (1958): *The Infrared Spectra of Complex Molecules*. London, Methuen Co.

Benoit, H. (1966): *J. Chim Phys.* **63,** 1507.

Bernauer, F. (1929): *Gedrillte Kristalle in Forschungen zur Kristalkunde*. Heft 2. Berlin, Borntrager.

Bertaut, E. F. (1950): *C. R. Acad. Sci.* **228,** 187–189. (1949), 492–494. (1949); *Acta Cryst.* **3,** 14–18. (1950).

Bodor, G. (1969): *Chemickle Vlakna* **19,** 42.

Bodor, G. (1981): *Diffraction Studies of Elastomer–Plastomer Blends. Diffraction Studies on Non-Crystalline Substances*. Ed by Hargittai and Orville-Thomas. Budapest, Akadémiai Kiadó.

Bodor, G. and Hangos, M. (1954): Research Report.

Bodor, G. and Füzes, L. (1979): Kisszögü röntgendiffrakciós vizsgálatok méreteloszlás meghatározására polimerekben. (Small Angle X-ray Diffraction Studies for Determination of Size Distribution in Polymers) MTA. Anyagszerk. Mbiz. ülése Mátrafüred, Oct. 8.

Bodor, G. and Dalcolmo, H. J. (1987): *Iupac*. Merseburg.

Bodor, G., Grell M. and Kalló, A. (1964): *Faserforschung und Textiltechnik* **15,** 527.

Bodor, G., Holly, S. and Kalló, Dné. (1959): *IUPAC*. Wiesbaden.

Brandrup, J. and Immergut, E. H. (1975): *Polymer Handbook* New York, Wiley–Interscience, 2. ed.

Bro, M. I. and Sperati, C. A. (1956): *J. Polymer Sci.* **38**, 289.

Broza, G., Rieck, U., Kawahguchi, A. and Petermann, J. (1985): *J. Polym. Sci. Polym. Phys.* **23**, 2623–2627.

Buerger, M. J. (1960): *Crystal-Structure Analysis.* New York, John Wiley and Sons, p. 243.

Bullock, A. T. and Cameron, G. G. (1976): ESR Studies of Spinlabelled Synthetic Polymers. In: *Structure of Macromolecules by Spectroscopic Methods.* Ed by Ivin, K. J. London, Wiley–Interscience.

Compostella, M., Coen, A. and Bertinotti, F. (1963): *Angew. Chem.* **74**, 618.

Cormia, R. L., Price, F. P. and Turnbull, D. (1962): *J. Chem. Phys.* **37**, 1333.

De Bye, P. and Bueche, A. M. (1948): *Chem. Phys.*

Dismore, P. F. and Statton, W. O. (1966): *J. Polymer Sci.* **C–13**, 133.

Dlugos, J. and Miche, R. I. C. (1960): *Polymer.* **1**, 41.

Dole, M. and Hettinger, W. R. (1952): *J. Chem. Phys.* **20**, 781.

Dorman, D. E., Otoka, E. P. and Bovey, F. A. (1972): *Macromolecules* **5**, 574. *Ser. A.* **175**, 208.

Elliott, A. (1959): *Advances in Spectroscopy* **1**, 214.

Erdey-Gruz, T. and Schay, G. (1952): Elméleti fizikai kémia. (Theory of Physicochemistry) I. Budapest, Tankönyvkiadó, 335.

Ewald, P. P. (1921): *Z. Krist.* **56**, 129.

Farrow, F. (1960): *Polymer* **1**, 518.

Farrow, G. and Ward, I. M. (1960): *British J. Appl. Phys.* **11**, 543.

Fatou, J. G. (1979): *Differential Thermal Analysis and Thermogravimetry of Fibers. Applied Fibre Science.* (Ed. Happes F.) London, Academic Press.

Ferguson, R. C. (1967): *Polymer Reprints* **8**, 1026.

Field, J. E. (1941): *J. Appl. Phys.* **12**, 23.

Fischer, E. W. and Schmidt, G. F. (1962): *Angew Chem.* **74**, 551.

Fischer, I., Kinsinger, J. B. and Wilson, C. W. (1966): *J. Polym. Sci.* **B–4**, 379.

Flory, P. J. (1942): Thermodynamics of High Polymer Solutions *J. Chem. Phys.* **10**, 51.

Flory, J. P. (1953): *Principles of Polymer Chemistry.* Ithaca, N. Y., Connell Univ. Press, p. 512.

Flory, P. J. (1949): *J. Chem. Phys.* **17**, 303.

Flory, P. J. and Krigbaum, U. R. (1950): *J. Chem. Phys.* **18**, 1086.

Flory, P. J. and Fox, T. G. (1951): *J. Am. Chem. Soc.* **73**, 1904.

Fox, J. J. and Martin, A. E. (1940): *Proc. Roy. Soc. (London)* Ser. A., **175**, 208.

Frenkel, S. Ya. (1968): *Dopolnenie monokristalli.* Leningrad, Khimiya.

Frisch, H. L., Mallons, C. L., Heatley, F. and Bovey, F. A. (1968): *Macromolecules* **1**, 533.

Fujiwara, Y. (1968): *Koll. Z.* **226**, 135.

Fuller, C. S. and Baker, W. O. (1943): *J. Chem. Education* **26**, 3.

Fuoss, R. M. and Mead, D. J. (1943): *J. Phys. Chem.* **47**, 59.

Gee, G. (1947): *J. Chem. Soc.* 280–288.

Gee, G. and Orr, W. Y. C. (1946): *Trans Faraday Soc.* **42**, 507–517.

Geil, P. H. (1960): *Polym. Sci.* **44**, 449.

Geleji, F., Bodor, G. and Dutka Gy. (1962): Chemiefaser Symp. Weimar, p. 421.

Geleji, F., Kóczy, L., Fülöp, I. and Bodor, G. (1977): *Orientation effects in solid polymers*. J. Polymer Sci. Symp. No. 58, 253–273.

Gent, A. N. (1954): *Trans Inst. Rubber Ind.* **30,** 139.

Goldfinger, G., Mark, H. and Siggia, S. (1943): *Ind. Engng. Chem.* **35,** 1083.

Goppel, J. M. (1947): *Appl. Sci. Res.* **A–1,** 3.

Gubler, M., Rabesiaka, I., Kovács, A. I. and Geil, P. H. (1963): *Polymer Single Crystals*. New York, Wiley–Interscience.

Guggenheim, E. A. (1949): *Thermodynamics*. Amsterdam, North Holland Publishing Co.

Guinier, A. (1927): *Compt. Rend, Seances Acad. Sci.* **204,** 1115.

Happey, F. (1951): *Br. J. Appl. Phys.* **2,** 117.

Hedvig, P. (1975): *Experimental Quantum Chemistry*. Budapest, Akadémiai Kiadó.

Heinen, W. (1959): *J. Polymer Sci.* **38,** 545.

Henniker, J. C. (1967): *Infrared Spectrometry of Polymers*. London, Academic Press.

Henry, N. F. M., Lipson, H. and Wooster, W. A. (1951): *The Interpretation of X-Ray Diffraction Photography*. London, Macmillan.

Hermans, P. H. (1946): *Contribution of the Physics of Cellulose Fibres*. New York, Elsevier.

Hermans, P. H. (1951): *Koll. Zeitschrift.* **120,** 3.

Hermans, P. H. and Weidinger, A. (1948): *J. Appl. Chem. Phys.* **19,** 491.

Hermans, P. H. and Weidinger, A. (1961): *Macromol. Chem.* **50,** 8.

Hermans, P. H. and Weidinger, A. (1949): *J. Polymer Sci.* **4,** 709.

Herzog, R. O. and Jancke, W. (1920): *Z. Phys.* **3,** 196.

Hess, K. and Kiessig, M. (1944): *Z. Physik. Chem. (Leipzig)* **193,** 196.

Hill, R. (1953): *Fibres from Synthetic Polymers*. Amsterdam, North Holland Publishing Co. Chapter 10.

Hildebrand, J. H. and Scott, R. L. (1950, 1964): The Solubility of Nonelectrolites. 3rd. ed. New York, Reinhold, New York, Boker.

Hosemann, R. (1950): *Kolloid Zeitschrift* **117,** 13.

Hosemann, R. and Bagchi, S. N. (1962): *Direct Analysis of Diffraction by Matter*. Amsterdam, North Holland Publ. Co.

Hosemann, R., Bonart, R. and Schonknecht, G. (1956): *Z. Phys.* **146,** 588.

Hoslam, J. and Ellis, H. A. (1965): *Identification and Analysis of Plastics*. London, Iliffe Books Ltd.

Howsmon, J. A. in Berl, W. (1956): *Physical Methods in Chemical Analysis*. Vol. I. New York, Academic Press. p. 142.

Hummel, D. O. (1966, 1978): *Infrared Spectra of Polymers*. New York, Wiley–Interscience.

Hummel, D. O. (1978): *Atlas of Polymer and Plastics Analysis*. Vol. 1. 2nd. ed. Münnich, Carl Hanser Verlag.

Huggins, M. L. (1942): *Ann. N. Y. Acad. Sci.* **43,** 1.

Huggins, M. L. (1942): *J. Phys. Chem.* **46,** 151.

Huggins, M. L. (1942): *J. Am. Chem. Soc.* **64,** 1712.

Immergut, E. H., Kollmann, G. and Malatesta, A. (1961): *J. Polymer Sci.* **51,** 57.

Ingersoll, M. G. (1946): *J. Appl. Phys.* **17,** 924.

Inoue, M. (1963): *J. Polymer Sci.* **A–1,** 2013.

Inoue, Y., Nishioka, A. and Chujo, R. (1971): *Polymer J. (Japan)* **2,** 535.

Johnson, L. F., Heatley, F. and Bovey, F. A. (1970): *Macromolecules* **3**, 175.

Jones, F. M. (1938): *Proc. Royal Soc.* **A–166**, 16.

Kakudo, M. and Kassai, N. (1972): *X-Ray Diffraction by Polymers.* Tokyo, Amsterdam, Kodanska–Elsevier.

Kalló Dénesné (1979): *Műanyag és Gumi* **16/10**, 285–291.

Kargin, V. A. and Mihailov, N. V. (1940): *Zh. F. H.* **14**, 195.

Kargin, V. A., Kitaigorodsky, A. I. and Slonimsky, G. I. (1957): *Koll. Zh.* **19**, 131.

Kast, W. (1953): *Feinstruktur-Untersuchungen an Künstlichen Zellulosefasern verschiedener Herstellungsverfahren.* Köln, Westdeutscher Verlag, 17. p.

Katz, I. R. (1925): *Chemi Kerztgt.* **49**, 353.

Kammerer, H., Rocaboy, F., Steinfort, G. and Kesu, W. (1962): *Macromol. Chem.* **53**, 80.

Keith, H. D. and Padden, F. I. (1963): *J. Appl. Phys.* **34**, 2409.

Keith, H. D. and Padden, F. I. (1964): *J. Appl. Phys.* **35**, 1270.

Keller, A. (1959): *J. Polymer Sci.* **36**, 361.

Keller, A. (1963): *Fiber Structure*, London, Butterworth.

Keller, A., Lester, G. R. and Morgan, L. B. (1954): *Phil. Trans.* **A–247**, 1.

Kirkwood, J. G. and Riseman, J. (1948): *J. Chem. Phys.* **16**, 567.

Kitaigorodsky, A. I. (1952): *Rentgenostrukturniy Analiz Melkokristallicheskikh i Amorfnykh Tel.* Moscow, Gostekhizdat.

Kocskina, A. (1968): *Kémiai Közlemények* **29**, 107.

Kozler, M. (1960): *Visokom. Soed.* **2**, 444.

Kinsinger, J. B., Fischer, T. and Wilson, C. W. (1966, 1967): *J. Polym. Sci.* Part. B. **4**, 379; Part. B. **5**, 285.

Krimm, S. and Tobolsky, A. V. (1951): *J. Polymer Sci.* **17**, 57.

Krimm, S., Miang, C. and Sutherland, G. (1955): *J. Chem. Physics* **25**, 549.

Kuribayashi, S. and Nakai, A. (1962): *J. Text. Ind. Jap.* **18**, 64.

Langford, J. I. (1978): *J. Appl. Cryst.* **11**, 10.

Laue, M. and Tank, F. (1913): *Ann. Physik.* **41**, 1003.

Liang, C. Y. and Krimm, S. (1956): *J. Chem. Phys.* **25**, 663.

Liang, C. Y. and Krimm, S. (1958): *J. Polymer Sci.* **27**, 241.

Luongo, J. P. (1960): *J. Appl. Polymer Sci.* **3**, 302.

Magill, I. H. (1962): *Polymer*, **3**, 655.

Mathas, A. and Meller, N. (1966): *Anal. Chem.* **38**, 473.

Matsuoka, S. (1960): *J. Polymer Sci.* **42**, 511.

Matsuoka, S. (1961): *J. Appl. Phys.* **32**, 2334.

McDonald, M. P. and Ward, I. M. (1961): *Polymer* **2**, 341.

Mandelkern, L. (1964): *Crystallization of Polymers.* McGraw-Hill, New York.

Mathot, V. B. F. and Pijpers, M. F. J. (1986): *In: Integration of Fundamental Polymer Science and Technology.* Ed by Kleintjens, L. A. and Lemstra, P. J., London, Elsevier Applied Science.

Meisel, W. (1971): *Exp. Tech. der Physik.* **IXX**, 23.

Meridith, R. (1944): *Shirley Inst. Mem.* **19**, 29.

Miller, R. L. and Nielsen, L. E. (1961): *J. Polymer Sci.* **55**, 643.

Moore, J. C. (1964): *J. Polym. Sci.* **A–2**, 835.

Möller, M. Cantow, H. S. and Ctloff, H. (1986): *Macrom. Chem.* **187**. 1237–1252.

Mullin, I. W. (1961): *Crystallization.* London, Butterworth.

Munden, A. R. and Palmer, H. J. (1950): *J. Text. Inst.* **41**, 609.

Nakanishi, K. (1964): *Infrared Absorption Spectroscopy Practical.* San Francisco, Holden-Day Inc.

Natta, G. (1955): *Polymer Sci.* **16**, 143.

Natta, G. and Corradini, P. (1955): *Atti. Accad. Naz.* Lincei Rend., Classe Sci. Fis., Mat. Nat. **18**, 19.

Natta, G. and Corradini, P. (1960): *Nuovo Cimento* **15**, Suppl. 1. 40.

Natta, G., Pino, P. and Corradini, P. (1955): *J. Am. Chem. Soc.* **77**, 1708.

Natta, G., Corradini, P. and Cesari, M. (1957): *Atti della Accademia Nazionale dei Linzei* Serie 8. **22**, Fac. 1. 11.

Natta, G., Porri, L. and Mazzei, A. (1959): *Chim. Ind. (Milan)* **41**, 116.

Nelson, M. L. and Conrad, C. G. (1948): *Text. Res. J.* **18**, 149.

Newing, M. J. (1950): *Trans. Faraday Soc.* **46**, 613–620.

Nyguist, R. A. (1960): *Infrared Spectra of Plastics and Resins.* USA, Dow Chemical Co.

Nikitin, V. N. and Pokrovsky, E. I. (1954): *Izv. Ak. N. S.S.S.R. Ser. Fiz.* **6**, 736.

Onogi, S. (1962): *J. Polymer Sci.* **58**, 1. 157–164.

Palmer, R. P. and Cobbold, A. J. (1964): *Die Makromol. Chem.* **74**, 174.

Pauli, W. (1924): *Naturwissenschaften* **12**, 741.

Pelzbauer, Z. (1967): *Straub-Reinhalt. Luft.* **27**, 233.

Petermann, J. (1987): *Eigenverstärkung in Blends durch Epitaxie.* Kolloquium Technologie der Eigenverstärkung von Thermoplasten 1987 Febr. Kassel, BRD.

Pokrovszky, E. I. and Kotova, I. P. (1956): *Zs. Tech. Izv.* **26**, 1456.

Powles, J. G., Strange, H. J. and Sandiford, D. J. H. (1963): *Polymer* **4**, 401.

Powles, J. G., Hunt, B. J. and Sandiford, D. J. H. (1964): *Polymer* **5**, 585.

Prietschk, A. (1958): *Koll. Zeitschrift.* **156**, 8.

Rabesiaka, I. and Kovács A. I. (1961): *J. Appl. Phys.* **32**, 2314.

Ranby, P. C. (1951): *Disc. Faraday Soc.* **11**, 158.

Reding, F. P. and Lovell, C. M. (1956): *J. Polymer Sci.* **21**, 157.

Reneker, D. M. (1962): *J. Polymer Sci.* **59**, 39.

Ruland, W. (1961): *Acta Cryst.* **14**, 1180.

Ruscher, Ch. (1958): *Faserforschung und Textiltechnik.* **9**, 486.

Samay, G. (1979): *Acta Chim. Acad. Sci. Hung.* **102**, 157.

Samuels, R. J. (1974): *Structured Polymer Properties.* New York, Wiley, p. 89–141.

Samuels, R. J. (1965): *J. Polymer Sci.* A–3, 1741.

Scherrer, P. and Zsigmondy, R. (1920): *Kolloidchemie* p. 339.

Scherrer, P., Herczog, R. O. and Jancke, W. (1920): *Z. Phys.* **3**, 196.

Schmidt, J. A. S. (1955): *Disc. Faraday Soc.* **19**, 207.

Sharples, A. (1966): *Introduction to Polymer Crystallization.* London, Edmond Arnold Publ.

Shirane, G. and Cox, D. E. (1962): *Phys. Rev.* **125**, 1158.

Shyluk, S. (1962): *J. Polymer Sci.* **62**, 318.

Sission, W. A. and Clark, G. L. (1933): *Ind. Eng. Chem.* **5**, 296.

Slichter, W. P. (1957): *J. Polymer Sci.* **24**, 173.

Slichter, W. P. (1958): *Fortschr. Hochpolym. Forsch.* **1**, 35.

Slichter, W. P. (1961): *Rubber Chem. Technol.* **34**, 1574.

Slichter, W. P. and McCall, D. W. (1956): *J. Polym. Sci.* **19**, 485.

Slichter, W. P. and Davis, D. D. (1964): *J. Appl. Phys.* **35**, 3103.

Small, P. A. (1953): *J. Appl. Chem.* **3**, 71.

Springer, J. (1970): *Einführung in die Theorie der Lichtsteruung verdünnter Lösungen grosser Moleküle.* Applied Research Laboratories.

Starkweather, H. W. and Moynihan, R. E. (1956): *J. Polymer Sci.* **22**, 363.

Starkweather, H. W. and Brooks, R. E. (1959): *J. Appl. Polymer Sci.* **1**, 236.

Statton, W. O. (1964): *Small-Angle X-ray Studies of Polymers in KE, B.: Newer Methods of Polymer Characterization.* New York, Wiley–Interscience.

Statton, W. O. (1967): *J. Polymer Sci.* **C–18**, 33.

Staudinger, H. and Signer, R. (1929): *Z. Crist*, **70**, 193.

Staverman, A. J. (1951, 1952): *Rec. Trav. Chim.* **70**, 344; **71**, 623.

Staverman, A. J., Pals, D. F. T. and Kruissink, A. Ch. (1957): *J. Polym. Sci.* **23**, 57.

Stehling, F. C. (1964): *J. Polymer Sci.* **A2**, 1815.

Stein, R. S. (1959): *J. Polymer Sci.* **34**, 711.

Stein, R. S. and Rhodes, M. M. (1960): *J. Appl. Phys.* **31**, 1873.

Stejskal, E. O., Schafer, J., Sefcik, M. D., Jakob, G. S. and Mackay, R. A. (1982): *Pure Appl. Chem.* **54**, 461–466.

Stokes, A. R. (1948): *Proc. Phys. Soc.* **61**, 382–391.

Strobl, G. R. (1970): *Acta Cryst.* **A 26**, 367–375.

Szymanski, H. A. (1963): *Theory and Practice of Infrared Spectroscopy.* New York, Plenum Press.

Takayanagi, M. (1957): *Mem. Fac. Eng. Kyushu Univ.* **16**, 112.

Takeda, M., Iimura, K., Yamada, A. and Imamura, Y. (1959, 1960): *Bull. Chem. Soc. Japan* **32**, 1150; **33**, 1219.

Tormala, P. and Lindberg, J. J. (1976): Spin Labels and Probes in Dynamic and Structural Studies of Synthetic and Modified Polymers. In: *Structural Studies of Macromolecules by Spectroscopic Methods.* London, Wiley–Interscience.

Tung, L. H. (1956): *J. Polymer Sci.* **2**, 144.

Tung, L. H., Moore, J. C. and Knight, W. (1966): *J. Appl. Polymer Sci,* **10**, 1261.

Turner-Jones, A., Aizlewood, J. M. and Beckett, D. R. (1964): *Die Makromol. Chem.* **75**, 134.

Vajnstein, B. K. (1964): *Diffraction of X-Rays by Chain Molecules.* Amsterdam, Elsevier.

Vancsóné Szmercsányi I. and Regős I. (1980): *Műanyag és Gumi* **17**.

Vándor, J. (1952): Kémiai Termodinamika. I. (*Chemical Thermodynamics*) Budapest, Akadémiai Kiadó.

Vonk, C. G. and Kortleve, G. (1967): *Kolloid Z. Z. Polymer* **220**, 19–24.

Vonk, C. G. (1982): *Synthetic Polymers in the Solid State in Small Angle X-ray.* Ed by O. Glatter and O. Kratky. London–New York, Academic Press.

Vonk, C. G. (1987): *The Small Angle Scattering and Related Methods.* 10th Discussion Conference, Prague.

Walker, E. E. (1952): *J. Appl. Chem.* **2**, 470.

Ward, I. M. (1962): *Proc. Phys. Soc. London,* **80**, 1176.

Ward, I. M. (1964): *Textile Res. J.* **34**, 806.

Waltermann, H. A. (1963): *Kolloid Z. Z. Polymere* **192**, 1.

Wildchut, A. J. (1946): *J. Appl. Phys.* **17**, 51.

Wiley, R. E. (1962): *Plastics Technology* **8**, 31.

Wilhoit, R. C. and Dole, M. (1953): *J. Phys. Chem.* **57**, 14.

Willbourn, A. H. (1959): *J. Polymer Sci.* **34,** 569.

Wilson, C. W. and Santee, E. R. (1965): *J. Polymer Sci.* **C–8,** 97.

Wlochowitz, A. and Eder, M. (1984): *Polymer* **25,** 1268.

Wood, L. A. and Bakkedahl, N. (1946): *J. Appl. Phys.* **17,** 362.

Wood, L. A. and Luongo, J. P. (1961): *Mod. Plastics* **38,** 132.

Woodbrey, J. C. and Trementozzi, Q. A. (1965): *J. Polymer Sci.* **C–8,** 113.

Wunderlich, B. (1963): *J. Polymer Sci.* **A–1.** 1245.

Wunderlich, B. (1973): *Macromolecular Physics.* Vol. 1. New York–London, Academic Press.

Yoshino, T. Kikuchi, Y. and Komiyama, J. (1966): *J. Phys. Chem.* **70,** 1059.

Zimm, B. H. and Myerson, I. (1946): *J. Amer. Chem. Soc.* **68,** 911.

Subject index